Asia in Transition

Volume 29

This book series, indexed in Scopus, is an initiative in conjunction with Springer under the auspices of the Universiti Brunei Darussalam – Institute of Asian Studies (http://ias.ubd.edu.bn/). It addresses the interplay of local, national, regional and global influences in Southeast, South and East Asia and the processes of translation and exchange across boundaries and borders. The series explores a variety of disciplinary and interdisciplinary perspectives.

Submission and Peer Review:

Proposal submissions are to be sent to the Series Editor-in-Chief, Prof Dr Hoon Chang Yau: changyau.hoon@ubd.edu.bn and Springer Publishing Editor Alex Westcott Campbell: alexandra.campbell@springernature.com. Please contact the publishing editor to obtain the correct proposal form. All proposals will undergo peer review by the editorial board members. If accepted, the final manuscript will be peer reviewed internally by the editorial board and may also be externally reviewed (single blind) by Springer ahead of acceptance and publication.

With special thanks to Bruno Jetin, outgoing Series Editor.

Hannah Ming Yit Ho · Ying-kit Chan
Editors

Transnational Southeast Asia

Communities, Contestations and Cultures

ubd INSTITUTE OF ASIAN STUDIES

Springer

Editors
Hannah Ming Yit Ho
Universiti Brunei Darussalam
Faculty of Arts and Social Sciences
Gadong, Brunei Darussalam

Ying-kit Chan
National University of Singapore
Singapore, Singapore

ISSN 2364-8252 ISSN 2364-8260 (electronic)
Asia in Transition
ISBN 978-981-96-3607-5 ISBN 978-981-96-3608-2 (eBook)
https://doi.org/10.1007/978-981-96-3608-2

This Springer imprint is published by the registered company Springer Nature Singapore Pte Ltd.
The registered company address is: 152 Beach Road, #21-01/04 Gateway East, Singapore 189721, Singapore

Preface

In an international symposium organised at the Faculty of Arts and Social Sciences (FASS) at Universiti Brunei Darussalam under the research grant UBD/RSCH/1.2/FICBF(b)/2018/008, the idea of putting together a book on the use of transnational as an operative framework to analyse the cultures and communities of Southeast Asia was mooted. We would like to thank both ex-Dean of FASS Noor Azam Othman and Dean of FASS Kathrina Mohd Daud for their unwavering support rendered to Debbie Ho and Hannah Ho, as co-organisers of this symposium.

Kuah Khun Eng and Ying-kit Chan were keynote speakers at this symposium that was held in November 2022. To the former, we are thankful for her engaging concepts and feedback that stimulated discussions. For all the exciting papers that were presented at this academic exchange, we are grateful to local and international scholars from their different disciplines and respective institutions across (Southeast) Asia who came together to participate and moderate the panels.

This book was made possible through the research grant UBD/RSCH/1.2/FICBF(b)/2023/039 awarded by Universiti Brunei Darussalam. Sincere thanks to Yazid Surieshtino and Nasri Tahir for their excellent and timely assistance in preparing the full manuscript for its publication. They have been reliable and diligent in helping to copyedit the chapters in this voluminous book.

We would like to extend our heartfelt gratitude to all chapter authors for their hard work, enthusiasm and passion. They have been receptive to comments and tireless in their drafting and redrafting of their chapters. The interdisciplinary range of academic work that is offered in this book is credited to their rich insights and fascinating research that they continue to conduct.

Our kind thanks are directed to the previous editor-in-chief of Springer's Asia-in-Transition Series, Professor Bruno Jetin, for his valuable direction and support especially in the early days of the book's proposal. Special thanks to Professor Chang-Yau Hoon, editor-in-chief of Springer's Asia-in-Transition Series for his excellent guidance and encouragement throughout the process of planning and writing the book from its inception and production to final publication.

Springer's commissioning editor Alex Westcott Campbell has been an absolute pleasure to work with as we brought this book to fruition. Her visit to Universiti

Brunei Darussalam in 2023 was a delightful opportunity to meet her in person. She has been extremely generous with her patience.

As a mentor, confidant and role model, Professor Chang-Yau Hoon deserves another mention. This book is dedicated to him and all who strive for research excellence. Finally, we hope that this book will inspire a new generation of thinkers, actors and researchers in the field of Southeast Asian studies.

Gadong, Brunei Darussalam

Hannah Ming Yit Ho
Ying-kit Chan

Contents

Part II Contestations

Part III Cultures

Editors and Contributors

About the Editors

Hannah Ming Yit Ho is Assistant Professor at the Faculty of Arts and Social Sciences at Universiti Brunei Darussalam. She is also a research associate at the Asia Research Institute, National University of Singapore. She was a postdoctoral fellow at King's College London and formerly a research fellow at the University of California, Berkeley. She is currently an associate editor of *Southeast Asia: A Multidisciplinary Journal*. She also co-edited *Engaging Modern Brunei: Research on Language, Literature and Culture* (Springer, 2021).

Ying-kit Chan is Assistant Professor of Chinese Studies at the National University of Singapore. He was formerly a research fellow at the International Institute for Asian Studies, Leiden University. He is now the deputy chief editor of the *International Journal of China Studies* and sits on the editorial boards of *Ming Qing Studies*, *Malaysian Journal of Chinese Studies* and *Southeast Asia: A Multidisciplinary Journal*. His latest book is *Southeast Asia in China: Historical Entanglements and Contemporary Engagements* (with Chang-Yau Hoon).

Contributors

Meredian Alam Faculty of Arts and Social Sciences, Universiti Brunei Darussalam, Gadong, Brunei Darussalam

Sharifah Nurul Huda Alkaff Faculty of Arts and Social Sciences, Universiti Brunei Darussalam, Gadong, Brunei Darussalam

Muhammad Arafat Faculty of Arts and Social Sciences, Universiti Brunei Darussalam, Gadong, Brunei Darussalam

Wafiqah Asnola Faculty of Arts and Social Sciences, Universiti Brunei Darussalam, Gadong, Brunei Darussalam

Ying-kit Chan National University of Singapore, Singapore, Singapore

Zhijian Cui School of Management, University of Science and Technology of China, Hefei, China

Rommel A. Curaming Universiti Brunei Darussalam, Gadong, Brunei Darussalam

Ashrol Rahimy Damit Faculty of Arts and Social Sciences, Universiti Brunei Darussalam, Gadong, Brunei Darussalam

Jely A. Galang University of the Philippines-Diliman, Quezon City, Philippines

Pagon Gatchalee Business School, Chiang Mai University, Chiang Mai, Thailand

Hana Ghani Faculty of Arts and Social Sciences, Universiti Brunei Darussalam, Gadong, Brunei Darussalam

Hannah Ming Yit Ho Faculty of Arts and Social Sciences, Universiti Brunei Darussalam, Gadong, Brunei Darussalam

Muhammad Adil Iqbal NUST Business School, National University of Sciences and Technology, Islamabad, Pakistan

Hema Kiruppalini Asia Research Institute, National University of Singapore, Singapore, Singapore

Chin-Fang Kuo Department of Sport Management, Aletheia University, New Taipei City, Taiwan

Hsien-Wei Kuo Center for General Education, National Tainan Institute of Nursing, Tainan, Taiwan

Ailin Lai Faculty of Arts and Social Sciences, Universiti Brunei Darussalam, Gadong, Brunei Darussalam

Kian Cheng Lee Faculty of Political Science and Public Administration, Chiang Mai University, Chiang Mai, Thailand

Rui Oliveira Lopes Museum of Convergences, Porto, Portugal

Elmy Maswandi Faculty of Arts and Social Sciences, Universiti Brunei Darussalam, Gadong, Brunei Darussalam

Kathrina Mohd Daud Universiti Brunei Darussalam, Bandar Seri Begawan, Brunei Darussalam

Shafinur Nahar Faculty of Arts and Social Sciences, Universiti Brunei Darussalam, Gadong, Brunei Darussalam

Najib Noorashid Faculty of Arts and Social Sciences, Universiti Brunei Darussalam, Gadong, Brunei Darussalam

Morragotwong Phumplab Faculty of Liberal Arts, Thammasat University, Bangkok, Thailand

Aranya Siriphon Faculty of Social Sciences, Chiang Mai University, Chiang Mai, Thailand

Matsumura Toshio Graduate School of Humanities, Osaka University, Osaka, Japan

Badriyah Yusof Faculty of Arts and Social Sciences, Universiti Brunei Darussalam, Gadong, Brunei Darussalam

Zulfadzlee Zulkiflee Faculty of Arts and Social Sciences, Universiti Brunei Darussalam, Gadong, Brunei Darussalam

Abbreviations

ABIM	Muslim Youth Movement of Malaysia
ASEAN	Association of Southeast Asian Nations
BNBHOG	British North Borneo Herald and Official Gazette
BB	Borneo Bulletin
B-B	Business-to-business
BH	Berita Harian
BI	Bahasa Indonesia
BIPA	Bahasa Indonesia for Foreign Speakers
BM	Bahasa Melayu
CBEC	Cross-border e-commerce
CWAA	Chinese Women Chin Woo Athletic Association
CWAF	Chin Woo Athletic Federation
C2C	Consumer-to-consumer
C-C	Customer-to-customer
DBP	Dewan Bahasa dan Pustaka
DRV	Democratic Republic of Vietnam
ESD	Education for Sustainabale Development
EU	European Union
EYD	Ejaan Bahasa Indonesia yang Disempurnakan
FDA	Food and Drug Administration
GMP	Goods Manufacturing Practice
GPSEA	Greenpeace Southeast Asia
HACCP	Hazard Analysis Critical Control Point
IMF	International Monetary Fund
KMT	Kuomintang
MABBIM	Majlis Bahasa Brunei- Indonesia- Malaysia
MASTERA	Majlis Sastera Asia Tenggara
MBIM	Majlis Bahasa Indonesia-Malaysia
MCN	Multichannel network
MIB	Melayu Islam Beraja (Malay Muslim Monarchy)
MHI	Mental health issues

MP	Media Permata
NAP	National Archives of the Philippines
NaPoWriMo	National Poetry Writing Month
NATO	North Atlantic Treaty Organization
NRCT	National Research Council of Thailand
NST	New Straits Times
NSTP	New Straits Times Press
OFW	Overseas Filipinos Works
PRC	People's Republic of China
QNPH	Qasrun Nasfis Publishing House
RVN	Republic of Vietnam
SEAPoWriMo	Southeast Asian Poetry Writing Month
SEASREP	Southeast Asian Studies Regional Exchange Program
SPN21	National Education System for the 21st Century
STEM	Science, Technology, Engineering, and Mathematics
TFB	Transnational Filipino-Bruneians
UNESCO	United Nations Educational, Scientific and Cultural Organization
WHO	World Health Organization

List of Figures

List of Tables

Chapter 1
Interrogating the Transnational Paradigm in Southeast Asia: Communities, Contestations and Cultures

Hannah Ming Yit Ho and Ying-kit Chan

Abstract This introductory chapter argues that Southeast Asia, a region whose boundaries are delimited more by politics than by other markers, is a conceptually meaningful site for interrogating the paradigm of transnationalism. The transnational paradigm is defined as a spatial area model that operates from the outer perimeters to study the way that the transnational has been deployed in various disciplines to examine historical connections and contemporary interactions across borders. This chapter suggests that transnationalism remains relevant for analysing cultural flows predicated on historical connections, but not without modifications or nuanced applications. By deploying the transnational paradigm as an operative framework, the introduction highlights various social networks at work, illuminated by the prevailing typologies of transnationalism. The volume is divided into three broad themes to underscore communities, contestations and cultures in Southeast Asia. The first section examines migrant and diasporic communities, their experiences of assimilating into host societies and their lingering connections with cultural homelands. The second section explores the construction and representation of identities by institutions and cultural others in novels, newspapers and literary writings, considering how such identities are contested by individuals and institutions. The third section evaluates the practice and production of culture, encapsulated in artefacts and literary discourses. Ultimately, the volume discusses adaptations of the transnational paradigm in Southeast Asia, breaking new ground with its transdisciplinary approach and dialogic criticism to interrogate the transnational paradigm.

Keywords Communities · Contestations · Cultures · Identities · Southeast Asia · Transnational paradigm

H. M. Y. Ho (✉)
Universiti Brunei Darussalam, Gadong, Brunei Darussalam
e-mail: hannah.ho@ubd.edu.bn

Y. Chan
National University of Singapore, Singapore, Singapore
e-mail: chscyk@nus.edu.sg

H. M. Y. Ho and Y. Chan (eds.), *Transnational Southeast Asia*, Asia in Transition 29,
https://doi.org/10.1007/978-981-96-3608-2_1

1.1 Introduction

In featuring research from and across different nations in Southeast Asia, this volume seeks to address the following questions: why is Southeast Asia an interesting and conceptually meaningful site to interrogate the transnational paradigm? How does Southeast Asia lend itself to nuanced applications of transnationalism, and what may be the wider cultural and collective implications? Instead of viewing the past and the present as oppositional concepts of time, a temporal continuum is applied to a time–space compression that is fundamental to the workings of the transnational paradigm in the region that we call Southeast Asia. The transnational paradigm is a conceptual tool encompassing various configurations of transnationalism across disciplines, then, becomes relevant for analysing global cultural flows, but not without due consideration of the nuances shaped by spatio-temporal trends. A paradigm shift in transnationalism from historical connections to contemporary connectivity is afforded by increased mobility and accelerated cultural flows, which have given rise to unprecedented economic productivity in the past century (IMF 2000) and digital connectivity in the new millennium.

1.2 Transnationalism: Engaging Past and Present Configurations

As a term, "transnational" may be traced back to the nineteenth century, but it was American writer Randolph Bourne (1886–1918) who popularised it. In his article "Trans-national America," Bourne (1992, pp. 248–264) called for a cosmopolitan rather than melting-pot United States, preferring cultural diversity to forced assimilation amid widespread fear of immigrants among native-born Americans. After a long hiatus, "transnational" reappeared in the 1960s as economists and particularly political scientists attempted to highlight interdependence between industrialised countries (Faist 2010), which coincided with the process of decolonisation in Southeast Asia. Political scientists Robert O. Keohane and Joseph S. Nye, Jr., truly resurrected the term with their edited volume *Transnational Relations and World Politics* (1972), which emphasised the unprecedented mobility, or ease of movement, of labour, capital and commodities that altered the power dynamics of state relations with civil society and market forces (Green 2019, pp. 34–35). In the 1970s and 1980s, "transnational" was conflated with "multinational," referring to nonstate, private corporations with cross-border operations that fostered a "global political economy based on hierarchy and inequality" (Camilleri 1976, p. 112). This brief genealogy of "transnationalism" suggests that the term was critically applied to describe capitalism gone awry; it had been used with considerable negative connotations than it is in the present (Green 2019, p. 36).

With the fall of the Berlin Wall, the implosion of the Soviet Union and the expansion of Europe by law and geographical scope to renewed multicultural identities, "transnationalism" began to assume more positive meanings. Compared to the older term "cosmopolitanism," which came to denote an elite social class, and "internationalism," which has transformed from a political slogan for cross-border solidarity into a byword for diplomatic or interstate relations, "transnationalism" more broadly and easily implied the transcendence of states and institutional borders by goods, people and even ideas (Green 2019, p. 36). Coupled with the poststructuralist turn in Western epistemological trends in the late twentieth century, the scholarly preference for "transnationalism" for analysing individual agency, hybrid identities and cross-border movement, which defied the notion of stasis associated with structures, is unsurprising. Since the 1990s, disciplines other than political science, most notably migration studies, have adopted the term. In 1992, anthropologists Nina Glick Schiller, Linda Basch and Cristina Blanc-Szanton published an edited volume titled *Towards a Transnational Perspective on Migration: Race, Class, Ethnicity, and Nationalism Reconsidered (Schiller et al.* 1992*).* Their volume proposed transnationalism as a new approach to migration studies, superseding the analytic models of either assimilation or ethnicity. Dissatisfied with assimilation that constructed the myth of social mobility and ethnicity, which reduces complex communities to discrete, tightly bounded groups, anthropologists anchored transnationalism in migration studies to tackle the scourge of "methodological nationalism" in the post-world war period (Schiller 2005).

Since then, transnationalism has been doubly conceived as a theoretical stance and an empirical phenomenon (Boccagni 2012). The term has been applied to theorise "sustained and continuous pluri-local transactions crossing state borders" (Faist 2010, p. 1673), and it can also describe everyday practices of mobility and productions of culture within migrant and diasporic communities. Transnationalism has been recurrently configured as a field (Carlson and Schneickert 2021; Portes 2003) and as space (Bruneau 2010; Fuchs 2007; Jackson et al. 2004; Roudometof 2005), be it physical or metaphorical. At present, transnationalism is understood as an idea (Lavers and Hickey 2016), discourse (Mitchell 1997) and optic (Levitt and Khagram 2007). While capacious to a fault, transnationalism has evolved into a *paradigm*, which not only invigorated migration studies but also inspired new conceptions of time and space.

Like anthropologists, sociologists have critically studied the phenomenon of transnationalism. In his book *Transnationalism* (2009), social anthropologist Steven Vertovec coins the term "superdiversity" to describe societies that are culturally complex and diverse for its inclusion of migrants and ethnic minorities who are agents and participants of cross-border processes. Continuous global migration has heightened the phenomenon of transnationalism. Among others, the transnational encompasses a "type of consciousness," "site of political engagement" and "mode of cultural reproduction" (Vertovec 2009, p. 4). This volume responds to these conceptual premises of transnationalism in the tripartite themes of community, contestation and culture. Accentuating an identity continuum, the transnational paradigm embraces a collective means to make sense of historical and contemporary social production and transformation.

The theorisation of the transnational phenomenon has been helpful for considering identity transformations in multi-localities, which shape cultural practices and productions. For one, the transnational offers "an imaginary coherence" (Hall 1990) that binds communities to social networks. This coherence is aligned with the related experience of transnationality, which describes an "awareness" (Faist 2010, p. 1666) of cross-border processes. Furthermore, such activities result in malleable identities that produce cultural hybridity or "new ethnicities" (Hall 1991). Rejecting the established idea of a homogenous ethnicity, new ethnicities reflect the forging of new relations with host societies but also a preservation of contact with cultures of origin. As a term, then, transnationalism in migration studies also denotes "super-ethnicity," (Green 2019, p. 47), even though it is rarely declared as such. Migrants are seldom entirely uprooted nor transplanted, and the transnational paradigm strikes a meaningful compromise between the binary opposition of assimilation and ethnicity of the migration spectrum.

In this volume, communities are discussed in terms of their transnational awareness of their migration patterns within chapters in the first section. The second section on contestations engages critical discussions of the cultural hybridisation of Brunei-Filipinos (Chapter 8) and Thai Yunnanese (Chapter 10), thereby manifesting new ethnicities. Rather than assimilating, these communities embrace their transnational existence to articulate super ethnicity in their language use, food choices and economic activities across state borders that are influenced by cross-border exchanges, communications and interactions. In the third section on cultures, the ambivalent treatment of an imaginary coherence of social networks becomes evident in an analysis of transnational Malay identity that is delineated in historical fiction (Chapter 15) as well as through cultural and historical artefacts of cultural contact (Chapter 16). Included in the final section on cultures, the two chapters highlight cultural production and transformation across traversal intersections of navigation during the European colonial rule of historical Malaya.

It is becoming clearer that the transnational can no longer be tied exclusively to historical processes of international migration, as was the case when the term was first deployed to refer to cohorts of immigrant entering industrialised countries (Roudometof 2005, p. 115). In today's digital age, the media is a vast resource that has diversified the meanings of transnationalism, which entails not only mobility, but also a "life-world" (Roudometof 2005, p. 116) that is increasingly shaped by media communications (see Beck 2000, pp. 96–97). In this volume, media reports on mental illness are critically analysed as a transnational issue that shapes public attitudes and cultural perceptions of mental health stigma (Chapter 17). Social media has also been utilised as a platform to build a transnational awareness of issues such as environmental crises and for promoting social activism across nation-states (Chapter 19).

Crucially, the transnational paradigm is an inclusive term that incorporates nuances, adaptations and modifications of the transnational to examine flows across communities and cultures. Although it is a singular paradigm, it is also flexible to encompass multiple understandings, theorisations and ideas about cross-border processes, resources and ways of life. As a point of departure, Thomas Kuhn's

(2012) notion of a scientific paradigm informs our present strategy of the transnational paradigm that has been conceived as a guide for scholars working on the theoretical and empirical instantiations of transnationalism. As it evolves from a term into a paradigm, transnationalism is conceived as a conceptual tool for research "across numerous disciplines [where] there is a widespread interest in economic, social and political linkages between people, places and institutions crossing nation-state borders and spanning the world" (Vertovec 2009, p. 1). There has been a further critical focus on the "edges" of the transnational process in efforts to understand its "core" (Yeoh et al. 2003). Likewise, this present study's proposition of the transnational paradigm as a spatial area model operates from the outer perimeters to survey and study the way that the transnational has been deployed in its various configurations across disciplines. Functioning inclusively as a tool to make sense of the ongoing formations of communities, identities and cultures, the transnational paradigm is a mode of aggregating the different ways in which these flows have developed across time and space.

1.3 Interrogating the Transnational Paradigm in Southeast Asia

This volume on Southeast Asian studies deploys the transnational paradigm to achieve three objectives. First, it aims to compare both historical and contemporary experiences of migrant communities, particularly those of assimilation and integration, in order to explore individuals' cultural retention, exercise of agency, and the impact of technological transformations on such agency. Historical connections are highlighted to contest both the rupture between past and present for migrants and the rigidity of current ethnic categories in modern nation-states. Second, the volume seeks to demonstrate the construction of identities by groups, individuals, institutions and cultural others, as represented in novels, newspapers and other modes of literature, illuminating how such identities have assumed hybrid, regional or multicultural forms over the past centuries. Third, the volume hopes to clarify the practice and production of culture, encapsulated in tangible artefacts and intangible media forms such as linguistics and works of fiction. Our historical approach emphasises the temporally comparative view that cultural production of the present shares many similarities with that of the past. This volume is thus a corrective to the "newness debate" of the transnational paradigm, arguing that transnationalism in the Southeast Asian context is not a new phenomenon as many anthropologists and sociologists have suggested (Green 2019, pp. 44–47). Rather, ideas, individuals and institutions of the past have always been transnational, even if they have not been named or recognised as such until recent decades. In this respect, the chapters constituting this volume modify the transnational paradigm in various ways to highlight communities, contestations and cultures in their respective case studies of Southeast Asia. The breadth and usefulness of the transnational paradigm are demonstrated across disciplines, including anthropology, art history, history, linguistics, literature, media studies and sociology, which broaden our inquiry beyond migration and diaspora studies. The diverse topics and disciplines discussed in this volume are necessary for reflecting the different and unequal ways in which people of both the past and the present experience transnationality.

The ambiguity surrounding the critical definition of transnationalism has meant that there is neither theoretical concurrence nor a definitive signification of the term, let alone of it as a paradigm (Boccagni 2012, p. 119). That said, scholars acknowledge its multiple indicators, divergent takes and competing perspectives (Curthoys and Lake 2005; Ietto-Gillies 2022). Understanding transnationalism to be a manifestation of globalisation, Steven Vertovec explains that "[transnationalism's] constituent processes and outcomes are multiple and messy" (2009, p. 2). Its manifold configurations across disciplines offer an opportunity to map the transnational onto a paradigm to evaluate both historical and contemporary practices of cross-border mobility and agency, both individual and communal, in the recently delimited region of Southeast Asia.

Given the inextricable entanglements between globalisation and transnationalism (Jay 2011), a historical precedent of cultural flows across geographical terrains has paved the way for present-day interactions that produce communities with "not consistent kinds of social formations and practices" (Vertovec 2009, p. 3). Consequently, new patterns of migration coupled with virtual connectivity have contributed to further identity formation and agency at individual and group levels. In the twenty-first century, sociocultural mobility and cross-border movement have persisted to be facilitated by the relentless process of globalisation. In relational practices of transnationalism, an international intensification and proliferation of social networks are evident:

> Transnationalism describes a condition in which, despite great distances and notwithstanding the presence of international borders (and all the laws, regulations and national narratives they represent), certain kinds of relationships have been globally intensified and now take place paradoxically in a planet-spanning yet common – however virtual – arena of activity.
>
> (Vertovec 2009, p. 3)

As a historical crossroads of culture and human migration, Southeast Asia is a conceptually meaningful site to interrogate the transnational paradigm, updating the uses, benefits and challenges of the paradigm in light of new trends and phenomena such as the rise of an assertive China, the advent of the digital age, the gradual loss of affective ties to cultural homelands and the intensification of racial politics in parts of Southeast Asia. For one, critical studies on Southeast Asia have emerged with scholarly works highlighting the "*political* and *intellectual* dimensions" of Southeast Asia as a "regional concept, [rather than a] geographical expression" (Fifield 1976, p. 151, original emphasis), which underscore its potential as a strategic region.

Today, Southeast Asia is a burgeoning region for generating critical scholarship on cross-border connections, social networks and economic transactions. Cultural flows into Southeast Asia have continued unabated due to migrant and diasporic movements motivated by business, educational and resettlement opportunities. Significantly, these patterns of human and cultural flow give rise to transnational families whose communities straddle cultural, linguistic and religious boundaries that shape their fluid identities. As a term, "Southeast Asia" barely came into being decades ago, largely as a political shorthand and as a result of geopolitical considerations by great powers (Emmerson 1984). Researchers studying non-political subjects began to take notice of Southeast Asia only later, also at the turn of the twentieth century when the transnational paradigm was taking shape. Even then, they tended to study Southeast Asia sub-nationally rather than as a region. The transnational paradigm can help to reconceptualise Southeast Asia as a transnational cultural zone of historical connections and contemporary interactions, as long as we are mindful of its limitations in painting a full picture of the past and in explaining the challenges of political interference and difficult logistics to transnational mobility.

1.4 Key Themes

This volume has not only three objectives but also three themes to organise the seemingly disparate chapters, which nonetheless share the similarities of speaking to its theoretical inquiry and applying the transnational paradigm to Southeast Asian studies. The first theme highlights communities that create or have been created by transnationalism. The transnational paradigm acknowledges the multilayers of communal spaces and both local and global networks without viewing them as mutually exclusive. In his article titled "Transnationalism, Cosmopolitanism and Glocalization," cultural sociologist Victor Roudometof contends that the transnational process entails "several layers ranging from the construction of transnational social spaces to the formation of transnational communities" (2005, p. 113). In addressing such communities, Roudometof explicates the nuances of "glocalization" that gives rise to an internal globalisation, which transforms the everyday lives of citizen-subjects without barring subjects of transnationalism. A consequential awareness of a "pluralization of borders" (Beck 2000, p. 17) with and without nation-states, in which an imaginary coherence forms within local and global spaces, informs the transnational paradigm that embodies a profound sense of multiplicity, malleability and adaptability that are accommodated, as this volume avers, in Southeast Asia.

The second theme is identities. A paradigm for studying transnationalism continues to highlight fluid and flexible identities. Despite their mutability, particular times and specific places shape the (re)constructions and (re)presentations of identities as articulated in various parts of Southeast Asia. In their study on diasporic subjects, Huang et al. accentuate transnationalism's "interconnectedness across borders [to] disavow essentialist and unchanging notions of identity" (2000, p. 392). In fact, transnational identities are informed by a time–space continuum, which feeds into an impression of a multiple existence culminating in a sense of "placelessness" (Yeoh et al. 2003, p. 213). With place as an empty signifier, the concept of multiple identities and multifaceted transnational identities become evident. Adding to their representations in contemporary literary works (Ho 2021), transnational self and identities are explored in chapters within this volume that engage the transnational paradigm to elicit hybrid, regional and multicultural forms of identities via analyses of literary, historical and cultural imaginations.

The third theme of cultural (re)production underscores not only ethnic and religious mobilities but also the use of technology to facilitate the flow of ideas, perspectives and social activism across historically cultural and currently ethno-national borders. While transnational mobility is not a recent process, historical flows in the "ethnoscape" and "ideoscape" (Appadurai 2000) were limited to physical cross-border journeys without the aid of advanced technology that is widespread in the present. In response to the "mobility turn" (Urry 2000), transnational studies adopted a "new mobilities paradigm" (Sheller and Urry 2006) at the dawn of the twenty-first century. This transnational paradigm gave additional impetus to transnational migration and diaspora studies. As a cultural paradigm today, transnationalism continues to shape identities, which affect familial and social structures that become "sites of negotiation" (Ho and Amran 2021) across national borders and digital spaces. Consequently, this volume adopts, modifies and updates the transnational paradigm to make sense of new epistemological ways of interpreting historical transformations and contemporary trends in Southeast Asia that is a revelatory region.

This volume is composed of three sections, each addressing one of its three themes using and interrogating the transnational paradigm. The first section, on the broad theme of "communities," opens with Rommel Curaming's chapter that reexamines the utility of the transnational paradigm in migration studies. It suggests that "anapolitics," a term that encapsulates both analysis and politics, is key to an understanding of transnational migration. Chapter 3, by Jely A. Galang, problematises the transnational migration of Chinese diaspora to Philippines who were neither socially nor institutionally accepted. It interrogates the transnational paradigm by discussing the emergence of an illegal community of Chinese traders who engaged in transnational activities in the nineteenth century. Transnational mobility was problematic to Spanish colonial interests in Southeast Asia, but it also presented opportunities of livelihood for settlers and sojourners alike in the colony.

Chapter 4 features Hsien-Wei Kuo and Chin Fang Kuo's examination of Chinese female immigrants with their various degrees of dependence on transnational networks between China and British colonised Malaya. Their inquiry into the transnational paradigm against a historical timeline illuminates the dichotomy of cultural tradition and feminist progress that shaped the transnational reach of education. The chapter explores the potential inadequacies of the transnational paradigm by emphasising the local achievements of these women within their communities consolidated in athletic associations. Chapter 5, by Pagon Gatchalee, Aranya Siriphon and Zhijian Cui, analyses entrepreneurial strategies of Chinese migrants who engage in transnational activities of purchasing local Thai products on behalf of consumers who live outside of Thailand. This chapter highlights Chinese people's inventive and resourceful ways to navigate Artificial Intelligence (AI) for sustaining their transnational entrepreneurship, especially in light of a post-COVID escalation of e-commerce competition that has altered the landscape of transnational micro businesses. Chapter 6 features Morragotwong Phumplab's research on overseas Vietnamese in Thailand during the Indochina War when this community utilised religious spaces as a form of political mobility to circumvent restrictions that the Thai government imposed upon them. Tapping into archival materials and oral history, this chapter deploys the transnational paradigm to study a resilient community against the political landscape of state and colonial powers.

Chapter 7, by Muhammad Arafat and Ai Lin Mei Sien, examines a culturally and linguistically hybrid community of Filipino-Bruneians, who straddle multicultural influences in their food and language choices. The community's articulation of their subjectivities reflects their multiple networks, cross-border movements and intercultural connections that are examined using the transnational paradigm, which also illuminates the challenges to establish a sense of belonging. Chapter 8 highlights Ashrol Damit and Zulfadzlee's delineation of linguistic adaptive strategies among Brunei's Belait people, whose domestic and cross-state migrations exhibit their transnational awareness and existence. The transnational paradigm helps to probe the way that the Belait people have evolved to arrive at various sites to take up local roots and regional routes.

The second section, on "contestations," examines the construction, representation and contestation of identities. Chapter 9, by Hema Kiruppalini, deals with the historical memory of Singapore's female bondservants in fictional works to investigate their transnational stories in colonial Singapore. In reading their narratives of life experiences against historical discourses from the British and Singapore's National Archives, this chapter queries the transnational processes that enabled their domestic servitude and cultural cosmopolitanism. Chapter 10 presents Lee Kian Cheng's study of Thai Yunnanese entrepreneurs through their everyday life practices and performances for a remaking of their identities. The transnational paradigm is engaged as a tool to forge not only a transnational but also translocal Chinese (Taiwanese) identity while holding Thai citizenship. Their multiple identities result in positive outcomes that challenge assimilationist demands while illuminating the pitfalls of dualistic perspectives of Thai Yunnanese identity, which is constructed in interstitial spaces of transnationalism.

Chapter 11, by Matsumura Toshio, examines the impact of Indonesia's post-independence nationality law on Chinese settlers in Indonesia who previously held dual citizenship as Chinese nationals and Dutch East Indies subjects. Studying Chinese-language newspapers, it examines their struggles in navigating this new law, especially when faced with a choice of having to concede their transnational status to retain their identities as Indonesians. Here, the paradigm of transnationalism is used to demonstrate threats to their transnational existence, which are undermined by legislative powers of the independent Indonesian nation-state. Returning to the discipline of literature, the next two chapters explore the challenges of the transnational paradigm in mapping permutations of Muslim masculinity and in its deployment as a literary genre. Chapter 12, by Shafinur Nahar and Hannah M. Y. Ho, foregrounds complexities of Muslim masculinities in contemporary novels produced by Malaysian and

Bangladeshi diasporic authors. In interrogating Muslim male identities, the transnational paradigm is used to underscore the similarities and stark differences between Southeast and South Asian cultural instantiations of Muslim identities, which are shaped through negotiations between national prescriptions and progressive global mobility. Chapter 13, by Kathrina Mohd Daud, addresses the reception of Anglophone Bruneian novels in discussing its target international readership. It argues for a transnational genre of literature in which contemporary literary works in English by Bruneian authors can be situated. Chapter 14, by Badriyah Yusof, Elmy Maswandi and Wafiqah Asnola, explores language planning across Malay Muslim dominant nations of Malaysia, Indonesia and Brunei Darussalam to illuminate structures of an ongoing transnational linguistics association.

The third section, on "cultures," investigates the productions and practices of culture encapsulated in literature, artefacts and the media, whether it be traditional or new media. Chapter 15, by Hannah M. Y. Ho, examines literary representations of Malayness that is studied as a cultural product of centuries of ethnogenesis. Using the transnational paradigm, it provides insights into the historical and contemporary contestations of Malayness across Malaysia, Singapore and Philippines. Chapter 16 features Rui Oliviera Lopes' research on cultural transfers in colonised Malacca to illuminate the historical integration between Portuguese and Malay identities. Chapter 17, by Sharifah Huda Alkaff, presents a comparative discussion of mental health representations in local media in Bruneian and Malaysian newspapers. The transnational issue of mental illness and social stigma against those living with mental illness are analysed through media cultures that promote and shape attitudes towards a local and global public health crisis.

Chapter 18, by Hana Ghani, is attentive to social media as an innovative platform for poetic performances known as "instapoetry." It discusses online strategies of transnationalism to advance literary culture, thus changing the ecological literary landscape to accommodate the digital era of production and consumption. Chapter 19, by Meredian Alam and Muhammad Adil Iqbal, examines social activism on online platforms that help to mobilise transnational social networks for a collective international cause. The digital scene enriches the transnational paradigm to shed light on the workings of a new cultural platform adopted by social organisations. Finally, Chapter 20, by Najib Noorashid, reflects on past and current research practices in linguistics that continue to be informed by transnational cultural and scholarly networks within and beyond the Malay world in Southeast Asia.

1.5 Final Remarks

The introduction of this volume has argued that the transnational paradigm remains useful for studying Southeast Asia, despite its limitations. Transnationalism is neither a new empirical process nor a novel theoretical concept. It has been configured and reconfigured as an idea, discourse and space. It has also manifested itself in various typologies of families, diaspora, associations and organisations. To subsume different approaches and its multiple manifestations, the transnational paradigm has been conceived as a means to address the multiplicity and fluidity of cultures and identities that are produced through past and present cross-border movements, economic transactions and cultural connectivity. Given today's digital globalisation, the transnational paradigm remains a productive instrument to analyse the continuous changes and developments of contemporary society. Both historical connections and contemporary connectivity are pertinent as sites of investigation through the use of the transnational paradigm, which can illuminate the adaptations, modifications and nuances of transnationalism across time and space in local and global realms of social contact, cultural interaction and rapid internet communication.

Central to this present study's argument is that Southeast Asia is a conceptually meaningful site for interrogating the transnational paradigm. The case studies presented in this volume speak to the theoretical inquiry of the transnational paradigm and develop the case for Southeast Asia as a germane region to demonstrate the transnational workings of communities, identities and cultures. It is evident that Southeast Asia contains rich empirical data and offers new theoretical discussions of multicultural, hybrid and cosmopolitan identities across various disciplines to challenge binary oppositional processes of assimilation and othering that were previously produced and reproduced through historical narratives of transnationalism. The volume's three objectives are aligned with three themes of communities, contestations and cultures to illustrate processes of agency, transformation and hybridisation across nation-states. By deploying and interrogating the transnational paradigm, the chapters in this volume situate Southeast Asia as a feasible and productive region marked not only by historical colonisation but also by cultural contact zones across newly formed independent nation-states. It is hoped that this volume will stimulate and pave the way for further research on Southeast Asia's production of fluid identities beyond sociocultural and political fault lines that have historically demarcated transnational studies, to which the transnational paradigm offers a corrective.

References

Appadurai A (2000/1990) Disjucture and difference in the global cultural economy. In: Lechner FJ, Boli J (eds) The globalization reader. Blackwell Publishers, Oxford, pp 94–103

Beck U (2000) The cosmopolitan perspective: sociology of the second age of modernity. British Journal of Sociology 51(1):79–105

Boccagni P (2012) Rethinking transnational studies: transnational ties and the transnationalism of everyday life. European Journal of Social Theory 15(1):117–132. https://doi.org/10.1177/1368431011423600

Bourne RS (1992) The radical will: selected writings, 1911–1918, edited by Hansen O. University of California Press, Berkeley

Bruneau M (2010) Diasporas, transnational spaces and communities. In: Bauböck R, Faist T (eds) Diaspora and transnationalism: concepts, theories and methods. Amsterdam University Press, pp 35–50

Camilleri JA (1976) Civilization in crisis: human prospects in a changing world. Cambridge University Press, New York

Carlson S, Schneickert C (2021) Habitus in the context of transnationalization: from 'transnational habitus' to a configuration of dispositions and fields. The Sociological Review 69(5):1124–1140
Curthoys A, Lake M (eds) (2005) Connected worlds: history in transnational perspective. Australian National University Press, Canberra
Emmerson DK (1984) "Southeast Asia": what's in a name? Journal of Southeast Asian Studies 15(1):1–21
Faist T (2010) Towards transnational studies: world theories, transnationalisation and changing institutions. Journal of Ethnic and Migration Studies 36(10):1665–1687. https://doi.org/10.1080/1369183X.2010.489365
Fifield RH (1976) Southeast Asian studies: origins, development, future. Journal of Southeast Asian Studies 7(2):151–161
Fuchs C (2007) Transnational space and the 'network society.' Journal of the Academy of Social Sciences 2(1):49–78. https://doi.org/10.1080/17450140601101218
Green NL (2019) The limits of transnationalism. Chicago University Press, Chicago
Hall S (1990) Cultural identity and diaspora. In: Rutherford J (ed) Identity: community, culture, difference. Lawrence and Wishart, London, pp 222–237
Hall S (1991) Old and new identities, old and new ethnicities. In: King AD (ed) Culture, globalization and the world-system. Macmillan, Houndmills, pp 41–68
Ho HMY (2021) Chinese Bruneian identity: negotiating individual, familial and transnational selves in anglophone Bruneian literature. The Wenshan Review of Literature and Culture 14(2):1–34. https://doi.org/10.30395/WSR.202106_14(2).0001
Ho HMY, Amran RH (2021) Constructs of the modern home: negotiating identity in anglophone Bruneian literature. In: Ho HMY, Deterding D (eds) Engaging modern Brunei: research on language, literature and culture. Springer, Singapore, pp 163–175
Huang S, Teo P, Yeoh BSA (2000) Diasporic subjects and identity negotiations: women in and from Asia. Women's Studies International Forum 23(4):391–398
Ietto-Gillies G (2022) Transnationality in the XXI century. concept and indicators. Critical Perspectives on International Business 18(3):338–361
International Monetary Fund (IMF) (2000) The world economy in the twentieth century: striking developments and policy lessons. In: World economic outlook: a survey by the staff of the International Monetary Fund, pp 149–181
Jackson P, Crang P, Dwyer C (2004) Transnational spaces. Routledge, London and New York
Jay P (2011) Global matters: the transnational turn in literary studies. Cornell University Press, Cornell
Keohane RO, Nye JS (eds) (1972) Transnational relations and world politics. Harvard University Press, Cambridge
Kuhn TS (2012) 1962) The structure of scientific revolutions. The University of Chicago Press, London
Lavers T, Hickey S (2016) Conceptualising the politics of social protection expansion in low- income countries: the intersection of transnational ideas and domestic politics. International Journla of Social Welfare 25(4):388–398
Levitt P, Khagram S (eds) (2007) The transnational studies reader: intersections and innovations. Routledge Press, New York
Mitchell K (1997) Transnational discourse: bringing geography back in. Antipode 29(2):101–114
Portes A (2003) Conclusion: theoretical convergencies and empirical evidence in the study of immigrant transnationalism. International Migration Review 37(3):874–892
Roudometof V (2005) Transnationalism, cosmopolitanism and glocalization. Current Sociology 53(1):113–135
Schiller NG (2005) Transnational social fields and imperialism: bringing a theory of power to transnational studies. Anthropological Theory 5(4):439–461. https://doi.org/10.1177/1463499605059231
Schiller NG, Basch L, Blanc-Szanton C (1992) Towards a transnational perspective on migration: race, class, ethnicity, and nationalism reconsidered. New York Academy of Sciences, New York

Sheller M, Urry J (2006) The new mobilities paradigm. Environment and Planning 38(2):207–226
Urry J (2000) Sociology beyond societies: mobilities for the twenty-first century. Psychology Press, New York
Vertovec S (2009) Transnationalism. Routledge, London and New York
Yeoh BSA, Willis KD, Fakhri SMAK (2003) Introduction: transnationalism and its edges. Ethnic and Racial Studies 26(2):207–217. https://doi.org/10.1080/0141987032000054394

Hannah Ming Yit Ho is Assistant Professor at the Faculty of Arts and Social Sciences at Universiti Brunei Darussalam. She is also a research associate at the Asia Research Institute, National University of Singapore. She was a postdoctoral fellow at King's College London and formerly a research fellow at the University of California, Berkeley. She is currently an associate editor of *Southeast Asia: A Multidisciplinary Journal*. She also co-edited *Engaging Modern Brunei: Research on Language, Literature and Culture* (Springer, 2021).

Ying-kit Chan is Assistant Professor of Chinese Studies at the National University of Singapore. He was formerly a research fellow at the International Institute for Asian Studies, Leiden University. He is now the deputy chief editor of the *International Journal of China Studies* and sits on the editorial boards of *Ming Qing Studies*, *Malaysian Journal of Chinese Studies* and *Southeast Asia: A Multidisciplinary Journal*. His latest book is *Southeast Asia in China: Historical Entanglements and Contemporary Engagements* (with Chang-Yau Hoon).

Part I
Communities

Chapter 2
Transnationalism: 'Anapolitics' of Conceptualising the Migrant Experience

Rommel A. Curaming

Abstract This chapter examines the conceptualisation of transnationalism in migration studies, focusing on (1) meta-analytic frames, (2) underlying political implications, and (3) the interplay of analysis and politics, termed 'anapolitics.' It shows how anapolitics operates using a critique of the conceptualisation by Portes et al. (1999) of transnationalism. It argues that scholars' lack of awareness of anapolitics could lead to analytic confusion and make them unwitting accomplices in marginalising people they expressly wish to empower. It concludes with the ethical implications for scholarly practice.

Keywords Transnationalism · Migration · 'Anapolitics' · Ethics of scholarship

2.1 Introduction

A shift in the efforts to analyse the migrant experience has long been observed (Levitt and Jaworsky 2007). The once prevalent assimilationist view of a linear movement from the sending to the receiving country, with the assumed absorptive capacity of the host culture, has given way to 'agency-oriented theoretical perspectives' (Espiritu 2003, p. 2). These perspectives recognise the migrants' more significant role in shaping their identities and experiences. Concepts such as transnationalism and hybridity have gained popularity, displacing assimilationist terminologies. Salman Rushdie (Rushdie 1992, p. 124) noted this shift by referring to the creation of a new type of human being, those rooted in ideas rather than places.

The shift can be attributed to several factors. The classical assimilationist paradigm, drawn from early European migrants' experiences in the United States, assumed a seamless integration by the third generation (Portes and Zhou 1993). However, the post-1960s migrants' experiences differed, necessitating new concepts and approaches (Basch et al. 1994; Schiller et al. 1992). Faist (2000), for instance,

R. A. Curaming (✉)
Universiti Brunei Darussalam, Gadong, Brunei Darussalam
e-mail: rommel.curaming@ubd.edu.bn

H. M. Y. Ho and Y. Chan (eds.), *Transnational Southeast Asia*, Asia in Transition 29,
https://doi.org/10.1007/978-981-96-3608-2_2

rejects the 'container' notion of culture, advocating for a view of culture as fluid and characterised by multiple, in-between practices typical of transnational lives. There has also been increasing awareness of "methodological nationalism," which refers to the tendency among scholars to take the nation-state as a natural starting or end-point of analysis. Assuming the congruence between the state, nation and the society, this approach primes up scholars to miss or misunderstand the processes or activities that are inherently cross-border or in-between, such as migration and globalisation (Koos and Keulman 2019; Wimmer and Schiller 2003).

Politically, assimilationist policies supported and were pushed by the dominant groups in host societies, justifying the marginalisation of migrants (Rumbaut 1997; Rumbaut et al. 1999). While conceptual issues and politics are customarily taken as distinct and opposites, conceptual adjustments may reflect and flag political implications. Conversely, the shift in political interests may require or accompany conceptual adjustments. I call it here anapolitics, entwined analysis and politics. The term is new but the idea is not. It has been highlighted in various ways and through different concepts such as power/knowledge and discourse (Foucault 1980), symbolic power; cultural capital (Bourdieu 1977), politics of aesthetics (Rancière 2006), conceptual history (Koselleck 2002), hegemony (Gramsci 1992), and situated knowledge (Haraway 1988), among others. While the idea has been regarded as illuminating by many scholars, particularly in critical or progressive tradition, many others have ignored it or appropriated it only to a limited extent. Perhaps due to its radical implications, they refuse to push it to its logical conclusion—a situation I call analytic constipation. Discussed elsewhere (Curaming 2020), it refers to the tendency for scholars to hold back analysis within permissible boundaries to avoid controversial moral, political or scholarly implications.

The entwining of the analytical and political aspects in conceptualising migration is often overlooked, leading to analytical confusion and inadvertently supporting politically pernicious projects. Trained to be objective or impartial and conditioned to believe that it takes only rigorous scholarship to overcome politics, scholars are often unmindful or oblivious of the political underpinnings as well as the consequences of their analytic choices. A lack of awareness could lead well-meaning scholars to become unwitting accomplices in disempowering people they seek to help. This chapter seeks to contribute to addressing this problem.

This chapter maps out definitions of transnationalism as a step towards addressing the conceptual ambiguities that hamper the analysis of the migrant experience. It aims to uncover the political underpinnings or implications of particular definitions or conceptualisations. To demonstrate anapolitics, I will offer a critique of a popularly received re-definition of transnationalism and I will utilise this critique, alongside my suggested alternative conceptualisation of transnationalism, to demonstrate the anapolitical consequences of conceptual readjustment. I refer to the interviews I did in 2008 with 15 first-generation Filipino migrants and 8 second-generation Filipino-Australians in Melbourne to illustrate relevant points. These interviews were part of the participant-observation I undertook for four months in early-mid 2008 to explore the various facets of transnational identity among Filipinos in Australia. The chapter concludes with notes on the ethical implications for scholarly practice.

2.2 Fluidity and Disciplining of Definitions

A comprehensive review of the literature reveals that scholars use the term transnationalism to refer to many interrelated but different things: process, experience, practices or activities, relationship, linkage or act of linking, phenomenon, outcome or driving force of transnational acts, ideology or sentiment, state of subjection, and platform for resistance. For instance, Schiller et al. (1992, p. 1) have defined transnationalism as a "process by which immigrants build social fields that link together their country of origin and their country of settlement." Similarly, Espiritu (2003, p. 70) has defined it as "the process by which immigrant groups forge and sustain strong sentimental and material links with their countries of origin." Levitt (2001, p. 202), on the other hand, defines it as "the cultural, economic, and political linking of people and institutions [that] deemphasises the role of geography in the formation of identity and collectivity and creates new possibilities for membership across boundaries." For her part, Lee (2004, p. 236) uses the term to refer to the 'the multiple connections maintained with the country of origin after migration.' When Kennedy and Roudometof (2002, p. 13) declared that 'transnationalism is necessary, unavoidable and advantageous' and that '(i)t has become a built-in feature of the cultural, social, political and economic lives of many people everywhere,' they were talking of yet another way of seeing transnationalism: as a favourable way of life.

The proliferation of various typologies reflects efforts to establish order, reining in the fluidity of definitions. They facilitate analysis but also added another layer of complexity and challenges. These typologies include transnationalism 'from below' (Smith and Guarnizo 1998); 'symbolic' (Espiritu 2003) and 'emotional' (Wolf 2002) transnationalism; 'comprehensive' and 'selective' transnationalism (Levitt et al. 2003); 'ethnic' and 'civic' transnationalism (Wheeler 2004); 'narrow' and 'broad' (Itzigsohn et al. 1999) or 'core' and 'expanded' transnationalism (Guarnizo 1997).

Fortunately, the problem did not go unnoticed. For instance, Pries (2005, p. 168) has warned that transnationalism is in danger of becoming the cliché of our time for failing to "provide a tangible, additional power of explanation." For their part, Portes et al. (1999, p. 219) observed that "if all or most things that immigrants do are defined as 'transnationalism,' then none is because the term becomes synonymous with the total set of experiences of this population." To address this problem Portes et al. (1999) offered a parsimonious approach to conceptualisation by defining transnationalism as "occupations and activities that require regular and sustained social contacts over time across national borders for their implementation" (p. 129). They identified three requisites for activities to be included under this rubric: regularity or resilience over time, a significant number of actors involved, and 'newness'; previous concepts do not cover the activity or behaviour (Portes et al. 1999). What drove their approach was, among other factors, the need to justify the use of transnationalism as a new concept that captures a "new mode of social and economic adaptation by simultaneously living and participating in two or more communities in different nation-states" (Portes et al. 2017, p. 1486).

The conceptual readjustment done by Portes et al. helps in no small measure in making the field "a clearly defined and measurable object of research" (1999, p. 218). It enabled them and others to produce a good amount of quantifiable, verifiable, and perhaps replicable data that are necessary as building blocks of theory building. Dahinden (2017) claimed that this article marked a milestone in transnationalism studies and that many scholars have followed their lead. In so narrowing the definition to easily observable and measurable activities, however, they paved the way for behaviour-centric emphasis in the study of transnationalism. This development results in a number of both potential and actual problems, the elaboration of which I will utilise below as a springboard for demonstrating the apparent organic relationship between the analytical and the political.

A caveat is in order. By singling out Portes et al.'s definition as a focal point of my analysis, I do not mean to imply that this is the only important or most influential conceptualisation in the field. One cannot ignore the sizeable corpus of scholarship on migrant transnationalism that emphasises the sentimental and symbolic aspects and the empowering potentials of the transnational experience (Levitt and Jaworsky 2007; Munro 2015; Tan et al. 2018). The main reason for the focus on Portes's definition lies in the air of scientificity or political innocence that it carries with the use of behaviourally operationalised concepts and statistical tools. If a non-political act, such as statistical- or behaviour-oriented analysis, can be shown to be political or politicised, we would be better positioned to uncover and deal with the hidden transcript of politics that suffuses the analytic act.

2.3 Consequences and Implications

First, the behaviour-oriented approach that Portes et al. (1999) adopt poses the danger of reducing the complexity of transnationalism to what is observable and measurable. Such danger manifests when Mahler (1999, p. 80), for example, misconstrued the migrants being settled or their relative lack of mobility as 'not maintaining transnational links.' Portes (2003) is more categorical: taking stock of the research in the field, he concluded that one of the identifiable theoretical and empirical convergences indicates that there are not many migrants who qualify as transnationals. Guarnizo et al. (2003) suggest, based on their survey of Mexicans, Dominicans and El Salvadorans, that they probably account for only about 10–15% of the immigrant population. Transnationalism, Portes declared, is 'not the normative or dominant mode of adaptation of these immigrant groups' (2003, p. 884).

That there are only a tiny number of transnationals seems to have become the standard view, as evident in Levitt and Jaworksky's (2007) comprehensive review of the state-of-the-art in transnational migration studies. In addition, Portes (2003) claims that most of these transnationals, far from what was widely thought, were not the most exploited or marginalised but rather are the most educated, well-connected, and established in the host country. This is the reason, he warns, for the need to take a more sober view towards the empowering potential of transnationalism.

I find these assertions both analytically and politically questionable. By narrowly defining transnationalism to include only observable, recurrent, and new activities, Portes and his followers privilege the conceptual-analytic over the actual lived experiences of migrants. They identify decisively but rather prematurely who are the transnationals. While it is true that they based their reconceptualisation on various existing studies, their database appears not broad enough to represent the spectrum of possibilities. Rushing to theorise aiming for analytic novelty rather than the genuine understanding of the phenomenon, their narrow definition risks excluding poorer, less settled, and less educated segments of the migrant population simply because they are not seen participating regularly in transnational activities. Transnational experience, like any lived experience, is rich, multi-layered, multi-dimensional, and shifting through time. Allowing rigid definitions to decide the analytic pathway is, in my view, a misstep: concepts should capture, not precede, reality. One of the primary goals of scholarship in this field should be to explore and map out the diverse modalities and intensities of such experiences. The narrow, behaviour-centric definition proposed by Portes et al. does not adequately serve this purpose. Instead, the broader definition offered by, say, Basch et al. (1994) in *Nations Unbound*—a work included among those implicitly critiqued by Portes et al.—may be more helpful in capturing the wider spectrum of transnational experiences.

Goldring (1998) argues that the transnational field provides the transmigrants with a particular context within which upward-status formation may be generated or cultivated. She points to the web of meaning generated within a community as the enabling mechanism for upward-status formation. This point is important as a counterpoint to the narrow and behaviour-centric criteria set by Portes et al. My interviews with Filipino first-generation migrants in Melbourne, Australia, confirm Goldring's assertion. One typical pattern in these interviews was that despite having long been settled, my informants claim that their daily life for the past 10–20 years in Australia—aspects that relate to human relations, family life, the meaning of work, the definition of success, their own and their children's future—cannot but operate within or be infused by their understanding and appropriations of what they call 'Filipino values' as adapted in the Australian setting. They have adjusted reasonably well, so they claimed, but Filipino-ness is something that 'cannot be shaken off from (their) system.' Several of them admit, for instance, that despite doing jobs that are not seen as 'lowly' or 'demeaning' in Australia (cleaning, clerical work, hospitality, construction, and other blue-collar jobs) but are seen as menial in the Philippines, there is something in their being Filipinos that disable them from thoroughly cleansing the jobs of their stigma. Owing to the traditional preference for white-collar jobs in the Philippines—all first-generation informants are college graduates with a few of them even having a Master's degree—those who do menial or non-professional jobs feel somewhat embarrassed whenever they were asked by fellow Filipinos what jobs they do.

On the other hand, what enables them to endure these jobs is not just their de-stigmatised status in Australia or the fact that they need such jobs to survive, but also the thoughts that it allows them to have a life much more comfortable than whatever they could ever have had they stayed in the Philippines. Their memories of, or ties with, the Philippines, in other words, provide them with a favourable comparative platform necessary to feel good about the otherwise tricky, sometimes frustrating life in Australia. By locating themselves favourably vis-a-vis their families or compatriots back home, they can neutralise (at least psychologically), endure, or accept their marginalised position in the host society. In addition, their being a citizen of an advanced industrial society provides them the economic, social, and cultural capital that sets them 'above' their compatriots back home.

In short regardless of their socio-economic standing in Australia and the extent of their participation in transnational activities, the very fabric of their lifeworld, simultaneously or in-between the host and the homeland, operates within the parameters of transnationalism. Contrary to the claims by Portes and Guarnizo, therefore, there may be far more migrants (at least the first generation) who are transnational by necessity, if not also by choice. The transnational ties, especially their symbolic or emotional aspects, are maintained and even nurtured as a necessary adaptive tool or a compensatory mechanism to make the migrant experience more bearable or enjoyable. It provides the mechanism by which migrants satisfy the psychological need to neutralise or justify the sacrifices or losses (perceived or real) that they have incurred in deciding to migrate.

From a political standpoint, Portes et al.'s behaviour-centric definition poses the danger of skewing scholarly attention towards the activities of small, more educated, and visible groups of migrants. One possible consequence of this situation is that it may deprive others, those who need it, the benefits that emanates from the potential for empowerment via the validation of transnational identity that scholarship on transnationalism may support. Of course, regardless of what scholars do, the migrants shall try to make the most of their rich and varied transnational lived experience. Scholarship, to say the least, is not essential in the life of the transmigrants and other ordinary people. One thing, however, that well-meaning scholarship can do is to affirm the ontological primacy of their lived experience as the basis for epistemological and political empowerment. To put it simply: to live is to know, and to be recognised as the knower or agent, as opposed to the 'merely' known or represented, may be empowering. By holding the lived experience hostage to definitional tyranny, as what Portes et al. do, it can abort the emancipatory or empowering potential of scholarship on transnationalism. If that happens, it foregrounds the question of whether the scholars are the only ones who benefit from such a scholarship, not the people they claim to represent.

The importance of transnational identity becomes clear against the backdrop of the hegemonic grip of nation-normativity or methodological nationalism in migration discourses in the Philippines, among other countries. Often chided by certain groups for lack of patriotism or nationalism, or for abandoning rather than helping the country become developed, overseas Filipinos (especially the highly educated and highly skilled ones) are pressured to assume the burden of guilt for "abandoning" the country. This sentiment was particularly strong in the 1980s and 1990s, fueled by the economic crisis and the intensifying brain-drain. While the appreciation for the economic contributions of overseas Filipino works (OFWs) grew in recent decades, as evident, say, in the widening acceptance of the "Bagong Bayani" (New Heroes) epithet, the tensions remain between the "nationalists" who stayed and the "traitor" who left recur at the surface from time to time. The identity afforded by being transnational lessens or supplants the marginalising effect of being labelled as "less than a Filipino," as evocatively captured by one of my interviewees (Encinas-Franco 2015; Manalo 2022).

Another problem with Portes et al.'s behaviour-centric conceptualisation is that it poses the danger of confusing transnational acts or practices with transnationalism. This is like mistaking the trees for the forest. The distinction between the two needs to be upheld, as actions do not necessarily reflect intention or motivation. Actions and meaning may or may not coincide. It appears, moreover, that transnationalism, just by virtue of being an -ism, entails a considerable amount of emotion or sentiments as well as subjective meaning-formation, not just actions.

The distinction made by (Levitt and Schiller 2004) between what they call 'ways of being' and 'ways of belonging' is instructive. According to them, a transmigrant who actively participates in transnational activities (such as sending money and travelling between the homeland and new land) may not necessarily identify with the politics or identity associated with the social field in which such activities take place. In other words, they do not have a sense of belonging to the homeland, and thus, their transnational attachment is, at best, shallow. On the other hand, an individual who may not be active in visibly identifiable transnational activities (and thus do not exhibit 'ways of being') but nonetheless nurtures a sense of affinity to the homeland—a kind of symbolic or sentimental attachment to the homeland—is said to demonstrate 'ways of belonging.' Twelve of my first-generation interviewees seem to qualify under the first category. They are involved in Filipino migrant communities in Melbourne but do not care as much about politics in the Philippines, as well as identity politics associated with being Filipino migrants in Australia. On the other hand, two of my informants are under the second category; they are politically engaged in varying extents and different ways, either in Philippine politics or in the migrant community. The importance of the idea of 'ways of belonging,' as Levitt and Schiller claim, lies in its being a better predictor of the resilience of transnationalism in the future.

The behaviour-centric conceptualisation also runs the risk of conflating the process of transnationalisation with transnationalism. The conceptual distinction between the process, on the one hand, and the principle that underlies or governs it, on the other needs to be maintained. In a rather complex manner, Pries (2005, p. 179) defines transnationalisation as "the strengthening or emergence of pluri-local societal configuration that span above and between the traditional container spaces, and in which the concentric circles or local, micro-regional, national, macro-regional and global phenomena are played out." Simply put, it refers to the processes of, and openness to, the forming and strengthening of transnational ties and spaces beyond what are commonly known. Such a definition is practically the same as the earliest and most popular definition of transnationalism offered by Schiller et al. (1992) cited above. Pries' definition is particularly significant in not confusing the process (transnationalisation) with the underlying principle (transnationalism).

In the Filipino-dominated church congregation where I participant-observed in Melbourne, one of the regular activities was sending volunteers to do community-cum-mission work in underprivileged areas in the Philippines, among other countries. In one of the Sundays I attended the service, two second-generation Filipino-Australians in their twenties, shared their recent experience doing such work. The testimonial of one of them struck me as relevant to be noted here. The community immersion in the Philippines afforded him deeper insights into himself, his spirituality, and the homeland of his parents, which for so long was nebulous to him. Intrigued by what he said, I talked to him further, and he said that seeing how tough life was in the Philippine community he stayed in, he came to understand first-hand the overriding importance his parents and other Filipino friends and relatives put on money-related things. Socio-religious activities like this were carried out outside the frame of transnationalism, but it contributed to the transnationalisation of second-generation Filipino-Australians, exemplifying the importance of separating the principle from the process.

The overly behaviour-centric conceptualisation also poses the danger of losing what seems to me an important distinction between the transnational and the transnationalist, on the one hand, and between them and the principle of transnationalism, on the other. A transnational and transnationalist may both participate in a set of transnational activities, and because of this they both exemplify transnationalism, if we go by Portes et al.'s definition. Unless we factor in the subjective meaning that they attach to or create or recreate through their participation in these activities, however, we shall miss the difference and lose sight of its significance. The centrality of meaning and practice is aptly emphasised by Guarnizo and Smith (1998) when they declared that while the global economic structure and the technological revolution exert enormous influence in defining the broader context for the formation of transnational social fields, "it is the everyday practices of migrants that provide a *structure of meaning* to the acts of crossing borders, living in bi-national households, and reproducing transnational social relations" (emphasis supplied, pp. 18–19).

What is the difference? A transnationalist, driven by conviction, purposively acts with conscious short- and long-term objectives: to derive benefits (material, psychological, political, social) clearly identified or not, from their being transnational. A transnational, on the other hand, does not have a similarly intense attachment or sense of purpose. They may have particular aims that run in the short term, but their acts are not driven by the self-conscious desire to achieve some goals by shaping aspects of their life following a particular principle or set of principles. Most Filipino-Australians I encountered in Melbourne in 2008 may be classified as transnational; a significantly smaller proportion were transnationalist. The distinction made by Levitt and Schiller (2004) between 'ways of being' and 'ways of belonging' as explained above is relevant here. Transnationals may be those who exhibit 'ways of being' while transnationalists are those who evince 'ways of belonging.'

The transnationals and the transnationalists, being actors, should not be confused with transnationalism. What they are and what they do may or may not manifest the underlying principle of transnationalism but they are not transnationalism per se.

2.4 Transnationalism as a Principle

The previous discussion sets the stage for clarifying what transnationalism is not. Transnationalism is not an act, nor is it the people involved, a relationship, or a process in itself. Instead, it is the principle or set of principles that underlie or govern all these aspects. Therefore, I define transnationalism as the principles guiding scholars' and transnational migrants' efforts to organise, make sense of, negotiate with, and benefit (as well as lose) from the transnational experience or being in the homeland and host country at the same time. It is important to note two things about this definition: first, the principles guiding scholars and transnational migrants may or may not align, and second, these principles operate with or without the migrants' awareness and regardless of their conscious intentions to benefit from the experience.

Defining transnationalism as a principle aligns with appropriate semantics, as the suffix "-ism" indicates ideational properties such as organising thoughts, principles, or ideology. This definition is also strategic because it helps identify a primary source of conceptual ambiguities in multiple definitions of transnationalism—the conflation of underlying principles with their manifestations. By separating the principle of transnationalism from its various behavioural, affective, and cognitive manifestations, it becomes clear that these manifestations—processes, activities, people, relationships, social spaces—should not be confused with the principle. Consequently, the object of empirical research is not mainly transnationalism itself, but its various manifestations: transnationalisation, transnational practices, transnational social spaces, transnational ideologies, and transnational identities. Empirical research on transnationalism per se would be limited to the underlying logic in transnationals' thoughts and behaviour, as well as those that can be extrapolated from scholars' (of transnational studies) analytic practices.

2.4.1 'Principle' Clarified

In defining transnationalism as a principle, I am aware of the danger of putting it at the mercy of idealism. One may complain that it is too cognitive or too conceptual such that it runs the risk of reducing transnationalism to the realm of ideas. One may ask, what happens to the behaviour, lived experience, feelings, human emotion, and dynamic processes that are central to transnationalism? Such apprehension is understandable. Admittedly, idealism is the equally pernicious opposite of behaviourism. As an analytic experiment, this is strategic to demonstrate the analytic and political consequences of a conceptual shift.

The term 'principle' is used here in two distinct but related ways. Firstly, in the experiential sense, it refers to the organising thread that underlies or runs through the various practices, processes, emotions, and experiences that manifest transnationalism. In the primordial state, let us imagine, it develops as an epiphenomenon of the individual and collective thoughts, feelings, and behaviour and their underlying causes—psychological, biological, social, and political. This is not something that is imposed by cognitive ordering, as that emanates from scholarship, but it is that which individuals and the community at large create, 'discover,' and/or simply live within their day-to-day lives. For lack of a better term, this may be called the inner logic (or logics) of the everyday life.

Secondly, as more and more individuals and communities come to have a comparable experience, scholars take notice and begin to study it as a sort of a 'phenomenon.' They name it, identify its constituting elements, and offer postulates, hypotheses, and theories to explain its inner workings. All efforts, from naming to hypothesis and theory building, are underlain by analytic conventions that manifest particular principles or logic at work. This is the more conventional usage of the term 'principle,' which may be called the logic of analysis.

2.4.2 Principle of Analysis Versus Logic of the Lived Experience

It is important that a distinction be made between transnationalism as a principle of analysis or as an analytic construct and transnationalism as the logic of the transnationals' lived experience. One reason for this, as this study argues, is that much of the confusion in the analysis of transnationalism is rooted in the elision of such distinction. If the idea is to resolve issues or minimise ambiguities, it is necessary to clarify the implications of such differentiation. Another important reason is that the cryptically political character of analysing transnationalism cannot be unravelled without reference to such a distinction. It is precisely in conflating the lived experience and the analytic construct that the insidious power of scholars and scholarship lies.

As an analytic construct, transnationalism refers to a set of theoretical and methodological assumptions and practices that seeks to explore and explain the modes and consequences of the experience of living 'in-between,' or living simultaneously in two or more nation-states. Central to this effort is the conception of 'simultaneity' (Levitt and Schiller 2004) as well as that of spatiality that accommodates what Pries (2005) calls the absolutist and relativist notions of space. The congruence between the geographic (physical) and societal (symbolic) spaces constitutes the absolutist space whereas the disjuncture between the two reflects the relativist notion of space. Transnational social fields or social spaces are examples of a relativist space (Pries 2005).

Transnationalism as an analytic construct is utilised by scholars to organise a set of practices and ideas in (or about) the national or transnational spaces into a meaningful, analysable unit of analysis. It also serves as an enabling mechanism that allows a set of practices to be considered as constitutive of transnationalism or transnational space. When Portes (2003) declares that transnationalism is not a new phenomenon but a new perspective, he was referring most likely to transnationalism as an analytic construct, not to transnationalism as a lived experience.

As a lived experience, what takes centre stage in transnationalism is the existential condition of living in or through transnational spaces. Such a condition sets a frame or mechanism by which migrants make sense of their experience—living literally or symbolically 'in-between,' or in both countries. It is their being subjects in the transnational lives they live—as the creator of meanings—that serves as the engine of the formation of consciousness favourable in negotiating with, maintaining, and enhancement of their transnationality.

The dual persona of transnationalism does coincide, but there is no reason to suppose that they always or often do. The gap between the two could be wide. The reasons, as I suggest in Table 2.1, are multiple. Firstly, the existential character of the lived experience makes it necessarily ambiguous, fluid, and open-ended. On the other hand, the rationalist nature of the analytic construct means that ordering, organising, and classifying the experience into sensible, composite units are unavoidable. Clarity, order, boundedness, and purposiveness are often imposed on the otherwise free-flowing experience. The aspects or features of the experience that do not fit within a preferred frame may be discarded or ignored. Classical sociology calls this the ideal type.

Secondly, analysis is primarily a cognitive exercise by nature, whereas the experience is multi-sensory and multifaceted. Aside from its cognitive aspects, the lived experience is also affective and behavioural.

Thirdly, and perhaps this is more decisive, their goals vary. As a principle of lived experience and at the same time an adaptive mechanism, transnationalism provides a means by which transmigrants make sense of, negotiate with, and perhaps get the most out of transnational experience. As an analytic construct, on the other hand, it serves primarily to satisfy the interests of the scholars, policymakers, and

Table 2.1 Transnationalism as analytic construct vs transnationalism as lived experience

As analytic construct	As lived experience
Rationalist and empiricist	Existential
Cognitive	Behavioural, affective, and cognitive
New	Long-standing; continuing
Transmigrants or transnationals are both objects and subjects of analysis	Transmigrants or transnationals are the subjects of experience
Emphasis on patterns of behaviour/thoughts	Primacy of individuality and singularity of the existential experience
Tends to see acts or phenomena as meaningful and purposive; coherent; logical	Things happen both with or without meanings or purpose; maybe disordered, incoherent, and irrational
Goal: for scholars to uncover or understand the principles that underlie or govern transnational experience, for their own academic interest, for public consumption as well as in support of public policies - To understand; to explain; or to rationalise	Goal: to enable the transmigrants to understand, negotiate, and navigate successfully through the transnational experience - To live, and live well

other stakeholders. Simultaneously it purportedly seeks to help the transmigrants understand and benefit from their experience.

Finally, transnationalism as an analytic construct is new, whereas as a lived experience, it is long-standing. This point needs elaboration as the debate that developed around the question of novelty or continuity sheds light on important issues.

The pioneering scholars of migrant transnationalism (Basch et al. 1994; Schiller et al. 1992) have asserted that the experience of the post-1965 migrants to the US was significantly different from that of the earlier migrants. One major factor they noted was the technological progress in communication and transportation in recent decades that facilitated time and space compression. Other scholars such as Waldinger (2013); Kivisto (2001); and Kennedy and Roudometof (2002) are not convinced. Kennedy and Roudometof (2002, pp. 2–3), for instance, assert that transnationalism is a phenomenon that is not just fairly common among the 'old migrants' but that it is discernible among communities of non-migrants. They claim that transnational connections, cultures, and communities were the normal state of affairs 'concealed during the relatively recent age of the modernising nation-state' (Kennedy and Roudometof 2002, pp. 2–3).

Portes (2003) seems to have resolved the issue when he claims that transnationalism is a long-standing phenomenon and that what is new is the perspective utilised in analysing (Levitt and Schiller 2004) this phenomenon. Herein lies one of the benefits of the distinction between the lived experience and the analytic construct drawn above.

A question may be raised, however, as to how, if we follow Portes (2003), we may reconcile the novelty of perspective with practices that are long-standing. Portes's suggestion indicates the seemingly intractable nature of the problem. As indicated earlier, Portes includes novelty as one of the requisites for activities to fall under the rubric of transnationalism. By that, he means that the features or practices should have not been covered by extant concepts. If they are, then, they are not within the frame of transnationalism. He asserts, for example, that because the migrants have been sending money and visiting the homeland since time immemorial, these activities cannot by themselves warrant the formulation of an allegedly new concept that is transnationalism (Portes 2003). There must be something new in these practices, he insists, seemingly unaware that it subverts his own declaration that what is new about transnationalism is the perspective and not the experience.

Portes seems oblivious that his overriding emphasis on the 'new' in transnational practices precisely makes his behaviour-centric definition of transnationalism untenable. If all there is to transnationalism are new practices, then it is analytically anomalous, for it negates the historical and logical imperatives that make continuity and change the two sides of the same coin.

The fairly sharp distinction spelt out above carries a number of important analytic and methodological implications. First, the main task of analysis is to narrow, if not really close, the gap. Second, considering the extent of differences in the features as well as the main goals, it would not be easy to narrow such a gap. Scholars, nonetheless, should aspire to achieve this goal. Third, the smaller the unit of analysis, the greater the chance of narrowing the gap between the analytic and the experiential.

This suggests that, fourth, qualitative methods such as ethnography seem more appropriate. It also means that, fifth, inductive rather the deductive approach seems more applicable. That is, the analysis must be geared, first and foremost, towards mapping out the modalities and permutations of transnational experience rather than measuring the extent, intensity, or resilience over time with the view towards hypothesis-testing. It does not mean that the latter is not important. It only suggests that timing is essential: once the overall terrain is already adequately mapped out, it becomes clear what aspects to measure, how, and for what purpose. Doing otherwise, as what is happening now, results in conceptual ambiguities and analytic confusion.

2.4.3 Crypto-Politics of Analysis and Good Intention

Taking transnationalism as a principle—both of analysis and lived experience—foregrounds the possibly sharp disjuncture between representation and the represented, between scholarship and the reality that it purports to represent. Drawing such a distinction is necessary in exposing the hidden source of power of scholarship. That is, the extent to which scholarship conflates the analytical and the lived experience is one of the often unrecognised pillars of the corporate interests of the scholarly class, following Bourdieu (Bourdieu 1990a, b). It also highlights the question, for whom is scholarship on transnationalism being undertaken? Is it really to help the transmigrants, or is there another more pressing reason?

Taking transnationalism as a principle (analytic construct) implies the demand for scholars to be self-reflexive in their analysis. When transnationalism is equated to activities (as Portes et al. do) or defined as a process or a phenomenon, as most migration scholars do, they unwittingly commit a sin of omission by removing themselves from the analytic equation as if they do not matter. Through denial of the ontological or ontologicalised, and of course synthetic, difference between transnationalism as analytic construct and as lived experience, scholars naturalise analytic exercise as inherently non-political. Without knowing it, they empower themselves by privileging their position as disinterested knowledge producers or arbiters of knowledge production (Bourdieu 1990b). At the same time, they absolve themselves of the responsibility for whatever consequences their analysis might have, precisely on the basis of doing a rigorous scholarship. It is a quintessential case of politics by invisible means and a politics without accountability.

Most, if not all, scholarships rest on good intentions. Because scholars profess the nobility of intention, they are often unaware of the potential danger that they can inflict on people their scholarship is supposed to help or empower. The development of the social sciences, in particular, the modernisation theory during the Cold War, for example, has demonstrated that regardless of the intentions of the scholars, they can unwittingly support ideological positions that could have dire consequences for millions of people they sought out to help (Gilman 2003). Apparently, innocent fields such as sociology, economics, anthropology, community development, social work, agricultural sciences, community medicine, performing arts, sports science, and others have lent support to the pernicious interests of either side in the Cold War divide (Berman 1983; Fisher 1993; Richardson and Fisher 1999).

In migration studies, as mentioned above, the upsurge in interest in transnationalism has been partly on account of its political salience. There are, of course, scholars who, unconsciously or by choice, stay clear of the politics of migration. Given that knowledge assumes a life of its own as it circulates in society, however, even the most seemingly innocent knowledge claim may not be spared from politicisation, if individuals or groups find the need for it. It does not mean that everything is political. It only means that everything may be politicised, and that includes good intentions.

2.5 Conclusion

The case I have tried to build in this article against Portes et al.'s behaviour-centric conception of transnationalism rests primarily on a number of conceptual, analytic, and methodological justifications. There is also politics in it, as I have noted, but it lies in the politically adverse implications of Portes et al.'s definition. I have argued that their conception of transnationalism deserves to be critiqued less because it is politically questionable than because it is conceptually, analytically, empirically, and methodologically problematic.

Such a distinction, however, is, at best, artificial. When I identified the problems with their conceptualisation, I was simultaneously in the process of asserting my own conception as preferable, thus offering my proposed alternative in the marketplace of ideas and approaches. If others would prove amenable to my proposal, I would somehow be empowered, as this means gaining intellectual capital, which is quite precious in the academic context. If, on the other hand, my proposal proves unacceptable, it would be established on the basis of a rebuttal or a counter-proposal to my own or a re-justification of the status quo. Knowledge production, in other words, is a sort of combat or contest for 'constituencies' which may eventuate in gaining an intellectual upper hand. If by politics we mean the exercise of power for one's or other people's gains or empowerment, then we can say that politics is potentially built-in within analysis, and a critique is a realisation-in-action of such a potential. Analysis, in short, is not inherently political; it may be politicised. It is the context, intent, and implications of analysis that constitute its politicisations, as what I am doing here exemplifies.

I may be chided for the sharp differentiation that I have drawn between lived experience and analytic construct. Some would argue that at best such a distinction is synthetic or artificial and that in real transnational life, they co-mingle fairly closely. One may add that it is precisely the task of a rigorous and honest scholarship to stretch the overlap to as closest congruence as possible. If one fails to do it, it is their fault.

Admittedly, drawing such a distinction serves a strategic and hardly innocent purpose. It must be emphasised, however, that just as in the case of my critique of Portes et al.'s definition, the justifications I laid down for such a distinction are packaged as primarily conceptual, analytic or theoretical. Despite the absence of the visibly political in these justifications, the political underpinnings and implications cannot be denied, simply because the analytical act cannot be performed outside of power relations. The necessary precondition of analysis is the assumption of the stability and accuracy of concepts, which are no more than frozen images of otherwise complex and perpetually unfolding reality. Who are authorised to do this act of freezing should be obvious by now. We merely assume or pretend that the socio-political contexts can be bracketed out of scholarly practice. Scholars like Bourdieu, Foucault, Derrida, Rancière, Bauman, and many others, including many feminist and postcolonial/decolonial scholars, have long and repeatedly reminded us of their inseparability. The question is no longer whether they are correct and whether those reminders deserve our attention, but why do we still act as if they are wrong? If we concede they may be right and pay attention, why do we take the idea only halfway, perhaps hoping that we can still find exceptions and prove them wrong? It is a question of great importance that it deserves a separate paper or even a full-length book.

Funding The research that made possible the writing of this article was funded by Australia's Endeavour Award, Research Fellowship Research (Grant #ERF08_PDR_590_2008). I would also like to acknowledge the Philippines Australia Studies Centre, School of Sociology and Anthropology, La Trobe University, Bundoora, Victoria, then under the leadership of Dr. Trevor Hogan, for hosting me in Melbourne while doing research.

References

Basch LG, Schiller NG, Blanc-Szanton C (1994) Nations unbound: transnational projects, postcolonial predicaments, and deterritorialized nation-states. Gordon and Breach, New York

Berman EH (1983) The ideology of philanthropy: the influence of the Carnegie, Ford, and Rockefeller foundations on American foreign policy. State University of New York Press, Albany

Bourdieu P (1977) Outline of a theory of practice. Nachdr. Cambridge studies in social and cultural anthropology. Cambridge University Press, Cambridge

Bourdieu P (1990a) In other words: essays towards a reflexive sociology. (trans: Adamson A). Polity Press, Cambridge

Bourdieu P (1990b) The scholastic point of view. Cultural Anthropology 5(4):380–391 https://doi.org/10.1525/can.1990.5.4.02a00030

Curaming R (2020) Power and knowledge in Southeast Asia: state and scholars in Indonesia and the Philippines. Rethinking Southeast Asia. Routledge, Abingdon, Oxon, and New York

Dahinden J (2017) Transnationalism reloaded: the historical trajectory of a concept. Ethnic and Racial Studies 40(9):1474–1485. https://doi.org/10.1080/01419870.2017.1300298

Encinas-Franco J (2015) Overseas Filipino workers (OFWs) as heroes: discursive origins of the 'Bagong Bayani' in the era of labor export. Humanities Diliman 12(2):56–78

Espiritu YL (2003) Home bound: Filipino American lives across cultures, communities, and countries. University of California Press, Berkeley

Faist T (2000) Transnationalization in international migration: implications for the study of citizenship and culture. Ethnic and Racial Studies 23(2):189–222. https://doi.org/10.1080/014198700329024

Fisher D (1993) Fundamental development of the social sciences: Rockefeller philanthropy and the United States social science research council. University of Michigan Press, Ann Arbor

Foucault M (1980) Power/knowledge: selected interviews and other writings, 1972–1977. In Colin G (ed) Pantheon Books, New York

Gilman N (2003) Modernization theory, the highest stage of American intellectual history. In: Engerman D (ed) Staging growth: modernization, development and the global Cold War. University of Massachusetts Press, Amherst and Boston, pp 47–80

Goldring L (1998) Power and status in transnational social fields. In: Michael Smith M, Guarnizo L (eds) Transnationalism from below. Transaction Press, New Brunswick and London, pp 165–195

Gramsci A (1992) In: Buttigieg JA (ed) Prison notebooks, vol 1. (trans: Buttigieg JA, Callari A). Columbia University Press, New York

Guarnizo LE (1997) The emergence of a transnational social formation and the mirage of return migration among Dominican transmigrants. Identities 4(2):281–322. https://doi.org/10.1080/1070289X.1997.9962591

Guarnizo L, Smith M (1998) The locations of transnationalism. In: Smith M, Guarnizo L (eds) Transnationalism from below. Transaction Press, New Brunswick, New Jersey, pp 3–34

Guarnizo LE, Portes A, Haller W (2003) Assimilation and transnationalism: determinants of transnational political action among contemporary migrants. American Journal of Sociology 108(6):1211–1248. https://doi.org/10.1086/375195

Haraway D (1988) Situated knowledges: The science question in feminism and the privilege of partial perspective. Feminist Studies 14(3):575. https://doi.org/10.2307/3178066

Itzigsohn J, Cabral CD, Medina EH, Vazquez O (1999) Mapping Dominican transnationalism: narrow and broad transnational practices. Ethnic and Racial Studies 22(2):316–339. https://doi.org/10.1080/014198799329503

Kennedy P, Roudometof V (2002) Transnationalism in global age. In: Kennedy P, Roudometof V (eds) Communities across borders: new immigrants and transnational cultures. Routledge, London and New York, pp 1–26

Kivisto P (2001) Theorizing transnational immigration: a critical review of current efforts. Ethnic and Racial Studies 24(4):549–577. https://doi.org/10.1080/01419870120049789

Koos AK, Keulman K (2019) Methodological nationalism in global studies and beyond. Social Sciences 8(12):327. https://doi.org/10.3390/socsci8120327

Koselleck R (2002) The practice of conceptual history: timing history, spacing concepts. (trans: Presner T, Behnke K, Welge J). Stanford University Press, California

Lee H (2004) 'Second Generation' Tongan transnationalism: hope for the future? Asia Pacific Viewpoint 45(2):235–254. https://doi.org/10.1111/j.1467-8373.2004.00240.x

Levitt P (2001) Transnational migration: taking stock and future directions. Global Networks 1(3):195–216. https://doi.org/10.1111/1471-0374.00013

Levitt P, Jaworsky BN (2007) Transnational migration studies: past developments and future trends. Annual Review of Sociology 33(1):129–156. https://doi.org/10.1146/annurev.soc.33.040406.131816

Levitt P, Schiller NG (2004) Conceptualizing simultaneity: a transnational social field perspective on Society. The International Migration Review 38(3):1002–1039. https://doi.org/10.1111/j.1747-7379.2004.tb00227.x

Levitt P, DeWind J, Vertovec S (2003) International perspectives on transnational migration: an introduction. The International Migration Review 37(3):565–575

Mahler S (1999) Theoretical and empirical contributions towards a research agenda for transnationalism. In: Smith M, Guarnizo L (eds) Transnationalism from below. Transaction Press, New Brunswick and London, pp 64–100

Manalo AR (2022) Bagong Bayani: tracing the reverence and portrayal of Overseas Filipino workers (OFWs) in Philippine political discourse. Emory J Asian Stud 4. https://ejasonline.org/2022/05/14/bagong-bayani-tracing-the-reverence-and-portrayal-of-overseas-filipino-workers-ofws-in-philippine-political-discourse/

Munro G (2015) Transnationalism: A review of the literature. Studies on National Movements 15(3):1–36

Portes A (2003) Conclusion: Theoretical convergencies and empirical evidence in the study of immigrant transnationalism. The International Migration Review 37(3):874–892
Portes A, Zhou M (1993) The new second generation: Segmented assimilation and its variants. The ANNALS of the American Academy of Political and Social Science 530(1):74–96. https://doi.org/10.1177/0002716293530001006
Portes A, Guarnizo LE, Landolt P (1999) The study of transnationalism: pitfalls and promise of an emergent research field. Ethnic and Racial Studies 22(2):217–237. https://doi.org/10.1080/014198799329468
Portes A, Guarnizo LE, Landolt P (2017) Commentary on the study of transnationalism: pitfalls and promise of an emergent research field. Ethnic and Racial Studies 40(9):1486–1491. https://doi.org/10.1080/01419870.2017.1308528
Pries L (2005) Configurations of geographic and societal spaces: a sociological proposal between 'methodological nationalism' and the 'spaces of flows.' Global Networks 5(2):167–190. https://doi.org/10.1111/j.1471-0374.2005.00113.x
Rancière J (2006) The politics of aesthetics: the distribution of the sensible, Pbk. Continuum, London and New York
Richardson TR, Fisher D (eds) (1999) The development of the social sciences in the United States and Canada: the role of philanthropy. Contemporary studies in social and policy issues in Education. Ablex Publishing Corporation, Stamford, Connecticut
Rumbaut RG (1997) Assimilation and its discontents: between rhetoric and reality. The International Migration Review 31(4):923–960. https://doi.org/10.1177/019791839703100406
Rumbaut R, Kasinitz P, DeWind J (1999) Assimilation and its discontents: ironies and paradoxes. In: Hirschman C, Kasinitz P, DeWind J (eds) The handbook of international migration. Russel Sage Foundation, New York, pp 172–195
Rushdie S (1992) Imaginary homelands: essays and criticism 1981–1991. Granta Books, London Nachdr
Schiller NG, Basch L, Blanc-Szanton C (1992) Transnationalism: a new analytic framework for understanding migration. Annals of the New York Academy of Sciences 645(July):1–24
Smith MP, Guarnizo L (eds) (1998) Transnationalism from below. Comparative urban and community research, vol 6. Transaction Publishers, New Brunswick, New Jersey
Tan Y, Liu X, Rosser A, Yeoh B, and Guo F (2018) Transnationalism, diaspora, and development: a purposive review of the literature. Geography Compass 12(12):e12413. https://doi.org/10.1111/gec3.12413
Waldinger R (2013) Immigrant transnationalism. Current Sociology 61(5–6):756–777. https://doi.org/10.1177/0011392113498692
Wheeler N (2004) A civic trend within ethnic transnationalism? Some insights from classical social theory and the Chinese American experience. Global Networks 4(4):391–408. https://doi.org/10.1111/j.1471-0374.2004.00100.x
Wimmer A, Schiller NG (2003) Methodological nationalism, the social sciences, and the study of migration: an essay in historical epistemology. The International Migration Review 37(3):576–610
Wolf D (2002) There is no place like 'home': emotional transnationalism and the struggles of second-generation Filipinos. In: Levitt P, Waters M (eds) The changing face of home: the transnational lives of the second generation. Russell Sage Foundation, New York, pp 255–294

Rommel A. Curaming is Senior Assistant Professor at the History and International Studies Programme, Universiti Brunei Darussalam. He obtained a Ph.D. in Southeast Asian Studies at the Australian National University (ANU). Before joining UBD, he was Postdoctoral Fellow at La Trobe University and the National University of Singapore (NUS). The empirical and thematic areas of his research vary, but they cohere around the analysis of the politics of the (supposedly) non-political, as evident in knowledge production and consumption in Indonesia, Malaysia, and the Philippines. He published in journals such as *Critical Asian Studies, South East Asia*

Research, Time and Society, and *Philippine Studies*, among others. His book *Power and Knowledge in Southeast Asia: State and Scholars in Indonesia and the Philippines* was published by Routledge (2020).

Chapter 3
Illegal Immigration and Chinese Transnational Networks in Southern Philippines, 1850–1898

Jely A. Galang

Abstract This chapter explores the illegal immigration of Chinese in southern Philippines from 1850 to 1898. During the second half of the nineteenth century, the Spanish colonial government in the Philippines implemented policies aimed at controlling illegal Chinese immigration in the southern part of the colony. Although most Philippine Chinese would disembark at the port of the colonial capital of Manila, authorities were aware of the unregulated movement of the Chinese in Mindanao and Sulu that posed a threat to the colony's political, military and economic stability. Chinese immigrants from Singapore, Labuan, Sandakan and other islands in the region provided commodity goods, arms and ammunition to local inhabitants and leaders such as the Sulu Sultan. Set against colonial conditions, this chapter examines the way that transnational networks in maritime Southeast Asia played a role in this illicit movement of people. Utilising Philippine and Spanish archives, it describes and analyses actors, institutions and processes operating in the Philippines and elsewhere that made this scheme possible, and from the state's view, challenging to manage. This chapter contributes to an understanding of the creative ways of Chinese immigrants' evasion of state regulations to engage freely in local and transnational activities.

Keywords Chinese · Philippines · Illegal migration · Sulu · Mindanao

3.1 Introduction

During the second half of the nineteenth century, the Spanish colonial government in the Philippines became particularly keen to suppress, if not totally eradicate, illegal Chinese immigration into the colony. The Spanish authorities were well aware that while most Chinese would disembark at the port of Manila—the colonial capital—there were many who entered the Philippines through a "backdoor route" in the Sulu archipelago and Mindanao island. Without the necessary travel documents,

J. A. Galang (✉)
University of the Philippines-Diliman, Quezon City, Philippines
e-mail: jagalang@up.edu.ph

H. M. Y. Ho and Y. Chan (eds.), *Transnational Southeast Asia*, Asia in Transition 29,
https://doi.org/10.1007/978-981-96-3608-2_3

these illegal migrants posed a serious threat to the political, military and economic stability of the colony. Mostly from Singapore, Labuan, Sandakan, Borneo and other islands in the region, these Chinese provided not only commodity goods but also arms and ammunition to the local inhabitants and their leaders (such as the Sulu Sultan) which the latter used to continue challenging Spain's sovereignty over the area, hence the need for the authorities to restrict their entry.

This chapter explores the illicit immigration and mobility of the Chinese in southern Philippines during the period between 1850 and the end of the Spanish regime in 1898. Set against the various transitions and transformations in the colony, and the broader Asian region at the time, it examines the factors that contributed to this unlawful movement of people. In particular, it provides special emphasis on how the transnational networks of the Chinese in maritime Southeast Asia played a crucial role in the immigration process. Utilising materials from Philippine and Spanish archives, it describes and analyses the actors, institutions and processes operating in the Philippines and elsewhere that made this scheme possible, and from the state's view, almost impossible to stop. It intends to contribute to our understanding of how people—in this case, the Chinese—in the past (as well as the present), found and employed creative ways and means to evade state regulations in order to freely engage in various commercial activities.

3.2 Chinese Migration to and Residence in Spanish Philippines

Chinese migration to and residence in the Philippines has a long and interesting story. Centuries prior to the Spanish colonisation of the archipelago, early Filipinos and the Chinese had already established commercial and cultural relations which can be traced back to the tenth century or even earlier. Chinese traders came to the islands and exchanged their goods with what the Filipinos gathered and produced. Although many of these sojourning merchants would leave after conducting their business transactions, there were some who remained and lived among the islands' native inhabitants (Go and Sy 2000; Scott 1989). Their small number of around 50 in 1570, significantly increased after the Spaniards claimed the Philippines as part of the Spanish empire and founded the Manila–Acapulco trading system in 1565, which was also called the galleon trade. This trading system served as an important part of the "early globalisation" that primarily attracted Chinese migrants to conduct trade and reside in the colony (Chong 2016; Flynn and Giraldez 2008; Guerrero 2015; Schurz 1939).

After the abolition of the galleon trade in 1815, the Philippine economy shifted direction from a monopolistic trade of commodity goods to a laissez-faire one based on the exportation of agricultural crops to different parts of the world. This economic transition created new opportunities for the Chinese (Legarda 1999; Wickberg 2000). As the colony needed additional workers to cultivate its vast and arable lands to produce commodities for global demand, the colonial government thus encouraged Chinese immigration to the islands. Therefore, in 1839 and 1849, two decrees were issued that allowed them to reside in different parts of the colony and to take up occupations that best suited them (Claveria 1849; Empadronamiento 1839; Sobre empadronamiento 1839). These decrees and other favourable immigration and employment policies contributed to the increase of the Chinese population in the colony, their dominance in the economy and their presence in many areas of the archipelago as traders, middlemen, artisans and labourers (Chinos, Sus Reglamentos 1893; Comenge 1894; Wickberg 2000). In particular, in terms of population, in 1850 there were only 9334 registered Chinese in the colony. After 15 years, in 1865, their number doubled and reached 16,898. By 1890, their number was pegged at 44,900 (Chinos 1870–1898; Chinos 1890–1891; Chinos 1891–1892a; Diaz Arenas 1850; Unreferenced 1865).

Such socio-economic developments within the Chinese community, from the middle of the nineteenth century onwards, made the government wary. Hence, while they were permitted to come to the Philippines, the Chinese had to abide by stricter measures related to immigration and registration. These forms of social control were primarily aimed at promoting the colony's political and financial stability. In 1871, for instance, the recently arrived Governor General Rafael de Izquierdo ordered an investigation of the conditions of the Philippine Chinese to determine the exact number of officially registered individuals, those who were overstaying, and the undocumented ones who most probably arrived in the colony illegally. The investigation revealed that there were over 18,000 registered Chinese in the Philippines. Of this number, only approximately 12,000 had proper documents, which meant that around 6000 were either overstaying or undocumented. This dire situation had been causing the colonial coffers some 50,000 pesos annually as these "illegal foreigners" were not paying their taxes (Ultramar 5217). As a result, two years later, in May 1873, the government issued a general decree to address this prohibited immigration (Rodriguez Berriz 1887–1888).

Based on intelligence reports, there were at least two ways through which illegal Chinese migrants entered the Philippines during the period under investigation. The first one was through northern Luzon—in Ilocos Sur and Cagayan (see Fig. 3.1). From the 1850s onwards, the port of Santiago (some 350 kms. from Manila) in Ilocos Sur became an important disembarkation point of commodities and peoples from southern China, usually from the ports of Hong Kong, Amoy (Xiamen) and Macau, going to the commercial city of Vigan and elsewhere in the region. This port commonly catered to Spanish and foreign ships *en route* to Manila, Southeast Asia and the Americas (Ultramar 436). The other means of access in Luzon was through the port of Aparri (approx. 620 kms. from Manila) in Cagayan province. Originally used as a trading post of the Japanese during the pre-Hispanic period, Aparri was made a major port by the Spaniards in the late seventeenth century (Malumbres 1918). Various galleons from Manila carrying Chinese merchandises would stop at Aparri to obtain additional supplies for their long journey to Acapulco in Mexico, across the Pacific.

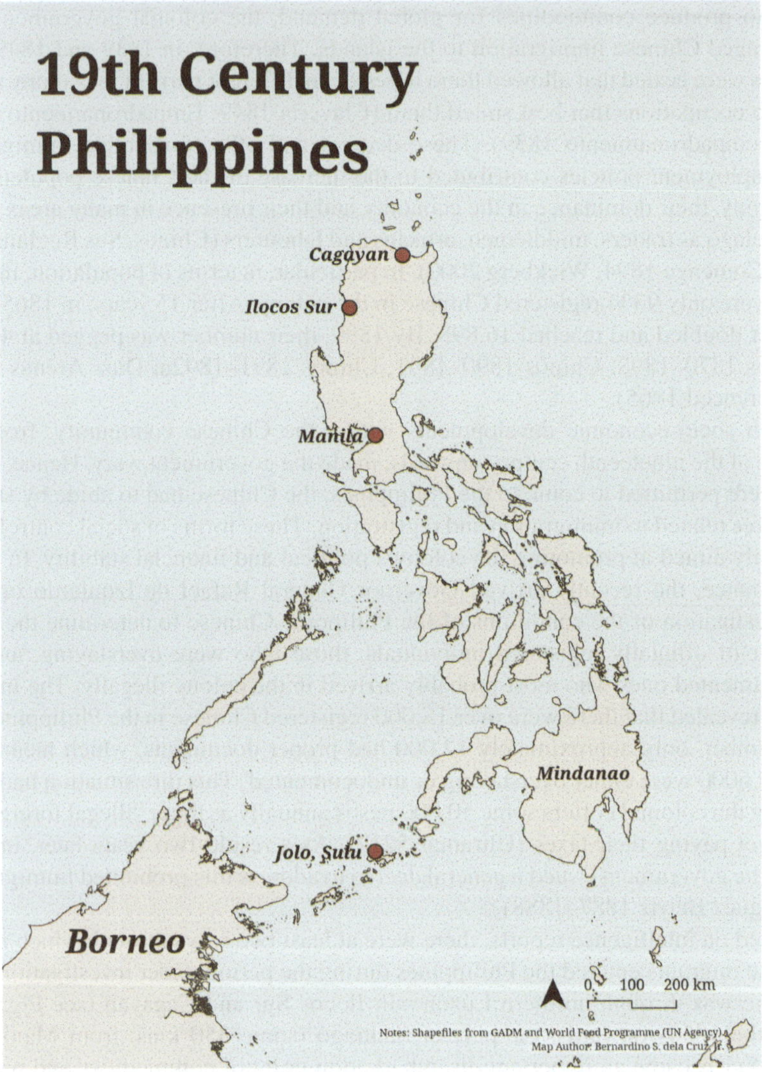

Fig. 3.1 The nineteenth-century Philippines showing points of legal and illegal Chinese

The second entry point—the one used more frequently than the first—was through the south, via Sulu and Mindanao. From the sixteenth century to the end of the Spanish regime, the Muslims (called "Moros") and Lumad peoples of Sulu and Mindanao had successfully resisted Spanish colonisation. Although the Spaniards were able to establish a foothold in Zamboanga in western Mindanao in 1635, military campaigns in the region generally had limited success. Muslim groups like the Sama–Balanguinguis were considered "pirates" and "marauders" who raided lowland Christian towns in Luzon and the Visayas, capturing slaves that were then sold in slave markets in various parts of maritime Southeast Asia. The Lumads, on the other hand, avoided direct confrontations. They evaded Spanish troops and missionaries by hiding in inaccessible parts of mountains. The Spaniards' weak presence in this region was one important factor that made illegal Chinese immigration rampant during this period (Majul 1973; Paredes 2013; Warren 1985).

3.3 Spanish and Chinese Presence in Southern Philippines

For the entire three centuries of Spanish regime in the Philippines, Spain's presence in Sulu and Mindanao was relatively weak as various groups continuously challenged and resisted Spanish hegemony in the region. Although Spain waged serious and costly military campaigns during the early modern period, these campaigns were more sustained in the 1840s and 1850s when the Spanish colonial government desperately looked for areas to open and develop for the Philippines' bourgeoning global commercial trade (Legarda 1999; Majul 1973; Warren 1985). After successfully occupying Balanguingui Island in the Sulu archipelago in 1848, Spain intended to subjugate Mindanao which, during this period, was divided into five districts under the island's politico-military comandancia or the "Government of Mindanao": 1) Zamboanga, 2) Misamis, 3) Davao, 4) Cotabato and 5) Surigao (see Fig. 3.2). The sixth district was the District of Basilan comprising the Sulu archipelago (Saleeby 1908). The Spaniards occupied Davao in 1849 and then two years later, Cotabato in the southwest part (Becket 1977). As Spanish troops advanced in these areas and the pacification process began, the ever-enterprising Chinese followed, establishing businesses in the districts' capitals (Wickberg 2000).

Apart from Spain's successful occupation of the region, there were other factors that contributed to Chinese immigration to Mindanao and Sulu. As noted, one factor was Governor General Narciso Claveria's 1849 decree that encouraged Chinese immigration to the Philippines. Another factor was the abolition of the *indulto de comercio* in 1844. This was the right given to provincial governors to engage in commercial activities that hindered the participation of Chinese, Chinese mestizo[1] and other players in economic activities in the provinces (Robles 1969; Wickberg 2000). The rise of Chinese mestizo elite families in urban commercial centres in central Philippines also led to the establishment of commercial and trading networks by these families in northern Mindanao and facilitated internal migration into the area (Cullinane 2022). The opening of the port of Zamboanga in 1855 also facilitated smoother traffic of commodities and peoples in this part of the colony (Aguilar 1994). In addition to these political, economic and infrastructure changes in the Philippines, there were also "push" factors in China at the time that made Chinese immigration possible. High population, natural calamities, social unrest and the history of migration of the Chinese in coastal region of China which included Fujian where the majority of the Chinese in the Philippines originated were but some of these factors (Yen 1985).

Fig. 3.2 Districts under the politico-military government of Mindanao, nineteenth century

Table 3.1 shows the population of the Chinese in Mindanao and Sulu. Note that these numbers are based on official records; not included in the data where those who illicitly came to and settled in the region. No data is available in certain areas especially in the mid-1800s due to either the unstable political and socio-economic conditions in these places or the lack of capable bureaucratic agencies to undertake the census or both. In Mindanao, for the forty-year period (1850–1890), the majority of legal Chinese residents were located in Misamis and Zamboanga. As noted, Zamboanga was a Spanish base founded in the seventeenth century (see Cañones–Agdeppa 2018) while Misamis was geographically close to the commercial cities of Cebu and Iloilo in the Visayas from which Chinese and Chinese mestizos originated (Alvero–Boncocan et al. 2023; Cullinane 2022). There were also companies that transported Chinese workers in Misamis to labour on abaca plantations there (Chinos 1891–1892b). In Sulu, the Chinese were located in the capital Jolo and in Isabela de Basilan. Their official number was recorded only in 1881 because of the fragile political and military situation in the archipelago despite Spanish presence (see Absari 2023).

The Chinese in Mindanao and Sulu during the second half of the nineteenth century engaged in various occupations. As elsewhere in the Philippines, many of them were merchants involved in wholesale and retail trade of goods from Manila and the Visayas (i.e. Cebu, Iloilo) as well as from Singapore and Malaya. Some of these merchants through their commercial networks within and outside the country also functioned as middlemen between local farmers and producers, and foreign

Table 3.1 Officially registered Chinese in southern Philippines, 1850–1890

Provinces	1850	1865	1881	1886	1890	Total
Mindanao Island						
Cotabato	–	–	166	133	158	457
Davao	–	10	17	36	45	108
Misamis	563	18	447	471	469	1968
Surigao	–	12	115	210	257	594
Zamboanga	40	138	557	606	465	1806
Sub–total	603	178	1202	1456	971	4933
Sulu Archipelago						
Isabela de Basilan	–	–	79	59	41	179
Jolo	–	–	–	–	599	599
Sub–total	–	–	79	59	640	778
OVERALLTOTAL (Mindanao and Sulu)	603	178	1281	1515	1611	5711

Sources National Archives of the Philippines (NAP), *Chinos (Manila, 1890–1891), SDS 13,066*, S 123–123b; NAP, *Chinos (Manila, 1891–1892), SDS 13,067*, S 766–766b; NAP, *Chinos (Paragua, Surigao, Zamboanga, Pampanga, Bulacan, Manila, Capiz, Cebu, 1870–1898), SDS 13,120*, S 46; Diaz Arenas (1950); Unreferenced; "Estado del numero de tributos de las islas Filipinas y de las cedulas de inscripcion para las provincias (1865), Papeles Relativos a las provincias de Ultramar colecionados por Eugenio Alonzo y Sanjurjo (Biblioteca Nacional de España (BNE))," p. 386

commercial houses located in urban commercial centres. There were also Chinese merchants who procured contracts with the government to bring into the region opium and to operate opium dens in provinces where there were numerous Chinese. In Spanish Philippines, only the Chinese were allowed to smoke opium. Opium-contracting was one of the government's important sources of revenue during the latter part of the colonial period (Anfion 1837–1898; Anfion 1847–1896; Anfion 1849–1869; Bamero 2006; Victoria 2023; Zarco 1969).

There were also small-scale vendors of various items in different parts of the region. These Chinese usually owned *tiendas de sari–sari* (general merchandise stores) from which locals bought their daily needs in exchange for goods they produced. They sometimes served as moneylenders, providing financial resources to Filipinos who were in dire need. There were also Chinese who worked in agricultural lands that produced commodities needed in the global market (Francia and Gonzalez 1898; Wickberg 2000). In the latter part of the nineteenth century, for example, the Spanish company Aldecoa y Compania brought hundreds of Chinese labourers to Surigao in northern Mindanao as well as in Samar and Leyte to work on the company's vast plantations of abaca. Although many of these contractual labourers would leave Mindanao after their usual three-year contracts, there were also those who remained, married local women and planted roots in the area (Chinos 1891–1892b).

3.4 Transnational Networks of Chinese in Southern Philippines

The commercial networks of the Chinese operating within and outside the Philippines played a crucial role in their immigration, residence and occupational conditions in Mindanao and Sulu during the period under investigation. Within the Philippines, their trade and community-based contacts and linkages with other Chinese groups and local communities—usually of fellow businessmen—became very crucial. As already mentioned, this was the case, for example, for many Chinese who settled in northern Mindanao (Cullinane 2022; Galang 2020; Wickberg 2000). Seeking economic opportunities in this "frontier" region, many of the Chinese migrants either came directly from Manila or from Iloilo and Cebu. In this migration pattern, the Chinese traders and middlemen effectively utilised their familial and kinship relations and the business connections they had established through the years. These connections proved necessary in procuring financial capital and inexpensive sources of supplies and labourers for their business and entrepreneurial activities in the region.

In addition to domestic connections, the Chinese also utilised their transnational networks in China and elsewhere in maritime Southeast Asia, particularly in the south of the Philippines, to facilitate, maintain and advance trading relations and successful immigration. Since the beginning of Spanish colonisation, Chinese immigration to the islands was based on their kinship, same village and same speech-dialect connections in China (Chen 2021; Chia 2006; Chin 2010; Wu 1977). Well-established Chinese businessmen in Manila, for instance, brought their relatives mostly from the Minnan region in Fujian province to the Philippines to help in their businesses as shop assistants and apprentices. Once the latter obtained enough experience and capital, they would start their own business ventures and invite others from the same clan or village to join them. Although at times affected by factors such as restrictive Spanish immigration and residence policies, such process and pattern continued for the entire Spanish regime and even beyond.

These transnational networks also proved important in the illegal entries of Chinese through Sulu and Mindanao as Spain began to assert her sovereignty, and as regional trade prospered and various modes of transportation improved during the second half of the nineteenth century. The main actors for these movements of people were Chinese merchants who took advantage of the economic opportunities in this southern region. These Chinese primarily served as commercial intermediaries that facilitated the movements and exchanges of goods and people between southern Philippines and the areas south of the "Sulu Zone".[2] Generally, there were two types of Chinese merchants involved in illegal Chinese immigration in this region and they can be classified based on their residence in the Philippines. The first type was the Chinese who had been permanently residing in Sulu and Mindanao for a very long time and had built vast trading networks within the Philippines and beyond the colony. The second type was the "itinerant" Chinese merchants who only remained in the Philippines on a seasonal basis (i.e. stayed only during busy trading periods) and frequently moved from one place to another to procure and sell trade commodities within the region. Both of these types of Chinese merchants oftentimes employed *personeros* or *dependientes* or agents that transacted business on their behalf. When the Chinese merchants could not do the business negotiations, it was these agents who did the work. From the standpoint of the Spanish colonial state, the resident Chinese merchants were easier to control than the itinerant ones as the former had documents of identification and travel permits and were paying their financial obligations to the state. The latter, on the other hand, although required to secure proper documents, could easily evade such policy as they constantly moved in different places.

Despite the difference in their residence status and how the colonial state viewed them, these resident and itinerant Chinese merchants had one important common characteristic: their vast commercial contacts and networks with other Chinese in neighbouring Singapore, Labuan, Sandakan, Borneo, Celebes, Sumatra and other areas in the region. These Chinese effectively utilised such linkages in the commodity exchange within this part of maritime Southeast Asia. From Sulu and Mindanao, they obtained marine products such as pearls, sea cucumbers, turtle shells as well as forest products such as beeswax and bird nests. In exchange, they brought to the region various commodities such as food stuff, textiles, luxury items and opium. Opium was legal in Spanish Philippines, but opium dealers were required to obtain contracts with the government. In southern Philippines, however, opium smuggling was rampant, which became a concern as the colonial coffers were losing immense revenue.

In addition to unregistered opium, these Chinese merchants also brought to the region other more dangerous items such as arms, ammunitions and saltpeter. As already mentioned, although the Spanish claimed to have conquered this southern region in the mid-nineteenth century, several Muslim leaders and groups continued to wage protracted war against Spain. One such leader was the Sulu Sultan who obtained arms from these Chinese merchants who brought these goods from Singapore and elsewhere in the region. Other Muslim groups who obtained these items used them in their continuous—albeit declining—raiding activities in Christian communities in Luzon and the Visayas (*BNBHOG* 1881, 1884; Montemayor 2005; Salazar 2000; see also Treacher 1891). In the 1890s, Spanish administrators who had a grim view of the situation in the south called this mutually beneficial commercial arrangement between the Sulu Sultan and Muslim groups, and Chinese merchants as a relationship between smugglers and pirates ("*relacion entre contrabandistas* y *piratas*") (Ultramar 5221; Ultramar 5339). A report written by a Spanish journalist that appeared in *Diario de Manila* in May 1889 stated that "The Chinese (in Mindanao and Sulu), instead of loving the Spaniard, hates him without the slightest motive, and makes close alliances against us even with the Moros" (author's translation, *China en Filipinas* 1889, p. 31).

These Chinese merchants–smugglers did not only purchase illegal goods and brought them to Sulu and Mindanao; they also facilitated the entry of undocumented Chinese into the Philippines. They and/or their agents would usually go to Singapore and coordinate with coolie agents there to obtain people who wanted to come to the Philippines. For would-be immigrants, this migration arrangement was considerably favourable as they did not have to seek permission, obtain passports or register with the Spanish authorities. As undocumented illegal migrants, they fell beyond the direct control of the Spanish colonial government, hence, need not pay taxes or abide by relevant policies imposed upon the Chinese and other foreigners in the Philippines at the time (Wickberg 2000).

Based on reports of Spanish officials deployed in the region, these undocumented Chinese freely moved within the Sulu Zone and Mindanao. The documentation of illegal entries became even more frequent during the last three decades of the nineteenth century. Governors General Emilio Terrero (term: 1885–1888) and Valeriano Weyler (term: 1888–1891) received numerous reports about such clandestine activities of the Chinese in southern Philippines. These officials were particularly intentional in knowing the conditions of the Chinese in the colony because of the heightening anti-Chinese campaigns in Luzon during the period, which added to the socio-economic problems that the state had to remedy. Governors General Terrero and Weyler wanted to obtain as much information on the ground as possible in order to come up with possible solutions to address this particular issue (Comenge 1894; Los Chinos 1886).

As noted, such reports highlighted the fact that such illegal movements of the Chinese were primarily aided by resident Chinese in Sulu and Mindanao as well as seasonal sojourning Chinese merchants from neighbouring islands. However, these Chinese merchants did not operate on their own, they utilised their networks with Muslim boat owners in the region. Boats and vintas operated by Muslims were used to bring into the colony illegal Chinese migrants. These Muslim vessels proved to be faster and more manoeuvrable compared to the steamships and other ships used by the Spaniards. In most cases, these Chinese migrants were secretly transported to Sulu and Zamboanga and, from there, they would go into the interior of Mindanao or to other parts of the Visayas. The illegal Chinese migrants would then contact and connect with the resident Chinese to conduct business there (Fonacier 1949; Wickberg 2000).

There are several reasons why illegal Chinese immigration to the Philippines commonly occurred via Sulu. The first of these was Sulu's geographical make-up, comprising hundreds of volcanic and coral islands and numerous rocks and reefs. It was, therefore, a challenging task for the Spanish navy to conduct regular patrols over Sulu's waters. Second, Sulu's proximity to British-controlled Singapore, Labuan and North Borneo made it easier for the Chinese from these areas to engage in Sulu's—and by extension, in Zamboanga's—trading activities. And finally, as mentioned, although Spain had claimed sovereignty over Sulu in the mid-nineteenth century, Spanish political military and economic clout on these islands were significantly weak (Warren 1985). In fact, there was no existing customs house in Sulu at that time. The nearest customs port was in Zamboanga, some ninety miles away from Jolo—Sulu's capital (Hurley 1936). Furthermore, in the 1870s and 1880s, trade in the Sulu zone was in the hands of British and German merchants who found ways to evade the Spanish blockade initiated in 1872. Foreign traders such as the British, German and Chinese were also engaged in gunrunning activities in the region (Ultramar 5221; Ultramar 5339; Wright 1972).

3.5 The 1889 Chinese Immigration Decree

In September 1888, Governor General Weyler issued a decree which aimed—among other pressing concerns in southern Philippines—to address the issue related to illegal Chinese immigration in this part of the colony. The decree required all Chinese immigrants arriving into the Philippines to disembark and register only at the port of Manila. Weyler specifically stated that this measure was absolutely necessary to "curb illegal entries" of the Chinese (Chinos. Sus Reglamentos 1893; Ultramar 5339). He further highlighted the fact that these undocumented and unregistered individuals were endangering the colony's financial stability and posed a threat to security as they did not pay their required financial obligations to the state, and they could move freely within the southern region of the colony. In the decree, the politico-military governors of the districts of Sulu and Zamboanga, with the help of the ill-equipped Spanish naval forces in the region were ordered to implement this decree in their respective jurisdictions.

While the decree was indeed necessary, at least from the point of view of the Spanish colonial government, it also led to unfavourable impacts on some economic activities in the region where Chinese capital and labour were relevant. A case in point was the Hacienda Gomantong, a tobacco plantation established by the German Borneo Company (*Deutsche Borneogessellshaft*) in 1884 (Galang 2020). Unlike other Spanish and foreign companies in Luzon and Visayas which employed Chinese workers from Manila, and Fujian and Guangdong in China, the German company obtained their Chinese coolies from Singapore. According to the hacienda's administrators, the September 1888 decree was detrimental to their operations for several reasons. First, Manila was farther away than Jolo from Singapore, the company's main source of labourers. Bringing the coolies to Manila to register there meant additional transportation and living expenses. Second, the company did not have agents based in Manila. It would be financially challenging to employ additional people in the capital just to supervise the collies' registration and then bring them to Jolo. The decree was one of the reasons for the abolition of the hacienda in late 1889.

Despite the Spanish officials' efforts, the September 1888 Chinese immigration decree proved to be ineffective as illegal Chinese migration continued. With regard to those arriving through Sulu, on 3 December 1892, more than two years after Weyler's decree, the politico-military governor of Sulu district claimed that suppressing illicit entries of undocumented Chinese and their movements and activities within the area was next to impossible. The district governor echoed the reasons cited by previous officials before the 1889 decree was issued. He claimed that the area's archipelagic make-up was the primary factor that led to the continuous unlawful entries of the Chinese[3].[4] Chartered boats from Singapore were used by these illegal Chinese immigrants to come to the islands. As in the earlier years, the official also mentioned that the islands' geographical proximity to Singapore and Borneo also hindered the proper implementation of the decree. Regular communication between the southern islands and the points of origins was also a factor (Chinos 1891–1892b). A regular steamer service which traversed between Sulu and Singapore was also available (Guia Oficial 1885). The district governors reported the same exact sentiments in their letter to Governor General Blanco on 9 March 1894 (Ultramar 5339). What these Spanish officials did not include was the transnational connections of the Chinese that made unlawful entries possible and even rampant.

When the Americans took over the Philippines from Spain in 1898, they were deliberate in taking full control of all the islands and islets of their newly acquired possession, especially the southern part which had been resisting Spanish domination for centuries. With regard to the "Chinese question," the Americans also had to be stricter about Chinese immigration to and residence in the Philippines as the Chinese Exclusion Law in America that was passed in 1882 was officially extended to the Philippines in 1903. This law prohibited the entry of Chinese labourers and other "types" of Chinese into the country. A few decades after this law was implemented, the American officials in southern Philippines continued to lament the illegal entry of the Chinese (Alejandrino 2003; Fonacier 1949).

3.6 Conclusion

The Spanish colonial government viewed the Chinese in the Philippines as "necessary outsiders." They were necessary because they provided essential services and commodity goods desired in the colony. In fact, some Spanish officials even claimed that without the Chinese, Spain could not have been able to effectively control and maintain the Philippines as part of the Spanish empire. On the other hand, the state also considered the Chinese as "outsiders" or foreigners whose loyalty to Spain was questionable. The tendency of many Chinese to resist assimilating culturally into the local population also made the Spaniards think in this manner. It is within this framework that the issue of Chinese immigration in Spanish Philippines has to be understood. The government, especially during the second half of the nineteenth century, needed additional labourers to augment the unreliable Filipino workers tasked to open and exploit the colony's vast agricultural lands to produce goods for global consumption. The state found the Chinese fit to perform this task. However, as many Chinese did not want to or could not pay the exorbitant taxes imposed upon them by the government, many of them entered and worked in the colony illegally. While some ports in Luzon were used for such illicit migration, the most effective route was through the southern Philippines, via Sulu and Mindanao.

To address this concern of illegal Chinese immigration, the Spanish colonial government issued the September 1888 decree requiring all Chinese migrants to register only at Manila port. Despite efforts of officials stationed in southern Philippines, this clandestine movement of the Chinese continued unhampered. There were several reasons why this was the case. These include the geographical make-up of Sulu, the proximity of Sulu and Mindanao to other islands outside the Spanish domain, the availability of regular steamship service in the region and the relatively weak presence of the Spaniards in this part of the colony. That which needs to be highlighted here is the domestic and transnational trade networks of the Chinese with Muslim boat owners and operators, as well as other Chinese merchants in Mindanao, Singapore, Celebes, Borneo and elsewhere in the region that proved crucial in their unlawful entries, residence and work in southern Philippines. These linkages were necessary for the Chinese not only to move from one place to another but also to obtain the capital and sources of labour that were important for their economic activities in Spanish Philippines and beyond.

Notes

1. Chinese mestizos or mestizos de chinos or mestizos de sangley were offspring of the intermarriages between Chinese fathers and Filipina mothers. For additional information on the Chinese mestizos, see Chu (2002), Cullinane (2022), Wickberg (2000), Wickberg (2001).
2. "Sulu Zone" is a term used by Warren (1985) (xxviii), to refer to a geographical area "comprising the Sulu archipelago, the northeast coast of Borneo, the foreland of southern Mindanao and the western coast of Celebes.".
3. This archipelagic nature of Sulu and Mindanao is also a geo-physical characteristic of the broader maritime Southeast Asia which contributed to illegal movements of people, goods, ideologies and technologies within the region (see Tagliacozzo 2005).

References

Absari D (2023) Sulu–China history from pre–Islamic to Spanish era. Chinese Studies Journal 17:227–242

Aguilar F (1994) Beyond inevitability: the opening of Philippine provincial ports in 1855. Journal of Southeast Asian Studies 25(1):70–90

Alejandrino C (2003) A History of the 1902 Chinese Exclusion Act. Kaisa Para sa Kaunlaran Inc., Manila

Alvero–Boncocan R et al (2023) Chinese immigration, settlement, agriculture, and commerce in Misamis in the nineteenth century. Chinese Studies Journal (17):40–74

Anfion (Opium) (1837–1898) Manila, books 1–3. National Archives of the Philippines (NAP), Manila

Anfion (Opium) (1847–1896) Tondo, books 1–7. National Archives of the Philippines (NAP), Manila

Anfion (Opium) (1849–1869) Binondo, books 1–2. National Archives of the Philippines (NAP), Manila

Bamero A (2006) Opium: evolution of policies, the tolerance of the vice, and the proliferation of contraband trade in the Philippines, 1843–1908. Social Science Diliman 3(1–2):49–83

Beckett J (1977) The Datus of the Rio Grande de Cotabato under colonial rule. Asian Studies: Journal of Critical Perspectives on Asia 15:46–64

British North Borneo Herald and Official Gazette (BNBHOG) (1 March 1881)

British North Borneo Herald and Official Gazette (BNBHOG) (1 March 1884)

Cañones–Agdeppa MC (2018) The Chinos were everywhere: mapping and locating the Chinese in Zamboanga, 1850–1898. Asia Mindanaw 4:1–32

Chen B (2021) The Hokkiens in early modern Hoi An, Batavia, and Manila: political agendas and selective adaptations. Journal of Southeast Asian Studies 52(1):67–89

Chia L (2006) The butcher, the baker, and the carpenter: Chinese sojourners in the Spanish Philippines and their impact on Southern Fujian (sixteenth–eighteenth centuries). Journal of the Economic and Social History of the Orient 49(4):509–534

Chin JK (2010) Junk trade, business networks and sojourning communities: Hokkien merchants in early Maritime Asia. Journal of Chinese Overseas 6(2):157–215

China en Filipinas: coleccion de articulos publicados en el Diario de Manila acerca de la inmigracion Asiatica en el archipielago (China in the Philippines: collection of articles published in the Diario de Manila about Asian immigration in the archipelago) (1889) Establecimiento Tipografico de Ramirez y Compania, Manila

Chinos (Chinese) (1870–1898) SDS 13120. NAP, Manila

Chinos (Chinese) (1890–1891) SDS 13066. NAP, Manila

Chinos (Chinese) (1891–1892a) SDS 13067. NAP, Manila

Chinos (Chinese) (1891–1892b) SDS 13044. NAP, Manila

Chinos. Sus reglamentos y sus contribuciones, (Chinese. Their regulations and contributions) (1893) comp. El Faro Administrativo Establecimiento Tipo–Litografico de Ramirez y Ca., Manila

Chong JL (2016) Los Indios Chinos del Galeon de Manila (The Indian Chinese of the Manila Galleon). In: Tornaviaje: La nao de China y el barroco en Mexico, 1565–1815 (Return trip: the ship from China and the baroque in Mexico). Gobierno del Estado de Puebla, Puebla, Mexico, pp 58–63

Chu R (2002) The 'Chinese' and the 'Mestizos' of the Philippines: towards a new interpretation. Philippine Studies 50(3):327–370

Claveria N (1849) Ultramar 5203, Expediente 14, no 4. (Decree of Governor Narciso Claveria) (Manila, 20 December 1849)

Comenge R (1894) Cuestiones Filipinas, 1a parte: los Chinos (Estudio social y politico) (Philippine questions, part 1: the Chinese (a social and political study)). Tipo-Litografia de Chofre y Compania, Manila

Cullinane M (2022) The Chinese Mestizos of Cebu city, 1750–1900. University of San Carlos Publishing House, Cebu

Diaz Arenas R (1850) Memorias historicas y estadisticas de Filipinas y particularmente de la grande isla de Luzon (Historical and statistical reports from the Philippines, particularly from the large island of Luzon). Imprenta del Diario de Manila, Manila

Empadronamiento (Registration) (1839) Ultramar 5203, Expediente 14, no 13. Empadronamiento de Chinos y sus tributos (Registration of the Chinese and their taxes) (Manila, 31 August 1839)

Flynn D, Giraldez A (2008) Born again: globalization's sixteenth century origins (Asian/Global Verus [sic] European Dynamics). Pacific Economic Review 13(3):359–387

Fonacier T (1949) The Chinese exclusion policy in the Philippines. The Philippine Social Sciences and Humanities Review 14(1):3–28

Francia B, Gonzalez J (1898) Las islas Filipinas. Mindanao (The Philippine islands. Mindanao). Imp. de la Subinspeccion de Infanteria, Habana

Galang J (2020) Hacienda Gomantong: the 1888 Chinese immigration decree, a German tobacco plantation, and Chinese laborers in Jolo, Sulu, Southern Philippines. Asian Studies: Journal of Critical Perspectives on Asia 56(2):1–32

Go BJ, Sy J (2000) The Philippines in ancient Chinese maps. Kaisa Para sa Kaunalaran Inc., Manila

Guerrero M (2015) The Chinese in the Philippines, 1570–1770. In: Chu R (ed) More Tsinoy than we admit. Vibal Foundation, Quezon City, pp 55–84

Guia officiál de Filipinas (Official guide of the Philippines) (1885) (1884) Establecimiento Tipog. de Ramírez y Giraudier, Manila

Hurley V (1936) Swish of the Kris. E.P. Dutton & Co. Inc, New York

Legarda B (1999) After the Galleons: foreign trade, economic change and entrepreneurship in the nineteenth–century Philippines. Center for Southeast Asian Studies, University of Wisconsin–Madison, Wisconsin

Los Chinos en Filipinas: Males que se experimentan actualmente y peligro de esa creciente inmigracion (The Chinese in the Philippines: evils that are currently being experienced and the danger of this growing immigration) (1886) Establecimiento tipografico de La Oceania Española, Manila
Majul CA (1973) Muslims in the Philippines. University of the Philippines Press, Quezon City
Malumbres J (1918) Historia de Cagayan (History of Cagayan). Tip. Linotype de Sto. Tomas, Manila
Montemayor MS (2005) Captain Herman Leopold Schück: the saga of a German sea captain in 19th Century Sulu-Sulawesi seas. University of the Philippines Press, Quezon City
Paredes O (2013) A mountain of difference: the Lumad in early colonial Mindanao. Cornell Southeast Asian Studies Program Publications, Ithaca
Robles EG (1969) The Philippines in the nineteenth century. Malaya Books Inc., Quezon City
Rodriguez Berriz M (1887–1888) Diccionario de la administracion en Filipinas (Dictionary of Philippine administration). Establecimiento Tipo–Lit. de Ma. Perez (Hijo), Manila
Salazar W (2000) German economic involvement in the Philippines, c. 1870–1918. Dissertation, University of London
Saleeby N (1908) The history of Sulu. Bureau of Science, Manila
Schurz WL (1939) The Manila Galleon. E.P. Dutton & Co. Inc, New York
Scott WH (1989) Filipinos in China before 1500. De La Salle University, Manila
Sobre empadronamiento (About registration) (1839) 205/2–c. Sobre el empadronamiento de Chinos (About the registration of the Chinese), 31 August 1839. Archivo de la Provincia Augustiniano de Filipinas (APAF), Valladolid, Spain
Tagliacozzo E (2005) Secret trades, porous borders: smuggling and states along a Southeast Asian frontier, 1865–1915. Yale University Press, New Haven, Connecticut
Treacher WH (1891) British Borneo: setches of Brunai [sic], Sarawak, Labuan and North Borneo. Government Printing Department, Singapore
Ultramar (436), Expediente 4. Conduccion de operarios para unas minas de cobre en Mancayan (Bringing workers for the copper mines in Mancayan). Archivo Historico Nacional (AHN) (National Historical Archives), Madrid, Spain
Ultramar (5217) Exp. 43, Sobre reforma general de la legislacion de chinos (About the general reforms in the legislation related to the Chinese) (Doc 3.2 Manila, Junio 1871). AHN, Madrid, Spain
Ultramar (5221) Expediente 57, Toma de medidas para evitar contraband chino en Mindanao (Measures taken to suppress the Chinese contraband trade in Mindanao). AHN, Madrid, Spain
Ultramar (5339) Expediente 13. El Subsecretario del Ministro de Estado traslada con R.O. fecha 4 de Febrero de 1890 una instancia del gremio de Sangleyes solicitando reformas en el Nuevo reglamento sobre capitacion personal de Chinos (The sub–secretary of the Minister of State forwarded with the Royal Order dated 4 February 1890 a petition from the Chinese Gremio soliciting reforms in the new regulation on the personal taxation from the Chinese). AHN, Madrid, Spain
Unreferenced (1865) Estado del numero de tributos de las islas Filipinas y de las cedulas de inscripcion para las provincias (1865), Papeles Relativos a las provincias de Ultramar colecionados por Eugenio Alonzo y Sanjurjo (State of the numerical tributes of the Philippine islands and from the receipts from the provinces (1865), papers related to the overseas provinces collected by Eugenio Alonzo y Sanjurjo). Biblioteca Nacional de España (BNE) (National Library of Spain), Madrid, Spain
Victoria F (2023) Opium, the ethnic Chinese, and the Philippine revolution: disruption and adaptation in the midst of regime change, 1896–1901. Chinese Studies Journal 17:140–183
Warren J (1985) The Sulu zone, 1768–1898: the dynamics of external trade, slavery, and ethnicity in the transformation of a Southeast Asian maritime state. New Day Publishing, Quezon City
Wickberg E (2000) Chinese in Philippine life, 1850–1898. Ateneo de Manila University, Quezon City
Wickberg E (2001) The Chinese Mestizo in Philippine history. Kaisa Para sa Kaunlaran, Manila

Wright LR (1972) The Anglo–Spanish–German treaty of 1885: A step in the development of British hegemony in North Borneo. The Australian Journal of Politics and History 18(1):62–75
Wu C (1977) Historico–economico aspects of Sino-Philippine relations, 1603–1762. University of the Philippines Library, Quezon City
Yen CH (1985) Coolies and Mandarins: China's protection of overseas Chinese during the Late Ch'ing Period (1851–1911). Singapore University Press, Singapore
Zarco R (1969) The Philippine Chinese and opium addiction. In: Felix A (ed) The Chinese in the Philippines, 1770–1898, vol 2, pp 96–109

Jely A. Galang is Associate Professor of History at the University of the Philippines (UP)—Diliman. He serves as Deputy Director of the UP Third World Studies Center and the Editor-in-Chief of the *Chinese Studies Journal*, an open-access, peer-reviewed journal of the Philippine Association for Chinese Studies. He has a Ph.D. Asian Studies (History) degree from Murdoch University, Western Australia. His recent works have appeared in *Journal of Chinese Overseas, Translocal Chinese: East Asian Perspectives* and *Southeast Asian Studies*, among others. His research interests include Philippines–China historical relations, Philippine social and economic histories, ethnic Chinese in Southeast Asia and modern history of China.

Chapter 4
An Emerging Feminine Force: The Malayan Women's Chin Woo Athletic Association (1920–1941)

Hsien-Wei Kuo and Chin-Fang Kuo

Abstract Chinese women immigrating to Malaya at the end of the nineteenth century constituted nearly one-fifth of the adult migrant population by the 1920s. This wave of Chinese female immigrants can be divided into active, dependent, and passive migration. Among them, a group of distinguished women founded the Selangor Chinese Women Chin Woo Athletic Association (CWAA) and Penang Ladies CWAA. Through a strategy that promoted 'enlightened women's education, robust female health, and a broadening of perspectives', the CWAA members broke free from the shackles of the Chinese traditional adage 'a woman's virtue lies in her lack of talent', thereby paving the way for women's rights in sports and nurturing an ideal of women versed in scholarly pursuits and martial arts. This symbolises the transnational reach of education. The two women's CWAAs offered Chinese women in Malaya the opportunity to integrate into the local community and participate in exchanges in China and Southeast Asia. This undoubtedly posed a challenge. The CWAAs' female pioneers and the members and learners represented a force in Malaya before the WWII outbreak at the end of 1941. The era stories of women's CWAAs demonstrate their historical roles in crossing borders and generations.

Keywords Chinese immigrant community · Feminine force · Malayan Chinese women · Martial arts · Transnational girl's education

H.-W. Kuo
Center for General Education, National Tainan Institute of Nursing, Tainan, Taiwan
e-mail: hsienwei@ntin.edu.tw

C.-F. Kuo (✉)
Department of Sport Management, Aletheia University, New Taipei City, Taiwan
e-mail: au4423kuo@gmail.com

H. M. Y. Ho and Y. Chan (eds.), *Transnational Southeast Asia*, Asia in Transition 29,
https://doi.org/10.1007/978-981-96-3608-2_4

4.1 Introduction

In the nascent Chinese immigrant society of Malaya (modern-day Malaysia), early immigrants were predominantly male, and initial community organisations were principally composed of men. These early Chinese immigrants established various associations driven by the need for connections, mutual support, and the protection of communal interests. These associations were typically founded on professional, kinship, and regional ties and secret societies (Chang et al. 2014), the latter of which was progressively outlawed by the British colonial government. The first three types of organisations became incubators for Chinese community leaders. They are trade unions based on "professional ties", kinship associations based on "blood ties", and regional guilds and halls based on "birthplace ties". Additionally, sacrificial organisations are also based on "spiritual (God) ties". Furthermore, secret societies were the main source of Chinese immigrants in Malaya in the early days. A link in the organisational structure of a community that is not unique to a certain dialect community. Secret societies can also be called private parties, the basis of the social order at that time. Especially when the colonial government wanted to obtain immediate economic benefits but had insufficient social knowledge, it had to rely on the operation of secret societies. However, when there were conflicts of interest between the colonial government and the Chinese society, and major divisions occurred within the Chinese community due to economic interests, the colonial government began to intervene in the management of social order. Thus, it began to ban secret societies in 1889 (Zhang et al. 2014).

Recognised leaders within the Chinese community (whom the colonial authorities also accepted) often arose from the Associated Chinese Chambers of Commerce and Industry, established in 1906 as an entity that transcended factional affiliations (Yong 1977). As the number of Chinese women in Malaya grew, demographics shifted; between 1891 and 1900, 91,047 Chinese women arrived, representing 5.19% of the adult migrant population. From 1911 to 1920, the annual average number of adult female immigrants reached approximately 20,000 individuals, accounting for 10–19% of the adult migrant population, indicating a rapid surge in Chinese female immigration. This wave of Chinese female immigrants can be roughly divided into three types, including active migration (referring to educated women who are willing to migrate to make a living), dependent migration (referring to women who follow their husbands or family members or looking for husbands), and passive migration (referring to a small number of involuntarily forced migrant women) (Fan 2005).

The establishment of Kuen Cheng Girls' High School in 1908 marked the advent of the first Chinese girls' school in Malaya, an era in which prevailing social mores stigmatised women for merely existing in the public eye. Initially attracting only 18 students, the school saw a gradual rise in enrolment (Nanyang Siang Pau, 14 October 1949). The school also indirectly catalysed the germination and subsequent growth of Chinese women's organisations in the region. In 1917, the Chinese Lady Association was founded to enhance the welfare of young Chinese women by providing them with social and mutual aid opportunities through educational classes (He and

Chen 2021). Beyond the abovementioned community structures, cultural and literary associations transcended the traditional confines of the clan, occupational, kinship, and regional organisations, reflecting an evolution in the organisation and expression of Chinese society post-1890 (Shi 2003). The Chin Woo Athletics Association (CWAA) exemplifies this category, including the Chinese Women CWAA in the state of Selangor (1921), the Chinese CWAA in Selangor (1921), the CWAA in the town of Kampar in the state of Perak (1921), the CWAA in Singapore (1922), the CWAA in the state of Negeri Sembilan (1922), the CWAA in Ipoh, the capital of Perak (1924), the CWAA in the state of Penang (1924), the Penang Ladies CWAA (1924), the CWAA in the town of Taiping in Perak (1930), and the CWAA in the state of Malacca (1932).

Most of these organisations are established and run by the children and grandchildren of transnational Chinese immigrants. Take the two Chin Woo Women's associations in Penang and Selangor as examples. The first president of Penang Ladies CWAA, Ms. Siew Yin Chung, is the daughter of wealthy businessman Kapitan Chung Keng Kuee. Siew Yin Chung's eldest daughter, Miaolan Lin (LAM Beow Lan), founded the Penang Ladies CWAA. The second president of Kuala Lumpur Selangor Chinese Women CWAA, Cuilan Lin (LAM Chooi Lan), is Chung Siew Yin's younger daughter. She is also one of the founders of the Penang Laddies CWAA. Cuilan Lin's husband is Alan Loke, the son of wealthy businessman Luk Yew of Kuala Lumpur, and his mother, Shugui Lin, is from a prominent Baba family (Wang 2019).

Distinguishing themselves within this network, Selangor and Penang featured not only the Selangor Chinese CWAA and the Penang CWAA but also the Selangor Chinese Women CWAA and the Penang Ladies CWAA. These sister associations, inspired by the original Shanghai CWAA, shared a commitment to promoting education and martial arts. The Penang Ladies CWAA thrived with notable achievements (Huang 1971), while the Selangor Chinese Women CWAA played a pivotal role in founding the Chin Woo Girls' School and offered services related to social welfare, striving to benefit society at large (Huang 1971). These two female athletics associations in the Chinese community engaged in martial arts, sports, and cultural and artistic pursuits but also actively participated in welfare and public service activities, contributing to the broader development of Malayan society and the Chinese community therein (Shi 2005). Here it refers more to ethnic societies. That is to say, in an era when "women's incompetence is a virtue", the Penang and the Kuala Lumpur Women's associations advocated both moral, intellectual, and physical fitness, which was of substantial significance to Chinese women and Baba Nyonya's sports and education. For example, Siew Yin Chung, the president of the Penang Ladies CWAA, emphasised that only with a strong body can one have a healthy mind to serve family and seek well-being for society (Penang Ladies CWAA 1970).

Meanwhile, the Chinese community in Malaya, amid the dominant Malay culture and other ethnic influences, established their presence in urban centres through regional concentration; in rural areas, they maintained their lifestyle and social ties through tribe settlement structures (Lin 2017). According to the immigration forms at that time, they could be divided into "non-free immigrants" and "free immigrants".

The former is "indentured labour", commonly known as "Zhu Zai (piglets)"; the economic status of the latter is relatively good. As the status of the Chinese evolved from migrant workers to key players in Malaya's mining, commerce, and agriculture sectors, along with an increase in their population, traditional private schools morphed into larger modern educational institutions. This shift facilitated the dissemination of political consciousness from China among teachers and students (Chen 2017). This illustrates that the transnational role of the Malayan Chinese and their transnational origins stem from this group of migrant workers and key players.

The Chinese diaspora harboured a variety of social and economic pledges to China, their ancestral homeland, fostering a sincere emotional and identarian connection (Ho 2018). In Malaya, Chinese women's associations and groups within Chinese associations were very active in society. Their members dedicated themselves to various activities to serve their associations and the Chinese community and advance Malaya's societal progress (Shi 2005). The outbreak of the Second Sino-Japanese War in 1937 galvanised Malayan Chinese women to become involved in relief efforts, inspired by the rallying cry that 'in the nation's peril, every woman has a duty'. Relief committees throughout the region harnessed public gatherings and commemorative occasions to raise awareness of their cause through speeches, songs, literary works, and dramatic performances (He and Chen 2021). This chapter investigates the organisational development of the Malayan women's CWAAs, focusing on their operational policies and strategic actions to elucidate their role and significance in that era. Furthermore, the transnational elements and implications of the relationship between the founders of the women's associations and their grandparents and the immigration experience can be made clearer from the above elaboration.

4.2 The Cradle of the Feminine Force: The Founding of the Women's CWAA

In the 1920s, Malaya's societal ethos was insular, strictly adhering to traditional rites and customs. Female participation in physical sports was rare and often met with disapproving glances unless a girl or woman had the benefit of progressive parents willing to allow entry into the realm of Chin Woo. It was widely believed among the contemporary intelligentsia that the vigour of a nation's population directly influenced its national power. Strengthening citizens' physical health was deemed essential to enhance the composite national strength of the state, with particular emphasis on the physical conditioning of women (Tan 1928, p. 9). The then-president, Jinglian Lu (1971), stated that one cannot help but be eternally grateful to the visionary founders of the CWAA branches who, with exceptional foresight and intellect, resolutely established a platform to foster the spirit of women's sports, bestowing enduring benefits upon future generations. The above passage means that the existence of the Chin Woo women's associations not only creates a place for the young generation born in Malaya's new homeland to improve their Chinese and

Western knowledge and physical skills but also creates an interpersonal network for the Chinese diaspora who have close ties with China.

4.2.1 The Selangor Chinese Women CWAA

Before 1920, the city of Kuala Lumpur (and today's capital of Malaysia) and, by extension, British Malaya, lacked grassroots organisations founded by women until the Shanghai CWAA dispatched five ambassadors of Chin Woo—namely Gongzhe Chen, Shichao Chen, Huisheng Li, Xiaoao Luo, and Shutian Ye—to Southeast Asia to promote the ethos of Chin Woo. During this campaign, the five ambassadors conveyed Chin Woo's philosophy through public speeches and martial arts demonstrations. Among them, Shichao Chen, the sole female martial artist, delivered impactful speeches on the physiological and psychological benefits of women practising martial arts and the significance of gender equality (Shanghai CWAF 2021). She performed martial arts, showcasing the agility and skill of women on par with their male counterparts, challenging the traditional image of feminine fragility, and adding a vibrant dimension to the societal perception of women (Federation of Chin Woo Associations Malaysia 2003). Afterwards, she also briefly stayed in Malaya for guidance and teaching of martial arts. The short exchange in Malaya of Shichao Chen from mainland China could be said to be a Chinese woman's effort to build a sense of community and identity across national borders.

The performance and efforts of the above-mentioned Shichao Chen quickly inspired a group of progressive women in Kuala Lumpur, including Caiyun Tan, Xuehua Wu, Danqiu Weng, Qiaomei Li, Zhiyi Yin, and Ruiqin Deng (Selangor Chinese Women CWAA 1971). In 1921, they founded the Selangor Chinese Women CWAA and successfully registered it with local authorities, officially establishing it on 1 July. The association became a vanguard for advocating for women's right to engage in sports and martial arts training and spearheading the gender equality movement (Yang 2005, p. 5). Caiyun Tan was elected as the first president, and in December, Zhixi Li was appointed secretary and coach (Jung 1963; Selangor Chinese Women CWAA 2005). In 1923, Cuilan Lin assumed the role of second president. Upon Miaolan Lin's invitation, Zhixi Li left to aid in the establishment of the Penang Ladies CWAA, with her previous duties taken over by the Shanghai CWAA's instructor, Qiongshan Feng (Selangor Chinese Women CWAA 2005).

As the number of members increased, the Selangor Chinese Women CWAA operations expanded, leading to purchasing land to construct a new clubhouse. Funds were raised through the concerted efforts of all members, and the new clubhouse was completed in 1929, allowing for the phased development of the association's expanded activities (Federation of Chin Woo Associations Malaysia 2009). In 1932, Qiongshan Feng resigned and returned to China, with Xiuan Cai succeeding as secretary and coach (Federation of Chin Woo Associations Malaysia 2003). The period from 1930 to the early 1940s marked an era of strategic expansion for the association, which was committed to nurturing a new generation of women, particularly through offering classes in the Chinese language, Chin Woo martial arts, and Western sports.

4.2.2 The Penang Ladies CWAA

In Penang's conservative climate of the 1920s, the traditional Chinese community held fast to the maxim that a woman's virtue was her lack of talent, leaving no room for women's organisations. Miaolan Lin (who was once the principal of Penang Pikru Girls' School) ignited the cause of the CWAA and rallied a group of women, including Siew Yin Chung (Kapitan Chung Keng Kuee's daughter, Miaolan Lin's mother), Shujun Lin, Sukuan Chen, Cuilan Lin (Miaolan Lin's sister), Xikun Xie, Zhenjin Zheng, and Ping Xie, to establish the Penang Ladies CWAA. The aim was to dispel the notion of female frailty and empower girls through literacy and physical strength (Li 1963).

In 1920, the local government enacted the Registration of Schools Ordinance to curtail gatherings that could disrupt the social order. Previously, the British government had adopted a laissez-faire approach to Chinese schools (Loh 1975), which meant that the colonial administration neglected Chinese language education (Roff 1974). This changed with the implementation of the Registration of Schools Ordinance, which empowered the director of education to shut down schools that did not meet stipulated requirements concerning the curriculum, administration, and hygiene. All schools and educational and administrative staff must register and comply with the director's mandates (Chen 2017). In 1923, the government appointed Europeans to be deputy directors of education in Chinese schools, and inspectors were designated a year later. The British government amended the ordinance in 1925, granting the director of education even more power to deny registration to any teacher involved in 'subversive' activities (Roff 1974). From this point onwards, the operation and registration of Chinese schools were tightly controlled.

The intentions of the Penang Ladies CWAA were pure, and the British authorities swiftly approved its registration, leading to its official formation in February 1924, with Xiuying Zheng serving as its inaugural president (Li 1963). Successive presidents (including Shujun Lin, Miaolan Lin, and Xikun Xie) passionately and strategically drove the association's progress. Educational initiatives, such as the establishment of "National Language Classes", promoted the national language movement and enabled education for older women who had missed schooling opportunities, thereby elevating the status of women (Li 1963). The curriculum for female members included Chinese classes on Monday and Friday mornings and Friday afternoons; literacy classes on Thursdays and Saturday afternoons at 2 pm; and martial arts training daily from 4 to 6 pm. The courses were primarily funded by donations from board and association members (Selangor CWAA 1979). Initially, the Penang Ladies CWAA operated on rented premises with fluctuating locations, courses, and schedules. During its growth, the association was active in relief efforts for China's resistance against Japanese aggression in 1931 and 1937 and in European war relief efforts for Britain, achieving notable fundraising success until the Japanese occupation of Penang in 1941 brought all activities to a standstill (Selangor Chinese CWAA 1949).

In sum, the Selangor Chinese Women CWAA and the Penang Ladies CWAA were led by their respective presidents with well-defined organisational structures emphasising Chinese education, martial arts, sports, and recreational activities. Beyond the fulfilment of their foundational tasks, the two women's associations proactively engaged with external entities from mainland China and other areas like Hong Kong to expand their social networks and participated in social relief efforts (see more details in the following parts).

4.3 Fostering Women's Education and Physical Fitness: The Action Strategies of the Women CWAA

A survey conducted in 1931 revealed that the literacy rate among Chinese women in Malay was 9.7%, surpassing the 3.9% of their Malay counterparts. One potential reason could be that Chinese women had already received some education before arriving in Malaya. Another significant contributing factor was the establishment of Chinese schools (Tan 2009, p. 310). Pictorials and magazines of the time depicted women stepping into modern educational settings and emphasising women's physical well-being and self-actualisation (Teoh 2021). The progress in these two domains can be attributed to the diligent efforts of the Selangor Chinese Women CWAA and the Penang Ladies CWAA, including establishing schools, dispatching representatives to participate in competitive events, organising first aid training courses, performing charitable acts, and engaging in close-knit exchanges with the Shanghai Chin Woo Athletic Federation (CWAF) and other CWAAs, school sports clubs, and the Nanjing Central Martial Arts (Guoshu) Institute in Mainland China, thereby contributing to social welfare and output.

4.3.1 Transnational Reach of Education: Establishing the Chin Woo Girls' School

Before the twentieth century, the Chinese in Malaya, aside from attending missionary schools, sought to perpetuate Chinese educational traditions through private tutoring, primarily in various Chinese dialects as a medium of instruction (Quek and Gao 2016). In 1898, the Qing Dynasty's Guangxu Emperor launched the Hundred Days' Reform, leading overseas private tutors to gradually build modern schools, which began offering an expanded curriculum including moral education, classics, the Chinese language, English, history, geography, math, physics, and gymnastics (Chen 2017).

The early twentieth century saw a significant influx of Chinese migrants into Malaya, consequentially boosting the Chinese population and leading to the proliferation of Chinese schools across towns and cities. The first of these modern educational institutions was the Confucian Chung Hwa School in Penang, founded in 1904; the first modern school for girls was Kuen Cheng Girls' High School, established in Kuala Lumpur in 1908. Following the fall of the Qing Dynasty in 1911, the Chinese community in Malaya embraced a new style of education, leading to an unprecedented boom in the opening of schools (Tan 2009). The nationalist government began to recognise education as a vital means of strengthening China and a tool for nation-building and modernisation. The nationalist government reorganised the education system and overhauled curricula. It replaced traditional teaching with ideologies of republicanism and modernism, further integrating overseas Chinese education into the mainstream of China's education system (Lu 2010). In their initial stages, many of these institutions were founded by scholars from mainland China. These literati, academics, officials, and writers traversed the Nanyang region, residing in or working there. Some chose to stay temporarily or permanently, turning to teaching as a livelihood (Teoh 2021).

The Chinese community in Malaya placed deep emphasis on secondary education in the wake of the May Fourth Movement of 1919. Community leaders recognised the role of education in raising literacy rates and the overall quality of local Chinese people and in preserving the cultural identity of Chinese people living overseas. Thus, they were willing to donate funds to establish Chinese secondary schools (Lu 2010). By 1920, the Federated Malay States had 181 Chinese schools, while the Straits Settlements had 313 (Chen 2017). The motivation behind the founding of Chinese girls' schools was to cultivate patriotic sentiment towards China, reinforce Chinese cultural knowledge, uphold social mores, and eliminate unsavoury customs (Wong 2004). By doing so, transnational ties of Chinese communities in Malaya back to Mainland China were fostered.

On 5 December 1921, Peking University President Yuanpei Cai made a brief stopover in Singapore and participated in a welcoming event organised by four local schools: Tao Nan School, Tuan Mong High School, Qifa Primary School, and Chinese High School. During his speech, he advocated for the four pillars of education—physical, intellectual, moral, and aesthetic, emphasising that physical education was about developing students' bodies and invigorating their spirits rather than seeking ephemeral fame through racing or sports (Khor and Teoh 2020). The 1920s marked a significant period of growth for Chinese communities in Malaya. Cultural enlightenment following the May Fourth Movement significantly influenced the promotion of modern Chinese education.

The operation of Chinese schools in pre-war Malaya remained deeply influenced by China, and these institutions essentially functioned as extensions of the Chinese education system abroad (Tay 1998). To facilitate the acquisition of the Chinese language for their members' children and to provide educational opportunities for women who had missed out on schooling earlier in their lives, the Selangor Chinese Women CWAA opened a women's half-day school in 1934, which was formalised as a primary school in 1936 and named the Chin Woo Girls' School. The establishment aimed to eradicate illiteracy among women, foster independence, and elevate their social status, marking the beginning of the educational empowerment of Chinese women overseas (Selangor CWAA 1979, p. 74). The school dedicated half an hour after class to teach calisthenics, including Tam Tui and Quan (martial arts, Chinese boxing) skills, and guided students in various ball sports. Moreover, the education provided often included discussions on the current status of women globally, including beyond Malaysia and other Southeast Asian nations in the vicinity, to nurture physically and mentally sound female citizens, thus enhancing women's status in cultural knowledge and socioeconomic aspects (*Nanyang Siang Pau* 19 October 1933).

Owing to the significant expenses involved in founding a school, members of the board of the Women's CWAA, who also oversaw the Chin Woo Girls' School, contributed both funds and efforts, and each board member sponsored the educational expenses of a student from an underprivileged background (Selangor Chinese Women CWAA 2005). In 1940, the board appointed Deng Zhicheng as the principal; simultaneously there was an increase in the number of teachers and classes, signalling a period of growth and improved educational facilities (S.J.K.C. Chen Moh 2016, p. 32). In 1941, Chin Woo Girls' School, Chin Woo School, Peide Girls' School, Baiping School, and Confucian Private Secondary School jointly hosted the first martial arts demonstration featuring both group and individual performances. This event facilitated mutual learning among schools and showcased students' talent to visitors, highlighting well-rounded and healthy female students (*Nanyang Siang Pau* 15 June 1941b). However, in December of that same year, the Japanese invasion of Malaya destroyed school buildings and educational equipment, causing Chin Woo Girls' School to cease operations. This cease of operation immediately impacted learning social and cultural knowledge and preserving a capital legacy.

4.3.2 Broadening Sports and Recreational Classes, Going Back to China for Games, and Engaging in Relief and Charity Activities

According to reports by *Nanyang Siang Pau*, the fourth Malayan Chinese Athletics Meet of 1937 included martial arts displays and participation in various sports competitions. Besides the Selangor Chinese Women CWAA and the Penang Ladies CWAA, which were purely female sports entities, other teams were primarily affiliated with associations or athletic clubs (Nanyang Siang Pau, 21 February 1939a). While the CWAA branches prioritised preserving and promoting martial arts, they also encouraged and developed other disciplines such as athletics, ball games, and swimming. Special reports on basketball, table tennis, lawn tennis, shot put, and swimming were featured in the Penang Ladies CWAA commemorative publication (1963).

The Penang Ladies CWAA actively promoted and facilitated basketball, volleyball, track and field, swimming, and martial arts training. Several outstanding players trained in basketball and volleyball were selected to participate in the second, third, and fourth Malayan Chinese athletics meets held in China (Li 1989, p. 51). Furthermore, the association organised swimming classes guided by staff members Xiuqiong Liang and Bingjian Lu on weekends. Athletes such as Yuying Zhang, Rihong Liang, and Manqiong Zheng, under the attentive coaching of committee members Huajin Chen and Zhenqing Chen, showed excellent results and were chosen to represent the team in national sports meetings in China (Li 1963).

Recognising the importance of martial arts proficiency in promoting the recruitment of new female members and enhancing their sense of achievement and self-identity, the Selangor Chinese Women CWAA administered a martial arts exam in 1939 to assess its members' skills. They invited Yaohui Cao and Taijie Chen, the founders of the Penang CWAA, as chief judges and Zhixi Li from the Penang Ladies CWAA and Xiuyuan Wu from the Singapore CWAA, who served as the main examiners (*Nanyang Siang Pau* 1 December 1939g). Mr. Yaohui Cao highlighted the increasing number of women practising martial arts, breaking the feudal mindset of women's weaknesses and forming a fundamental capability for wartime efforts. He also stressed this as self-defence (*Nanyang Siang Pau* 18 December 1939h). The twelve graduates inspired each other to further their studies and skills. During the fifteenth anniversary of the Penang CWAA and the graduation ceremony for martial arts students, Ms. Yaomei Liu encouraged graduates not to rest on their laurels but to continue striving for self-improvement, particularly during the crucial period of war and nation-building, calling for a total mobilisation of spirit from everyone (*Nanyang Siang Pau*, 20 June 1939d). It can be seen from this that the Second Sino-Japanese War also affected Southeast Asia and China and their transnational relations.

In addition to promoting martial arts and various sports, the Selangor Chinese Women CWAA and the Penang Ladies CWAA introduced recreational activities to relieve stress and fatigue, such as music, dance, and drama performances for social charity (Li 1963). For instance, they participated in commemorative assemblies and fundraising activities for social relief. In 1922, the Selangor Chinese Women CWAA celebrated its first anniversary at 'Xin Garden', featuring a five-act play, dance, and martial arts (Selangor Chinese Women CWAA 2005). In February 1930, the Penang Ladies CWAA held a tenth-anniversary celebration where the then-chairwoman Yaomei Liu addressed the gathering, stating: 'The establishment of the Penang Ladies CWAA has had a profound societal impact, pioneering the study of physical education among women and fostering its widespread adoption. This initiative has been instrumental in dispelling the erstwhile debilitating customs of women's frailty in our ancestral homeland of China, nurturing a robust constitution, and invigorating the spirit of women' (*Nanyang Siang Pau* 28 February 1934).

In May 1938, the Penang Ladies CWAA responded to Soong Meiling's call for a charity performance (Cantonese opera) to raise funds for disaster-stricken children in China, urging the overseas Chinese community to contribute to relief efforts (*Nanyang Siang Pau* 19 May 1938). In April 1939, as the local government intensified its air raids, the Penang Ladies CWAA decided to conduct first aid training classes to assist the government in a war (*Nanyang Siang Pau* 14 April 1939b). Within a month, 28 individuals, all proficient in English and possessing a reasonable level of education, participated in the training (*Nanyang Siang Pau* 26 May 1939c). In June of that same year, the Selangor Chinese Women CWAA, in response to the bombings by Japanese forces in China, made not only generous donations but also organised recreational fundraising for medical supplies to aid wounded soldiers and refugees in the motherland (*Nanyang Siang Pau* 26 June 1939e). Additionally, they fused elements of Chinese culture with patriotism and promoted these values to an international audience through specially curated performances and martial arts displays that embodied the spirit of female citizens (*Nanyang Siang Pau* 9 July 1939f).

In June 1940, the Penang Ladies CWAA prepared plays, carnivals, and martial arts demonstrations to raise money to support Britain (*Nanyang Siang Pau* 2 April 1940a; *Nanyang Siang Pau* 24 June 1940b). By March 1941, the Penang Ladies CWAA held charitable performances and sales during the Penang Chinese Relief Fund Carnival, once again dedicating its efforts to the cause of China (*Nanyang Siang Pau* 26 March 1941a). These initiatives illustrate how members of the Women's CWAAs applied their acquired skills to their utmost capacity during wartime, providing support to their ancestral homeland and the colonial government and, of course, assisting local communities with relief.

4.3.3 Transnational Exchanges Within Kinship and Lineage Organisations

Transnationalism, as a perspective that examines localities across multiple nations, involves the multiple ties and interactions that connect people or institutions across national borders (Nowicka 2020; Vertovec 1999). Migrants can establish and maintain connections with family members, communities, institutions, and governments in their countries of origin, serving as vehicles for social and cultural exchanges, such as those in the realms of education, sports, and research (IOM 2010).

The organisational evolution of the CWAAs in the Malayan region fundamentally mirrors a model of bloodline replication (Kuo and Kuo 2023). Coupled with the Chinese immigrant community's inclination towards clustering based on professional, familial, and regional ties, cross-border exchanges through various activities influenced the economy, society, culture, and politics of both the diaspora's new homes and ancestral land. In the 1920s and 1930s, martial arts instructors hired by Malayan CWAAs were often recruited from the Shanghai CWAF. Notable instructors included Keyi Luo, Zhixi Li, Jinglin Cai, Qiongshan Feng, Weichang Lu, Chanbo Yao, Shaolin Li, Ruibiao Li, Xiuyuan Wu, Liancheng Zhao, Zhixiang Liu, Qiying Chen, Yuqin Wang, Qinggui Liu, Chengzhang Wang, Fenqi Ye, Shuixiang Ye, Shutian Ye, Shushen Ye, Yuanfeng Wei, Dechun Zhang, and Peixian Li, who were tasked with teaching and guiding martial arts in numerous associations (Tsao 1963).

The two Women's CWAAs are no exception to this practice. For instance, in the early days of the Selangor Chinese Women CWAA, five individuals—Jinglian Lu,

Sulian Gao, Yuelan Deng, Dexin Mai, and Peiying Liu—were sent to the Shanghai CWAA to study martial arts for four years (1923–1926). They earned advanced diplomas and participated in martial arts performances at the 1924 National Sports Meet (Lu 1971). In 1927, during the third-anniversary celebration and middle school graduation at the Penang Ladies CWAA, coinciding with the visit of the Hong Kong Chin Woo touring team, which was promoting Chinese martial arts and national music, the Penang Ladies CWAA hired a team to run a two-night carnival at the Western Assembly Hall to launch a fundraising campaign for their clubhouse (Selangor Chinese CWAA 1949). In 1929, Penang Ladies CWAA President Siew Yin Chung travelled to China, visited the Shanghai CWAA, and subsequently invited Cantonese music instructor Shifu Gan to Penang to coach female members. At the time, the association saw an influx of 30–40 new members (Selangor Chinese CWAA 1949).

In August 1932, men's and women's basketball and volleyball teams in Guangdong came to Malaya for exchanges. Sports organisations and school societies in Penang convened a special meeting to deliberate on the details of the sporting exchange; the Penang Ladies CWAA was also invited, and Zhixi Li contributed to the discussions (*Nanyang Siang Pau* 13 August 1932a). In September of that same year, to universalise Chinese martial arts, the CWAA branches across Malaya collaborated in a performance at the Great World arcade, initiated by a martial arts speech by Weichang Lu, followed by diverse martial arts demonstrations by associations from Singapore, Penang, and Ipoh (*Nanyang Siang Pau* 22 September 1932b).

At the third Penang–Malaya Sports Meet held at the Chinese Recreation Club's field in June 1935, the Penang Ladies CWAA was invited to perform; 32 members dressed in Chin Woo uniforms performed Chinese martial arts in precise unison (*Nanyang Siang Pau* 26 June 1935b). Subsequently, the Selangor Chinese Women CWAA dispatched a basketball team for a friendly match with the visiting basketball team from the Shanghai Liangjiang Women's School of Physical Education (*Nanyang Siang Pau* 3 May 1935a). In 1936, Zhijiang Zhang, the director of the Nanjing Central Martial Arts (Guoshu) Institute, led his staff and students from the National Normal College of Martial Arts and Physical Education during promotional and exchange visits to Malaya. Before they arrived in Kuala Lumpur, the Associated Chinese Chambers of Commerce and Industry invited the CWAA, along with other community organisations and school representatives, to discuss and coordinate a welcome event and performance (*Nanyang Siang Pau* 12 February 1936a). Twenty days later, the Penang Ladies CWAA and the overseas Chinese community collaborated in hosting a special welcome tea party for the exchange (*Nanyang Siang Pau* 2 March 1936b), during which the importance of being versed in the literature and martial arts was emphasised, including instruction on martial arts and physical education techniques provided to students as well as nurturing knowledge and moral values, truly befitting the term 'intellectual'.

Before Malaya was overtaken by the Japanese invasion in December 1941, the two Women's CWAAs and other sister associations in the region frequently interacted with students regarding their operations, instructors, and membership. Moreover, they maintained connections with the Shanghai CWAA in China while exchanging with organisations from China, Hong Kong, and other territories. These transnational engagements served as public displays of the growth and development of Chinese women in and beyond Southeast Asia, thus providing significant arenas and opportunities for new thinking and the broadening of horizons, especially in cultural, language, sporting, and martial arts knowledge and skills.

4.4 Conclusion

In the 1920s, a group of advanced Chinese women in Malaya established two women's CWAAs to create platforms through these grassroots organisations that offered Chinese women opportunities for integration and in-depth engagement with Malaya's social, cultural, and economic spheres. This historical record presents opportunities and challenges for future research. These women represented an emerging force within the Malayan Chinese community during that time.

Given the challenges in determining the exact number of Malayan Chinese women and their children who benefited from these two women's CWAAs (due to the scarcity or incomplete nature of records following the Japanese occupation), it is difficult to ascertain the precise increase in literacy rates attributed to these associations or to present definitive indices of health improvements as a direct outcome of martial arts and physical training. However, the linguistic agility of the second and third generations—who, since the 1980s onwards, have adeptly switched between Malay, Chinese (including regional dialects), and English—is evident. Their global forays in pursuing education, business ventures, transient sojourns, and long-term settlements, testify to these associations' intangible yet authentic and ongoing legacy, whose impact has been unmistakably profound and enduring, although not easily quantified.

The CWAAs in Malaya, including the two women's CWAAs illustrated in this article, started their establishment and development through the cloned model passed down from the bloodline of the Shanghai CWAF in China (Kuo and Kuo 2023). Through education, martial arts, and sports, the two CWAAs in Kuala Lumpur, Selangor, and Penang have improved the perception of women in the Chinese community and Malayan society while working hard to establish themselves and thrive. In addition, through cross-border exchanges, maintain umbilical cord relations with China, maintain friendly relations with Southeast Asia, and influence each other. The contemporary stories and challenges of these two women's CWAAs demonstrate a historical role that crosses national borders and generations.

References

Chang HP, Chang WA, Lee LS (2014) The belief in god, human relations and social organization: Penang tanjong tokong thai pak koong and its ritual organizations. Global Hakka Studies 3:111–138

Chen CN (2017) A study on Chinese education and Chinese self–identity in Malaysia. The Elementary Education Journal 64(1):67–81. https://doi.org/10.6701/TEEJ.201703_64(1).0005
Fan RL (2005) Immigration, gender, and the Chinese community: a study on Malayan Chinese women. Zhongguo Hua qiao, Beijing
Federation of Chin Woo Associations Malaysia (2003) Federation of Chin Woo Associations Malaysia special edition. Malaysia
Federation of Chin Woo Associations Malaysia (2009) Special edition for the 16th National Chin Woo Carnival. Malaysia
He YH, Chen TY (2021) Malaysian Chinese women's social participation in democratization: a case study from two Taiwan alumni. Thought and Words: Journal of the Humanities and Social Science 59(3):107–175
Ho KC (2018) Virtual and reality of national identity: a preliminary study of the national congress of delegates of Malayan Chinese registered societies' struggle for citizenship. Taiwan Journal of Southeast Asian Studies 13(1):9–28
Huang CYT (1971) Unity and mutual assistance in Chin Woo as one family. In: Selangor Chinese Women's CWAA (ed) Golden jubilee commemorative special issue of Selangor Chinese Women's CWAA 1921–1971. Malaysia
IOM (2010) Migration and transnationalism: opportunities and challenges. International Organization for Migration. https://www.iom.int/migration-and-transnationalism-opportunities-and-challenges-2010
Jung J (1963) A brief biography of Ms. Zhixi Li. In: Penang Ladies CWAA (ed) 40th anniversary commemorative special issue of Penang Ladies CWAA (1924–1963). Malaysia
Khor TH, Teoh HS (2020) Cai Yuanpei's cross–border experience in Nanyang and overseas Chinese educational context. Bulletin of Educational Research 66(2):1–36. https://doi.org/10.3966/102887082020066602001
Kuo HW, Kuo CF (2023) Bloodline organization of disseminating Chin Woo Athletics to Nanyang: from cultural matrix China to sea. The International Journal of the History of Sport 40(6–7):514–532. https://doi.org/10.1080/09523367.2023.2216653
Li CH (1963) The history of the association. In: Penang Ladies CWAA (ed) 40th anniversary commemorative special issue of Penang Ladies CWAA (1924–1963). Malaysia
Li CH (1989) A brief history of Penang Ladies CWAA. In: Selangor CWAA (ed) 60th anniversary commemorative issue and fundraising for school building of Kuala Lumpur Pudu SJK(C) CHIN WOO 1929–1989. Malaysia, p 51
Lin HY (2017) The intertwined and integrated imaginations of "tradition": some observations on the repertoire and performance of kî–lîn (麒麟) professional hoklo (Min) opera troupe in Malaysia. Journal of Chinese Ritual, Theatre and Folklore 198:63–126
Loh FSP (1975) Seed of separatism: educational policy in Malaya 1876–1940. United Kingdom, Oxford
Lu CL (1971) Chin Woo memoirs. In: Selangor Chinese Women CWAA (ed) Golden jubilee commemorative special issue of Selangor Chinese Women's CWAA, 1921–1971. Malaysia
Lu CS (2010) A study of the dongjiaozong and development of Chinese education in Malaysia, 1951–2000. Dissertation, National Chi Nan University
Nanyang SP (13 August 1932a) Penang sports organisations convene to decide on welcoming the tour of South China men's and women's basketball and volleyball teams; dates set for matches, number of games still under discussion. Nanyang Siang Pau, p 9
Nanyang SP (22 September 1932b) The All–Malaya CWAAs unite for a Guoshu performance at the great world city; the final night's exhibition is moved to the new boxing venue, attracting throngs of spectators in a lively atmosphere. Nanyang Siang Pau, p 7
Nanyang SP (19 October 1933) A survey of Chinese education in Selangor: Kuala Lumpur Women's half–day school. Nanyang Siang Pau, p 8
Nanyang SP (28 February 1934) Penang Ladies CWAA celebrates its tenth anniversary in grand style; nine students graduate from junior and senior middle school, striving to spread education

widely among women and eradicate the frail and feeble customs of the past. Nanyang Siang Pau, p 9

Nanyang SP (3 May 1935a) The Chinese sports community in Selangor has decided to welcome the visiting Shanghai Liangjiang women's basketball team; the event is scheduled for this afternoon at 7 o'clock at the Beauty World Dance Hall, with the first night's match against the Confucian private secondary school boys' team, and the second night against the Selangor women's team. Nanyang Siang Pau, p 10

Nanyang SP (26 June 1935b) The third Penang Chinese sports meet concludes triumphantly with outstanding achievements; 16 records broken, CWAA and Fujian girls' school win general team championships, with Yinjiao Deng as the female champion and Suian Ye as the male champion. Nanyang Siang Pau, p 9

Nanyang SP (12 February 1936a) Following the arrival of the Guoshu group in Kuala Lumpur, the Associated Chinese Chambers of Commerce and Industry held a representative meeting of organisations and schools to establish a welcoming committee for the national sports performance. In: Jincong H, Zhongji H, Man Q, Zhanmei C, Kunling Z (ed) were appointed as chairs and vice–chairs, with the Chinese Assembly Hall selected as the welcome venue, and the timing to be determined by the chair. Nanyang Siang Pau, p 9

Nanyang SP (2 March 1936b) SMJK Chung Ling Pulau Pinang welcomes the Guoshu touring group, with Director Zhang delivering a speech encouraging students to cultivate both martial and scholarly pursuits and moral and intellectual development. Nanyang Siang Pau, p 8

Nanyang SP (19 May 1938) The Penang Ladies CWAA prepares to stage a performance to provide relief for disaster–stricken children. Nanyang Siang Pau, p 15

Nanyang SP (21 February 1939a) The third report reviews overseas Chinese sports in Malaya. Nanyang Siang Pau, p 35

Nanyang SP (14 April 1939b) The Penang Ladies CWAA decided to offer a first aid training class and is about to recruit students and welcome women from the community for participation. Nanyang Siang Pau, p 14

Nanyang SP (26 May 1939c) The Penang Ladies CWAA's first aid class includes regular training with 28 enrolled participants. Nanyang Siang Pau, p 14

Nanyang SP (20 June 1939d) The Penang Ladies CWAA celebrates its fifteenth anniversary with great fanfare and distributes certificates to the graduates of the martial arts class; the chairperson exhorts women to assume wartime responsibilities. Nanyang Siang Pau, p 16

Nanyang SP (26 June 1939e) The Selangor overseas Chinese Women's Athletic Association schedules a carnival to fundraise relief on the seventeenth of the following month. Nanyang Siang Pau, p 15

Nanyang SP (9 July 1939f) The Kuala Lumpur overseas Chinese Women's Athletic Association organises a carnival for relief efforts, featuring patriotic stories in a programme rich in international propaganda. Nanyang Siang Pau, p 11

Nanyang SP (1 December 1939g) The Kuala Lumpur Women CWAA advocates for martial arts; martial arts exams are scheduled for the next month, and a special martial arts publication will be issued for promotion. Nanyang Siang Pau, p 15

Nanyang SP (18 December 1939h) A strong nation must first strengthen its people. The Selangor Chinese Women CWAA holds martial arts exams, filling the venue with an atmosphere of promoting martial virtues, and 12 graduates encourage each other to further their studies. Nanyang Siang Pau, p 16

Nanyang SP (2 April 1940a) The Penang Ladies CWAA plans to play the game in June to aid Britain. Nanyang Siang Pau, p 16

Nanyang SP (24 June 1940b) The Penang Ladies CWAA holds a fundraising carnival to support Britain; speaker Ling Ye attends and delivers a speech hoping for a victory for Britain, with an event worth over 2000 dollars each night. Nanyang Siang Pau, p 15

Nanyang SP (26 March 1941a) The Penang Ladies CWAA decides to hold a charity sale on 29 March, issuing a statement calling for the overseas community to donate. Nanyang Siang Pau, p 14

Nanyang SP (15 June 1941b) The Selangor CWAA's martial arts performance dazzles with swordplay and blade skills and features 48 spectacular events and an enthusiastic audience. Nanyang Siang Pau, p 7

Nanyang SP (14 October 1949) Remembering the pioneer of Chinese education in Kuala Lumpur, Ms. Xuehua Wu. Nanyang Siang Pau, p 7

Nowicka M (2020) (Dis)connecting migration transnationalism and nationalism beyond connectivity. Comparative Migration Studies 8(20):1–13

Penang Ladies CWAA (1963) 40th anniversary commemorative special issue of Penang Ladies CWAA (1924–1963). Malaysia

Penang Ladies CWAA (1970) Souvenir Commemorating the Official Opening of the New Building of Penang Ladies Chin Woo Athletic Association. Malaysia

Quek SL, Gao HH (2016) The Chinese language in Singapore and Malaysia: a brief history and current perspective. Taiwan Journal of Southeast Asian Studies 11(2):45–82

Roff WR (1974) The origins of Malay nationalism. University of Malaya Press, Malaysia

SJKC Chen Moh (2016) 80th anniversary and Building B teaching block inauguration commemorative special issue of S. J. (C) Chen Moh. Malaysia, p 32

Selangor Chinese CWAA (1949) Selangor Chinese CWAA, Special edition. Malaysia

Selangor CWAA (1979) Carnival special issue of the Selangor CWAA 1979. Malaysia

Selangor Chinese Women CWAA (1971) Golden jubilee commemorative special issue, 1921–1971 of the Selangor Chinese Women CWAA. Malaysia

Selangor Chinese Women CWAA (2005) 24th National Chin Woo Carnival and the 84th–anniversary commemorative special issue of the Selangor Chinese Women CWAA. Malaysia

Shanghai CWAA (2021). Chin Woo Zhi. Wenhui, Shanghai

Shi CJ (2003) A study on the history of the associations of overseas Chinese in Malaysia. Dissertation, Jinan University

Shi CJ (2005) Female Chinese overseas societies in Malaysia. Southeast Asian Studies 2005(1):79–84

Tan AB (2009) Gender roles and nation–building: the transformation of social and economic roles of Chinese women. In: Voon PK (ed) Malaysian Chinese and nation–building: before Merdeka and fifty years after, vol 1. Centre for Malaysian Chinese Studies, Malaysia, pp 301–324

Tan YC (1928) The necessity of women learning martial arts at Chin Woo. Selangor Chinese Chin Woo, p 9

Tay LS (1998) The history of Chinese education development in Malaysia, vol 1. Gabungan Persatuan Guru-Guru Sekolah Cina Malaysia, Malaysia

Teoh HS (2021) The imprints of women's education in Nanyang: graphics of Nanyang female intellectuals in early modern Chinese pictorials and magazines and their significance. Chung Cheng Chinese Studies 37:137–154

Tsao YH (1963) Chin Woo: Nostalgia. In: Penang Ladies CWAA (ed) 40th anniversary commemorative special issue of Penang Ladies CWAA. Malaysia, pp 1924–1963

Vertovec S (1999) Conceiving and researching transnationalism. Ethnic and Racial Studies 22(2):447–462

Wang KC (2019) The afterglow of the prosperous age: Lu Youxing and Lu You's descendants. In: Zheng H, Zhou W (eds) Mid–level Elevator: Soaring to Qingyun Road. Zhonghua, Hongkon, p 267

Wong SK (2004) Issues of Chinese women in Penang: centred on the discourse of women's education in Penang Sin Poe. Journal of Malaysian Chinese studies 7:43–66

Yang PC (2005) Foreword. In: Among Selangor Chinese Women CWAA (ed) 24th National Chin Woo carnival and 84th anniversary commemorative special issue of the Selangor Chinese Women CWAA. Malaysia, p 5

Yong CF (1977) A preliminary study on the social structure and leadership of the pre–war Singapore Chinese Community. The South Seas Society, Singapore

Zhang H, Zhang WA, Li LS (2014) Belief in gods, human relations and social organization: Tua Pek Gong and his sacrificial organization in Haizhu Island, Penang. Global Hakka Studies 3:127–135

Hsien-Wei Kuo is Professor at the Center for General Education of the National Tainan Institute of Nursing, Taiwan. His research focuses on the history of sports, qualitative research, martial arts, sports diplomacy, and body culture.

Chin-Fang Kuo is Professor in the Department of Sport Management at Aletheia University, Taiwan. Her research focuses on qualitative research and oral history, exploring topics such as sports diplomacy, women's sports experiences, and the history and culture of sports.

Chapter 5
"Daigou" Practices, Chinese Entrepreneurial Strategies, and Transnational Business Cultures: Case Study in Thailand

Pagon Gatchalee, Aranya Siriphon, and Zhijian Cui

Abstract This article explores contemporary mobility and the new wave of Chinese emigrants in Thailand engaging in "Daigou", a micro-entrepreneurial activity where individuals purchase local goods for customers in mainland China or elsewhere. This research employs qualitative approaches and methodology through a case study of Chinese residents in Thailand. It examines how these Daigou operators strategically navigate challenges in the post-COVID-19 landscape and amidst shifting global dynamics, especially in response to new Chinese e-commerce laws and competitive markets. Despite facing challenges posed by new regulations and changing market preferences, Daigou operators have demonstrated remarkable resilience and adaptability with three strategies: (1) maintaining a low profile or discontinuing practices, (2) using inventive language to evade AI detection, and (3) adapting professionally within legal frameworks. These adaptive strategies reflect responses to contemporary challenges and contribute significantly to a deeper understanding of transnational entrepreneurial subjects and the evolving landscape of transnational business cultures.

Keywords Daigou · Cross-border commerce · Transnational entrepreneurship · Chinese migrant · Chinese entrepreneurs in Thailand

P. Gatchalee (✉)
Business School, Chiang Mai University, Chiang Mai, Thailand
e-mail: pagon.g@cmu.ac.th

A. Siriphon
Faculty of Social Sciences, Chiang Mai University, Chiang Mai, Thailand
e-mail: aranya.s@cmu.ac.th

Z. Cui
School of Management, University of Science and Technology of China, Hefei, China

H. M. Y. Ho and Y. Chan (eds.), *Transnational Southeast Asia*, Asia in Transition 29,
https://doi.org/10.1007/978-981-96-3608-2_5

5.1 Introduction

The global movement of Chinese migrants has attracted scholarly attention in recent decades (e.g. Chan and Koh 2018; Nyiri 2007; Nyiri and Saveliev 2002; Siriphon 2015, 2017, pp. 79–96). Some scholars have investigated the surge in contemporary Chinese migration, which has led to the expansion of Chinese influence, partly driven by the Chinese state (Nyiri 2002, pp. 208–241). This movement potentially re-establishes transnational connections between China and other regions (Barabantseva 2011, pp. 8–9; Liu 2008, pp. 118–148; Nyiri 2002, pp. 208–241).

Recent studies have delved deeper into the developing trends in transnational Chinese migration that have emerged since 2010. Guo (2022a, b, pp. 847–872) and authors in a special issue of the *Journal of Ethnic and Migration Studies*[1] have captured the essence of contemporary Chinese migration trends, suggesting a shift characterised by unprecedented levels of hypermobility, hyper-diversity, and hyper-connectivity. The Chinese migration today exhibits extensive global engagement through multiple and circular sojourns, diverging from the conventional unidirectional movements or settlements in a single location. Furthermore, the recent Chinese diasporas, often referred to as "transient migrants" (Xiang 2017, pp. 3–5), comprise affluent, well-educated, and highly skilled professionals and entrepreneurs. They display flexible mobility driven by the pursuit of a preferred lifestyle and greater socioeconomic opportunities elsewhere (Chan and Koh 2018). Fortunately, the integration of technology with migration infrastructure improvements, advancements in transportation and logistics systems, and the rapid evolution of the digital world—marked by the proliferation of mobile devices and internet connectivity for social media and e-commerce—has substantially facilitated the increasingly borderless movement across the globe. Consequently, diverse forms of transnational mobilities, including sociocultural networks, economic trade and capital flows, and symbolic exchange of ideas, information, and images, evidently demonstrate the dynamic nature of Chinese diasporas in the era of transnational migration (Guo 2022a, b, pp. 847–872).

This chapter explores the recent Chinese emigrants engaging in "Daigou" (代购), a micro-entrepreneurial practice involving full-time or part-time "buying on behalf" while utilising transient migration in Thailand. It primarily focuses on this specific group, contextualising the Daigou practice—a micro business activity—within the framework of "transnational business cultures" (Jaime 2018, pp. 647–665). This chapter defines transnational business as more than just an economic entity with cross-border management impact. It also encompasses the dynamic relationship between consumer actions, the marketplace, and cultural significance (Jordan 2010, pp. 15–25) in which cultural patterns, languages, values and beliefs, codes of conduct, and rules have been instrumental in guiding business practices. For this article, we examine how Daigou deploys transnational entrepreneurial strategies in navigating evolving business landscapes, encompassing the use of transnational digital entrepreneurship (Mishra et al. 2023, pp. 1–29), to ensure business adaptability in light of emerging challenges of post-COVID-19, specifically considering the influence of the new Chinese e-commerce laws and the global competitive dynamics among traditional retailers and contemporary companies.

The Daigou participants in this chapter are perceived as actively engaged in transnational entrepreneurial subjects (Flisfeder 2015, pp. 553–570; Ong 2005, pp. 1–34) or entrepreneurial self (Bröckling 2016). They contribute to shaping modern knowledge and governance, fostering neoliberal rationality, and catalysing shifts in economic behaviour from a culture of dependency to one of self-reliance. In this regard, the transnational entrepreneurial subjects represented by Daigou operators exemplify how they strategically navigate, identify opportunities, and devise solutions within specific contexts.

Methodologically, we employed a case study approach focusing on recent Chinese migrants engaged in Daigou services, both full- time and part-time, residing temporarily in Thailand, specifically in the Chiang Mai Province. All interviewees featured in this chapter are referred to by anonymous names. Our research utilised qualitative methods involving interviews and text mining from relevant Chinese social media (from January 2022 to November 2023) using the Baidu Index tool. This combined approach facilitated a more comprehensive gathering of information and enhanced understanding within our study. This chapter constitutes an extensive study conducted as part of the research project titled "Impact of China's Cross–Border E–Commerce on Thailand: Transnational Practices of Chinese Entrepreneurs Residing in Chiang Mai Province (2020–2021)", funded by the National Research Council of Thailand (NRCT).

This study highlights the distinctive patterns observed in the dynamic evolution of transnational micro-businesses facing emerging challenges. We have identified at least three observable patterns within various strategies and adaptability measures: first, the decision-making between abandoning practices or maintaining a low profile, second, the use of inventive languages for effective digital communication with customers to evade automated Artificial Intelligence (AI) detection algorithms, and third, the professional adaptability of Daigou practices within legal frameworks.

Hence, the Daigou practices exemplified in the Thailand case study encompass not only conventional business skills but also technological proficiency, fostering flexible arrangements, diverse entrepreneurial strategies, and the intelligent utilisation of symbolic languages on digital platforms, thereby enabling the evolution of transnational business culture. Within these practices, while the maintenance of mutual trust, morality, and credibility remains pivotal for nurturing relationships, digital communication styles through symbols, non-verbal signs, and images are creatively employed to exert influential effects. Daigou practices and their transnational entrepreneurial strategy today no longer resemble the traditional model, but it is a reconfiguration of a transnational business culture at a micro business scale, but significantly contributing to a substantial volume of international trade.

5.2 Literatures on Daigou and Transnational Entrepreneurship

The existing literature on Daigou issues has contributed to an understanding of its dynamics and complexities. Some studies highlight how daigou business operation have shown distinctive entrepreneurial behaviours manifested in non-traditional contexts, identifying Daigou as important middlemen who behave entrepreneurially in heterodox settings where the legal boundaries of businesses are ambiguous (Glavas et al. 2023). In addition to the typical entrepreneurial orientation-based behaviours such as autonomy, proactiveness, innovativeness, competitiveness, and risk-taking propensity, four other behaviours have been observed within the specific unorthodox settings of Daigou, which are tolerance for ambiguity, opportunism, ability to pivot, and non-system-bound orientation.

This literature on entrepreneurial behaviour links with the concept of transnational entrepreneurship, which contributes to economic transnationalism, particularly among immigrants involved in various forms of entrepreneurship. Chinese entrepreneurs in Southeast Asia, whether operating on a small or medium scale, actively pursue business opportunities across national borders. Their engagement signifies an economic footprint serving diverse purposes in cross-border mobility and international travel such as exploring new markets and facilitating international trade (Chen and Tan 2009).

By investigating consumer aspects associated with Daigou entrepreneurship, a study (Li et al. 2022) has unveiled the factors influencing purchase intentions of Daigou buyers. A key finding highlights the significance of self-efficacy and the strength of social ties between buyers and shoppers, wherein a buyer's trust in the shopper plays a crucial role (Li et al. 2022). This aligns with relevant literature, such as Sandal (2023), which contends that overseas Daigou networks assist Chinese consumers in accessing reliable foreign-made products, often validated through visual evidence for credibility. Scholars such as Chih-Hsien Hughes and team (Hughes et al. 2019) have indicated that despite its rapid growth in transnational business, the Daigou business model remains rooted in trust, service quality, and social capital. Operational systems also show a lean quality, cost-efficiency, and a primary orientation towards income generation, in which social media serves as the primary channel for customer engagement within their personal business or enterprises.

There are several research studies that examine the utilisation of digital media space by Daigou operators. Xie (2018, pp. 4052–4065) contributes to the discourse on "mobility politics", highlighting how Daigou practices illustrate imbalanced power dynamics within hierarchical classifications of people's corporeal mobilities. Jaime (2018, pp. 647–675) identified the use of common languages on digital platforms, technology management, and diverse entrepreneurial strategies as integral elements in business management among Daigou operators. Zhao (2021, pp. 2554–2574) explores the engagement of Chinese international students in Daigou activities through digital media, shedding light on the various forms of digital labour and unproductive work inherent in their online endeavours.

The literature on Chinese outbound tourism studies has extended its focus to encompass the "Daigou phenomenon", which contributes to the emerging concept of "Daigou tourism" (Zhang et al. 2021, pp. 549–569). The concept of Daigou tourism captures the associated practices of tourists engaging in Daigou activities while travelling outbound. It arises from the blurred boundaries within the Daigou space, redefining tourists' classifications from those seeking leisure experiences to individuals financing ventures as brokers. It represents a diverse market shaped by factors, such as product quality, profit margins, and associations with the country of origin. Daigou tourists also undertake risks including product unavailability, price fluctuations, damages, item losses during transportation, and encountering unreasonable buyers (Zhao et al. 2016).

5.3 Contextualising Daigou Within Transnational Business Culture and Mobility Regulation Background on Daigou

Daigou, a micro-business activity involving "buying on behalf", refers to both the activities and individuals functioning as intermediaries in cross-border e-commerce under a consumer-to-consumer (C2C) relationship (Huang and Nan 2021; Wang 2018). Fundamentally, Daigou operators purchase local products overseas and subsequently resell them to Chinese customers in mainland China or elsewhere through Chinese social media. The exact number of Daigou digital business operators and entrepreneurs remains ambiguous due to the varied nature of this micro business type, encompassing both million-dollar-order-based trades and what is commonly termed as "suitcase trade" (Sandal 2023).

The beginning of the Daigou transnational business stemmed from the period around 2008 when Chinese consumers began losing confidence in domestically manufactured products, notably following the "melamine scandal", involving a Chinese company producing counterfeit infant formula, which resulted in the deaths of 6 infants and over 10,000 falling ill. Subsequently, a substantial number of Chinese consumers shifted their preference towards purchasing foreign products over domestically produced ones (Sun and Yu 2023, pp. 155–181).

The heyday of Daigou transnational micro business emerged massively in China during the past decade (2010–2020). Gradually, a significant number of Chinese customers not only involved with seeking safe infant formula from abroad but also expanded to include various other imported goods, ranging from luxury items to fashion products from major foreign cities.

Typically, Daigou service providers utilise Chinese social media platforms such as WeChat (Weixin) and Weibo to interact with friends or clients by sharing their location. Subsequently, individuals within this network place orders for specific goods, which the agents then collect and transport back to China in their luggage. These items, considered for personal use, often yield profits two to three times their original purchase price. Daigou service providers are typically not subjected to customs inspection and import taxes if not flagged by customs, effectively evading taxes for this group of individuals. Marked by the rapid growth and development of China's online economy, e-commerce, and social commerce platforms, such as Taobao, Tmall, Alibaba, WeChat, Weibo, Xiaohongshu, or Douyin, became widely known and immensely flourished, often setting up personal online stores where customers can visit and purchase their items.

The emergence and proliferation of Daigou have appeared within a specific context. On the one hand, the improved economic status of Chinese individuals led to increased spending power and greater opportunities for international travel, whether as tourists, students, or expatriates. On the other hand, advancements in internet communication technology, mobile phone networks, online payment systems, and social media platforms equipped with live streaming capabilities facilitated direct products and services immediate customer queries, and product advertising on platforms such as WeChat's newsfeed or Moments. Furthermore, foreign private companies, manufacturers, and sellers of luxury goods have recognised Daigou operators as a gateway to access the Chinese consumer market. This was particularly crucial during the times when foreign private enterprises faced significant limitations in accessing Chinese consumers due to strict language barriers and trade regulations imposed by the Chinese government. Huang and Nan (2021) reported that approximately 200,000 Daigou agents operated in Australia in 2016, comprising of tourists, Chinese residents, and students who collectively sold Australian products, generating an average income of around $80,000 US dollars per person. Moreover, the Daigou Index report in 2020 as cited in Tang (2023) estimated the annual value of the Daigou market at 57 billion US dollars. Consequently, China experienced an estimated loss of tax revenue amounting to 800 million US dollars from Daigou transnational business activities (Huang and Nan 2021).

5.4 Mobility Regulations and Emerging Situation

Daigou, as a transnational micro business, however, is unlikely to reach its heyday market size when facing several challenges today. One of the primary factors pertains to the introduction of new laws and policies by the Chinese government, which came into effect on 1 January 2019 (Wong 2019). The policy aims to impose tighter regulations by subjecting cross-border e-commerce to stricter controls. This includes mandating Daigou to pay taxes, imposing heavier fines on social media platforms, while simultaneously supporting its growth by promoting more legitimate shopping channels and reducing taxation for sales made through cross-border e-commerce

(CBEC) channels. The second factor is in regards to the COVID-19 pandemic, which spanned three years, that has halted overseas travel, significantly hampering the capacity of Daigou to conduct transnational business. Thirdly, due to China's new policy, which offers more flexible measures, overseas, and international brand companies have taken an active role in operating their cross-border e-commerce ventures within the Chinese market. As a result, many Chinese consumers shifted towards more legitimate e-commerce channels or duty-free department stores, where an expanded selection of foreign products became readily available.

Although there has been a recovery from the impact of COVID-19 in 2023, China's stricter mobility regime and policy support favours formal cross-border e-commerce (CBEC). However, this chapter asserts that Daigou and their transnational digital entrepreneurship may not remain the same as it was in its former flourishing stage, as they adapt to new circumstances influenced by the Chinese e-commerce law and COVID-19, which impose restrictions such as mandatory business registration and full tax payments.

5.5 Daigou in Thailand: Transnational Entrepreneurial Strategies Under the Emerging Challenge with New Chinese E-commerce Law and COVID-19

5.5.1 Daigou in Thailand

Daigou in Thailand can generally be categorised into three groups: Chinese students, residents, and professional buyers. Chinese students leverage their social media connections to build trust and conduct trades, supplementing their income for educational and living expenses. Chinese residents typically engage directly with factories or foreign companies, overseeing product orders and warehouse management. Professional buyers, who are well-versed in the demands of Chinese consumers, specialise in sourcing products and primarily focus on exporting them back to China. The boundaries between these three groups are not always clear-cut, as some Chinese agents transitioned from tourists to students and eventually evolved into professional traders over time (Siriphon et al. 2022, pp. 46–48).

From 2020 onwards, the Daigou practices in Thailand have witnessed a decline. Xiaowu (interviewed in May 2021), who previously engaged in Daigou from 2013 to 2018 while studying at a Thai university, continued these activities after completing his studies. He established his own brand, "Chiang Mai Royal", focusing on rubber and other Thai products including cosmetic and luxurious items to serve Chinese customers' demand. From 2016 to 2018, he expanded his market by establishing multiple sub-agents for Thai products, facilitating direct sales to these agents or shipping products directly to their customer groups. This operation was conducted from either a warehouse in China (rented in Chongqing) or one in Chiang Mai, based on specific customer requirements.

Xiaowu observed that the new Chinese e-commerce law, effective from 1 January 2019, primarily requires market entities to register and obligates them to pay taxes. This means Daigou merchants must now obtain a business licence and pay full taxes, including import taxes, a requirement many previously avoided while operating as personal shoppers (Sosnovskikh 2021). He views this as the culmination of his Daigou activities. This is not only due to the new law imposing strict regulations on customs procedures and its crackdown on tax evasion related to illegal imports. Daigou operators will also experience reduced profits due to increased import taxes under the new regulations mentioned above, leading to escalated operational costs stemming from both logistics and tax fees. Moreover, the popularity of Thai products, especially cosmetics and skincare items, has waned among Chinese students and middle-class female consumers. Simultaneously, a growing number of Chinese customers have shifted to purchasing these items through registered stores and online platforms within the CBEC domain. This transition has resulted in a decline in Xiaowu's customer base and amplified competition in the market.

The fluctuations in Daigou practices and customer engagement were observed on various digital platforms. Our study focused on analysing the prevalent utilisation of Daigou practices on select popular platforms such as WeChat's "Weidian" (微店), Xiaohongshu, Douyin (Chinese TikTok), employing Baidu Index tools to search for "泰国代购" (Thailand Daigou).

This method revealed a pattern in online searches across digital platforms, showing increased interest in Xiaohongshu and Douyin, and a consumer preference shift with Weixin and Weidian's decline.

The period of 2014–2015 saw an upward trend in Weixin and Weidian searches, peaking in August 2015 with 800,000 searches per day, followed by a fluctuating decline from 2016. Currently, search volumes for Weixin have decreased to about 100,000 per day, while Weidian stands at approximately 2000 per day. WeChat Official Accounts began to exceed 10,000 daily searches in 2015, but during the COVID-19 pandemic, it reached a maximum of 50,000 searches per day and is currently averaging about 22,000 searches per day. This implies a period when Chinese public was keen on receiving information from the digital world. Weixin remained a popular online media platform, not specifically limited to online shopping, but offering a variety of content, marking a divergence from Weidian, which is an e-commerce service on Weixin and was a preferred tool among Daigou providers during its boom.

However, changes in e-commerce law and the impact of COVID-19, the search terms reflected a declining trend due to legal restrictions and international travel limitations and a shift in consumer behaviour towards Daigou services. On the Xiaohongshu platform, one of China's popular lifestyle social media platforms, consumer behaviour is reflected through search and online discussions. As tracked by the Baidu Index, it shows that Xiaohongshu has gained increasing popularity since 2019, peaking in 2023. This trend aligns with that of Douyin or Chinese TikTok, although Douyin's search volume began to increase from 2018, coinciding with the decline of Weixin and Weidian. Weidian's search volume has been below 10,000 per day since 2016, with the years 2014–2015 seeing the highest search volumes for Weidian, exceeding 10,000 per day. Concurrently, Xiaohongshu gained increased popularity as a social media platform enabling direct product sales.

An interview conducted in October 2023 with Zhang Lin, a 32-year-old online marketing entrepreneur from Shandong Province, revealed insights into the evolving trends within the e-commerce landscape. Lin highlighted that Weidian, a Consumer-to-Consumer (C2C) style shop on WeChat, historically fostered a closer relationship between Daigou service providers and their customers. However, with major e-commerce platforms in China such as Tmall and JD offering direct channels for imported products while ensuring product authenticity, the demand for purchasing via Daigou operating individual shops solely on WeChat has diminished. This decline is also attributed to growing apprehensions regarding counterfeit products and potential deception when buying through WeChat-based platforms.

Another case, Ms. Zu (interviewed in October 2023), a Chinese expatriate residing in Thailand and an influential online business owner and blogger on Xiaohongshu, shared her insight about platforms such as Weidian and Xiaohongshu. According to Ms. Zu, both platforms presented challenges for foreigners due to documentation limitations. However, she preferred Xiaohongshu due to its larger popularity and user base, especially for popular Thai products such as cosmetics and snacks. The platform primarily caters to a female demographic, aligning with her customer base. Ms. Zu also highlighted a prosperous period from 2018 to 2019, wherein she earned significantly from product reviews on Xiaohongshu, averaging over 100,000 baht (around 2800 US dollars) per month. This period coincided with the peak of Chinese tourism in Thailand, familiarising the Chinese audience more acquainted with Thai products.

However, she noticed the increased limitations on Xiaohongshu from 2020 onwards, particularly for foreign products. Advertising and selling products on the platform required formal company registration and product licencing, posing hurdles for unregistered foreign products or those unavailable in China since 2019 onwards. Following the implementation of the new Chinese e-commerce law, Xiaohongshu has launched its "brand collaboration platform". This initiative was aimed at controlling product advertising content and promoting the use of Xiaohongshu for marketing and selling products. The timing of this launch in 2019 coincided with China's new e-commerce regulations, reflecting a broader shift in the digital marketing landscape in the country. Consequently, Daigou marketing on Xiaohongshu has adapted to subtler strategies, such as avoiding overt commercial language. The situation particularly impacts Daigou operators known as Yesheng (野生), independent creators without multichannel network (MCN) company contracts (Guo 2022a, b).

5.5.2 Daigou: Adapting Strategies to Emerging Changes

In response to the transnational business landscape influenced by the new Chinese e-commerce law, the global changes triggered by COVID-19, and other concurrent shifts, this study identifies three distinctive patterns reflecting varied strategies and adaptability among the Daigou operators and their businesses by using case studies in Thailand.

Firstly, some Daigou service providers opted to cease their operations, exemplified by Xiaowu's decision as mentioned above. He chose to discontinue Daigou practice due to amplified risks associated with trading Thai products into China under the new laws. These regulations not only heightened the likelihood of legal violations but also led to increased operational expenses for businesses, including import taxes, elevated logistic costs, registration fees, accounting services, and potential legal consultations significantly augmented overhead costs. Furthermore, the market dynamics for imported goods in China also changed drastically. Daigou service providers needed to compete with registered Chinese or foreign businesses that were legally licensed and operated on legitimate platforms.

However, we also found some small-scale Daigou operators in Thailand, planning to continue their Daigou services, have shifted their approach by opting for smaller quantities. By reducing the size of their shipments, they aim to maintain a lower profile and decrease the likelihood of drawing attention from customs authorities. Opting for smaller transactions simplifies the tax calculation process and might lead to lower tax liabilities, presenting a more appealing option for Daigou sellers. To mitigate risks, Daigou sellers in Thailand now prefer smaller shipments, which carry less risk. In the unfortunate event of a package being detained or confiscated by customs, the financial loss is minimised compared to larger shipments. This approach also helps limit potential legal issues stemming from customs violations.

Yi (interviewed in October 2023), a Chinese student running a Daigou business in Thailand, encountered challenges due to the new Chinese e-commerce law, particularly Article 26. This law necessitates compliance with state import and export regulations, including specific laws such as the Chinese drug law. Consequently, platforms such as WeChat have implemented stricter policies, such as prohibiting the posting of medical products. WeChat noticed Yi when trying to post medical product info on WeChat moment. Additionally, the COVID-19 pandemic further disrupted shipping from Thailand, exacerbating these challenges.

Adapting, Yi transitioned from a customer-to-customer (C–C) model to acting as an intermediary for a wholesale company in business-to-business (B–B) transactions, managing bulk orders and focusing on smaller, selective orders.

Secondly, Daigou service providers have changed their communication strategies with customers on digital platforms to evade detection by AI algorithms. The role of social media in advertising and selling Daigou products has become more complex and subtle. Operators frequently employ coded languages and create private groups to interact with their customers, minimising the risk of detection while effectively reaching their intended audience.

The case of using the search term "Thai Daigou" (泰国代购) on Xiaohongshu refers to purchasing agents who specialise in buying Thai products for their clients, illustrates how digital users (both Daigou and customers) have adapted tactically to platform restrictions. Instead of using the direct term "Daigou", users have created content using alternative terms. A language example is 泰国代沟 (tàiguó dàigōu), where 代沟 sounds phonetically similar to "Daigou" but has a completely different meaning, referring to a generational gap. However, users mutually understand it to signify Daigou. Another term, 泰国代买 (tàiguó dàimǎi), translates to "purchasing on behalf of someone in Thailand", aligning closely with the concept of Daigou.

Although Xiaohongshu's system does not directly recommend searches for Daigou, content related to Daigou can still be posted in Chinese, including hashtags such as #代购, or #泰国代购 (Thai Daigou). Significantly, the Thai Daigou content on Xiaohongshu is uniquely generated with a tactic of creative content without directly selling Thai products. The Thai Daigou content, for example, is posted about well-known Thai products using only names and photos, without detailed descriptions for each item, and just describing visits to Thailand, including travel dates, framed as a shopping service for tourists. This includes how to avoid direct mentions of product prices and payment methods in posts, especially for general content not linked to a Xiaohongshu store. Additionally, using terms and methods to evade AI detection, especially for indicating contact through WeChat is crucial tactics today. This involves using a WeChat ID similar to the Xiaohongshu username and explaining it in Xiaohongshu account details with tactics such as homophones and emojis (e.g. 'V' for WeChat) to evade AI detection.

Daigou operators employ several strategies on Chinese online platforms that have also used images and videos to portray everyday life rather than solely focusing on selling a product or product explanation. Crucially, the strategy of building networks and communities plays a pivotal role in the marketing process. Nurturing relationships with like-minded individuals has significantly assisted in promoting and marketing products or services online. Furthermore, establishing a platform for exchanging ideas and information fosters trust and generates interest in offerings within a dependable consumer base. This implies the importance of WeChat Groups, among other platforms, as a marketing channel that facilitates interaction and communication with customers.

In addition to adopting communication strategies on digital platforms to evade AI detection, a changing market from luxury to local products is another issue among Daigou operators. Previously, high-end luxury items dominated this market, but there has been a discernible shift towards more niche, local products. These products often possess distinctive ties to travel destinations, encompassing a diverse range of local crafts, specialty foods, and beauty products. This change is driven not only by shifting consumer preferences but also as a strategic adaptation to increasingly strict customs regulations. Purchasing specialty local products from the country of residence or travel, often deemed for personal use rather than commercial purposes, is seen as a method to circumvent new e-commerce laws.

Lastly, certain Daigou operators are now obligated to register as formal business entities. This transition necessitates a deep understanding of both Chinese and international legal and tax systems, particularly in the countries where their operations are conducted. The shift towards formalisation signifies a substantial departure from the previously informal nature characterising many Daigou operations. This pattern involves adhering to the new e-commerce laws of China, in which Daigou operators have to be registered legally and partnered with foreign companies.

Xie Xie, a Chinese entrepreneur in Thailand, launched his Daigou venture in 2017 as Chinese tourists flocked to Thailand, driving demand for Thai products. As a Daigou operator, he swiftly expanded his customer base and established a strong network in both China and Thailand. Subsequently, he founded a marketing agency to assist Thai entrepreneurs in promoting their businesses in China, aligning it with his core Daigou operations. However, new Chinese e-commerce regulations in 2019, coupled with tax concerns and the disruption caused by the COVID-19 pandemic, prompted Xie Xie to transition away from Daigou. He ceased Daigou operations and shifted his focus to the marketing agency, which he registered in Thailand.

Similarly, Su, a Chinese resident in Thailand, seized the opportunity presented by the popularity of Thai products in China from 2016 to 2017. She established herself as a Daigou operator at that time and later registered companies both in China and Thailand in 2019 to ensure compliance with new regulations. Su continues to thrive by collaborating with top Thai product brands in Thailand, even in the face of COVID-19 challenges.

Professional perspectives of Daigou agents and their transition into formal business entrepreneurs such as Su have revealed a shift in their operations to comply with legalities and form new collaborations with foreign companies. According to Biondi (2021), it indicated that in 2021, products from five notable brands—Gucci, Prada, Loewe, Celine, and Balenciaga—amounting to 4.3 million US dollars, were ordered and sold by Daigou agents through Taobao platforms. These luxurious brand companies view Daigou service agents as an essential mechanism in expanding sales market through a new collaborative channel, recognised as the "New stream of Daigou". These luxurious brand companies have perceived Daigou service agents as official partners who serve as commercial mediators, tastemakers, and taste shapers.

5.6 Conclusion

This chapter highlights the transformative nature of contemporary Chinese diasporas, specifically focusing on those involved in the micro-entrepreneurial practice of "Daigou" in Thailand. It examines how these Daigou operators strategically navigate challenges in the post-COVID-19 landscape, and amidst shifting global dynamics, especially in response to new Chinese e-commerce laws and competitive market.

Despite facing challenges posed by new regulations and changing market preferences, Daigou operators have demonstrated remarkable resilience and adaptability. Employing qualitative approaches and a case study methodology, this study uncovers distinctive patterns within Daigou practices. Three discernible strategies emerge: the choice to maintain a low profile or discontinue practices, the use of inventive language to evade AI detection, and professional adaptability within legal frameworks. Their adaptive strategies not only reflect responses to contemporary challenges but also significantly contribute to a deeper understanding of transnational entrepreneurial subjects and the developed terrain of transnational business cultures. In the future, a shift in Daigou's business practices will likely undergo further changes, such as collaborating directly with brands and even transforming the business model to be a formal business.

Notes

1. Please see the Special Issue: Chinese Diasporas Studies in a Transnational World, Journal of Ethnic and Migration Studies, 48, no. 4 (2022).

References

Barabantseva E (2011) Overseas Chinese, ethnic minorities and nationalism: de–centering China. Routledge, New York

Biondi A (2021) China's Daigou personal shoppers aren't going away. https://www.voguebusiness.com/consumers/china–daigou–not–going–away. Accessed 4 October 2023

Bröckling U (2016) The entrepreneurial self: fabricating a new type of subject. Sage Books, London

Chan YW, Koh SY (2018) New Chinese migrations: mobility, home, and inspirations. Routledge, London

Chen W, Tan J (2009) Understanding transnational entrepreneurship through a network lens: theoretical and methodological considerations. Entrepreneurship Theory and Practice 33(5):1079–1091

Flisfeder M (2015) The entrepreneurial subject and the objectivization of the self in social media. South Atlantic Quarterly 114(3):553–570

Glavas C, Mortimer G, Ding H et al (2023) How entrepreneurial behaviors manifest in non–traditional, heterodox contexts: exploration of the Daigou phenomenon. Journal of Business Venturing Insights 19:e00385

Guo J (2022a) The postfeminist entrepreneurial self and the platformisation of labour: a case study of yesheng female lifestyle bloggers on Xiaohongshu. Global Media and China 7(3):303–318. https://doi.org/10.1177/20594364221095896

Guo S (2022b) Reimagining Chinese diasporas in a transnational world: toward a new research agenda. Journal of Ethnic and Migration Studies 48(4):847–872

Huang E, Nan X (2021) Daigou: cross–border digitalised hidden economy transactions are now detectable. https://www.austaxpolicy.com/daigou–cross–border–digitalised–hidden–economy–transactions–are–now–detectable. Accessed 4 October 2023

Hughes CH, Lucas G, Brown D (2019) The Daigou business model: explanation and implications. Academy of Management 2019(1). https://doi.org/10.5465/AMBPP.2019.15531

Jaime P (2018) Recycling the idea of race: socio–political agenda, transnational business culture, and diversity management in Brazil. Social Identities 24(5):647–665

Jordan AT (2010) The importance of business anthropology: its unique contributions. International Journal of Business Anthropology 1(1):15–25

Li L, Xu J, Kumar U (2022) Consumers purchase intention in consumer–to–consumer online transaction: the case of Daigou. Transnl Corp Rev 1–17

Liu H (2008) Immigrant transnational entrepreneurship and linkages with the state/network: Sino-Singaporean experience in a comparative perspective. In: Wong RSK (ed) Chinese entrepreneurship in a global era. Routledge, London, pp 118–148

Mishra AA, Maheshwari M, Pandey J et al (2023) Fempreneurship through digital platforms. Journal of Global Information Management 31(8):1–29. https://doi.org/10.4018/JGIM.329606

Nyiri P (2002) From class enemies to patriots: overseas Chinese and emigration policy and discourse in the People's Republic of China. In: Nyiri P, Saveliev I (eds) Globalizing Chinese migration: trends in Europe and Asia. Ashgate, Burlington, pp 208–241

Nyiri P (2007) Chinese in Russia and Eastern Europe: a middleman minority in a transnational era. Routledge, Abingdon
Nyiri P, Saveliev I (eds) (2002) Globalizing Chinese migration: perspectives from Europe and Asia. Ashgate, Aldershot
Ong A (2005) Anthropological concepts for the study of nationalism. In: Nyíri P, Breidenbach J (eds) China inside out. Central European Press, Budapest, pp 1–34
Sandal MJ (2023) Daigou: transnational flows, relatedness, and the reproduction of the Chinese family. Dissertation, Aarhus University
Siriphon A (2015) 'Xinyimin', new Chinese migrants and the influence of the People's Republic of China (PRC) and Taiwan on the northern Thai border. In: Santasombat Y (ed) Impact of China's rise on the Mekong region. Palgrave Macmillan, London, pp 147–166
Siriphon A (2017) Border guanxi: Xinyimin and transborder trade in northern Thailand. In: Nyíri P, Tan D (eds) Chinese encounters in Southeast Asia: how people, money, and ideas from China are changing a region. University of Washington Press, Seattle, pp 79–96
Siriphon A, Gatchalee P, Li J (2022) Impact of China's cross–border e–commerce to Thailand: transnational practices of Chinese entrepreneurs residing in Chiang Mai province: final report. The National Research Council of Thailand (NRCT)
Sosnovskikh S (2021) A new form of parallel trading within economics relations between Russia and China. Post–Communist Economies 33(1):94–118
Sun W, Yu H (2023) Mobility and micro–entrepreneurship: Daigou as transnational subjects. In: Sun W, Yu H (eds) Digital transnationalism Chinese-Language media in Australia. Brill, Leiden, pp 155–181
Tang F (2023) In China, the resale market (Daigou) is impacting luxury. https://www.luxurytribune.com/en/in–china–the–resale–market–daigou–is–impacting–luxury. Accessed 4 November 2023
Wang L (2018) What a new law could mean for China's 'Daigou'. https://www.sixthtone.com/news/1003372/what–a–new–law–could–mean–for–chinas–daigou. Accessed 4 October 2023
Wong FKH (2019) China's crackdown on Daigou, new cross–border e–commerce policies. https://www.china–briefing.com/news/chinas–crackdown–daigou–cross–border–e–commerce–policies/. Accessed 6 October 2023
Xiang B (2017) Hundreds of millions in suspension. Transitions: Journal of Transient Migration 1(1):3–5
Xie Z (2018) Im/materializing cross–border mobility: a study of mainland China–Hong Kong Daigou (cross–border shopping services on global consumer goods). International Journal of Communication 12:4052–4065
Zhang X, Tham A, Liu Y et al (2021) 'To market, to market': uncovering Daigou touristscapes within Chinese outbound tourism. Journal of China Tourism Research 17(4):549–569
Zhao K(2016) An empirical investigation of online gray markets. Journal of Retailing 92(4):397–410
Zhao X (2021) Digital labour in transnational mobility: Chinese international students' online boundary work in Daigou. New Media & Society 23(9):2554–2574. https://doi.org/10.1177/1461444820934096

Pagon Gatchalee is a lecturer in the Department of Marketing, Business School of Chiang Mai University (Faculty of Business Administration). He is also a PhD Candidate of Business Administration, School of Management, University of Science and Technology of China (USTC), China under the supervision of Prof. Cui Zhijian. He obtained his BSc in Computer Science (2011) from Chiang Mai University and MEng in Computer Software and Theory (2014) from Beihang University, China. His research interests include AI and algorithm aversion in enterprise, AI marketing, innovation strategy in business, cross-border commerce, and Chinese entrepreneurs in Southeast Asia. His e-mail is pagon.g@cmu.ac.th

Aranya Siriphon is Associate Professor at the Department of Sociology and Anthropology, Faculty of Social Sciences, Chiang Mai University, Northern Thailand. She obtained her MA (2002) and PhD (2008) from Chiang Mai University. She received a one-month exchange scholarship from Ernst Mach ASEA-UNINET, serving as a visiting fellow at the Faculty of Social Sciences, University of Vienna (2018). She also held a one-month visiting fellowship at the School of Ethnology and Sociology, Minzu University of China, Beijing, PRC (2019). Additionally, she was a visiting scholar at ISEAS, Singapore for 8 months (2021). Her research interests include Anthropology and Development Studies, Chinese influence and mobility in the Mekong region, transborder and transnational migration studies, and ethnicity in Southeast Asia. Her e-mail is aranya.s@cmu.ac.th

Zhijian Cui is Professor of Business Administration and Director of MBA Program at the School of Management, University of Science and Technology of China. He was selected for "Anhui Province Hundred Talents Plan" in 2020 and served the role as Chair of Technology, Innovation Management, and Entrepreneurship Society at INFORMS from 2017 to 2019. Before moving back to China in 2019, he was tenured Associate Professor at IE Business School in Spain, Associate Professor and Head of Marketing and Operations Management department at Nazarbayev University Graduate School of Business in Kazakhstan, and Visiting Assistant Professor at Hong Kong University of Science and Technology Business School. His work has appeared in leading journals including *Production and Operations Management, Decision Sciences,* and *IEEE Transaction on Engineering Management*. Zhijian holds PhD in Management from INSEAD (Fontainebleau), MBA from Pepperdine University (Malibu), MS in Management from Euromed Marseille Ecole de Management (Marseille), and Bachelor in Engineering from Tsinghua University (Beijing).

Chapter 6
Temple and Church: Political Mobility, Socioreligious Space and Transnational Lives of the Overseas Vietnamese (Việt Kiều) in Thailand

Morragotwong Phumplab

Abstract The emergence of colonialism and the Cold War's relocation of geopolitics had a profound effect not only on high politics but also on the populace on the ground. Vietnamese nationalists and communists utilised Thailand as a major geographic conduit in their anti-French colonialism and anti-American operations. During the Cold War, temples and churches became the political refuge for Overseas Vietnamese (*Việt Kiều*) and the arena of conflict between the various groups of supporters, namely Hanoi and Saigon supporters. As Thai authorities monitored these Vietnamese activities, religious space became a refuge for overseas Vietnamese who required protection from the Thai government. From the 1950s to the 1970s, the Cold War encompassed more than the proxy conflict between bipolar political ideologies. This was instead a fierce conflict between people on the ground. The Thai government was concerned about the Overseas Vietnamese in northeastern Thailand, who were predominantly perceived as communists and had ties to the anti-colonial struggle. Using Vietnamese and Thai archival sources as well as oral history, this chapter examines the history of the Cold War from below through the transnational lives of the Overseas Vietnamese, particularly the use of religious space as a sphere of the Overseas Vietnamese' political mobilities.

6.1 Introduction

The Southeast Asian people have maintained their homeland's connections and established new settlements through physical proximity and migration, even before the colonial era (Ricklefs 2010). Thailand has been a pivotal location in the geopolitics of mainland Southeast Asia for nationalists seeking asylum and conducting political movements from neighbouring countries including Vietnam, Cambodia, Laos, and Burma (Baker and Phongpaichit 2009; Wyatt 1984). Southeast Asia's political,

M. Phumplab (✉)
Faculty of Liberal Arts, Thammasat University, Bangkok, Thailand
e-mail: morragot@tu.ac.th

H. M. Y. Ho and Y. Chan (eds.), *Transnational Southeast Asia*, Asia in Transition 29,
https://doi.org/10.1007/978-981-96-3608-2_6

economic, and social aspects were impacted by the postcolonial struggle between capitalists and socialists for alternative political ideologies both locally and globally (Ang 2010; Goscha and Ostermann 2009; Westad 2005). In particular, Thailand and the Indochinese peninsula served as hotspots for political conflict. The *Việt Kiều* (Overseas Vietnamese), who had spent decades relocating between the Thai and Indochinese borders, were among those whose movements across national-state lines for political, social, and economic reasons were affected.

The *Việt Kiều* migrated to Siam/Thailand during the pre-colonial and colonial eras. Thailand's Ministry of Interior categorises the *Việt Kiều* into two groups: the old Overseas Vietnamese (*Yuan kao* or *Việt Kiều cũ*) who came to Siam before World War II, and the new Vietnamese (*Yuan mai* or *Việt Kiều mới*) who arrived after the 1946 counterinsurgency at Thakhaek in Laos. The *Việt Kiều*, which served as a base and a conduit for Vietnamese nationalists and communists going between Thailand, Indochina, and South China, was crucial to the anti-colonial movement's growth. As Christopher Goscha noted, "Thailand became the center of a larger Vietnamese revolutionary network in Southeast Asia during the colonial period" (Goscha 1999). The Vietnamese developed and enlarged their nationalist network by leveraging economic, family, and marital connections with the Thais, Overseas Chinese, and *Việt Kiều cũ*, who were effectively assimilated into the local society before colonisation, with the assistance of the Siamese state (Goscha 1999; Khamwan 2019; Phumplab 2023; Poole 1970; Sripana and Trinh 2005). Thailand was thus connected both internally and outside, not just in terms of military assistance but also in the provision of other essential supplies and resources. Despite a shift in policy towards the *Việt Kiều* by the Thai government at the onset of the Cold War, all overt political activities had to be clandestinely relocated.

The archival sources from Thailand and Vietnam offer a valuable understanding of the political and cultural strategies employed by the *Việt Kiều*, as well as the psychological competition that existed between Saigon and Hanoi. This chapter examines the history of *Việt Kiều* from the perspective of Cold War history, using Vietnamese and Thai archival materials, as well as oral history. The primary focus of the oral interview is on the *Việt Kiều* who lived in northeastern Thailand during the Cold War era. Ten individuals, who are aged between 80 and 98 years, were interviewed. The interviews, performed between December 2019 and April 2021, have a duration of two hours each. Their narratives unveiled the correlation between the political landscape and the Vietnamese populace, as well as the ways in which individuals from the lower echelons of society reacted to and navigated these governmental measures. The analysis also examines the transformation of Thailand into a disputed battleground for the two opposing factions of Vietnam and for the subsequent alterations in the lives and political movements of its inhabitants throughout the Cold War. This study aims to provide insights into the political transnational movement of Overseas Vietnamese during the Cold War era.

In the 1940s, the total number of Vietnamese individuals who crossed the borders of Cambodia and Laos to enter Thailand amounted to 48,702. The majority of Vietnamese on the Cambodian border were of southern ancestry and lived in the provinces of Battambang and Siem Reap, which were granted an alien certificate by the Thai government in 1942, during Thailand's rule over these territories. They lived in Prachinburi Province, which is located on the border between Thailand and Cambodia. Following the surrender of Japan in World War II, a substantial number of Vietnamese people relocated from Laos to the northeastern regions of Thailand. Subsequently, the leadership of the Việt Minh contemplated the possibility of venturing into Laos. In order to facilitate this endeavour, a clandestine organisation called the Overseas Vietnamese National Salvation Association (Hội Việt Kiều Cứu Quốc) was established as a subsidiary entity under the Party Committee of the Overseas Vietnamese (Đảng Bộ Việt Kiều) (Khamwan 2019). This organisation aimed to oversee the Vietnamese populations in Laos and Thailand, with the same hierarchical structures as the Communist Party. However, in 1946, the French authorities in Indochina brutally suppressed the Vietnamese, marking a significant turning point. As a consequence, the Vietnamese were forced to evacuate from Laos Khammouane (Thakhek), Savannakhet, and Pakse to Thailand by crossing the Mekong River (Panha Kiewkap Chao Yuan 1967).

The status and policy of the *Việt Kiều* were influenced by both diplomatic relations and Thailand's internal political power struggle. During the period after World War II, the Thai government provided significant assistance to the nationalist movement in Indochina under the leadership of Prime Minister Pridi Banomyong (March–August 1946) and Prime Minister Khuang Aphaiwong (November 1947–April 1948). Việt Minh received shelter and munitions assistance from the Thai government. To help the Vietnamese sustain themselves, the Thai government granted them land and financial aid. In addition, in Bangkok, the Democratic Republic of Vietnam (DRV) office of delegation for political and diplomatic movements was granted permission by the Thai government to the Việt Minh.

The status of *Việt Kiều* was changed during Phin Choonhawan's coup in 1947, enabling Phibulsongkram to regain control of the government. As a result, a system was developed to monitor the movement of Vietnamese immigrants in the northeastern region of Thailand. It is worth noting that Phibulsongkram had shown support, either directly or indirectly, for the Vietnamese nationalist struggle during World War II. One of the main reasons was due to the connections of Tiang Sirikhan, a senior Member of the House of Representatives and a close friend of Pridi Banomyong. Sirikhan had supported the nationalist movement in the Northeast for *Việt Kiều* (Goscha 1999). In 1949, the victory of the Communist Party of China had a significant impact on the Việt Minh Army's fight against the French. This success also caused concern for the Thai government regarding the growing influence of the Vietnamese communist movement, which operated from Thailand as a strategic location.

At the beginning of the 1950s, when the US solicited the aid of its allies in the Indochina struggle, Thailand became politically involved in the Second Indochina War (Algie et al. 2014; Chaloemtiarana 2007; Fineman 1997). Phibunsongkhram's formal acknowledgement of the Bảo Đài Government's authority over the State of Vietnam confirmed Thailand's pro-capitalist position. This marked the initiation of Thailand's anti-Communism drive, specifically targeting the *Việt Kiều*. As a result, the Thai government constructed a narrative in which the *Việt Kiều* were portrayed as communist agents, sympathisers, and rear guards, thereby creating a security risk for Thailand. Thai lawmakers regarded the *Việt Kiều* problem as a security threat. In response, the Thai government endeavoured to reduce the influx of Vietnamese immigrants by repatriating them to either the DRV or the Republic of Vietnam (RVN). The Thai government expressed apprehension regarding the Overseas Vietnamese communities in northeastern Thailand, as they were mostly inclined towards communism due to their strong affiliations with the anti-colonial movement and their roots in northern and central Vietnam (Naewkhit Thang Yutthakan Thi 114 1970).

Policies towards *Việt Kiều* were directly impacted by the coup and following political transition in Thailand between the 1950s and 1970s. Thai and RVN authorities established a relationship and pursued collaborations in multiple sectors, such as the anti-communism campaign, military assistance, diplomatic initiatives, and social cooperation. Conversely, relations between Thailand and the DRV have experienced fluctuations. Bangkok established a positive relationship with Saigon, but its elites were careful to avoid a full-scale conflict with Hanoi. The *Việt Kiều* community had a significant role in shaping the diplomatic relations between these three states (Jayanama 2017).

Despite this, the *Việt Kiều* implemented a diverse range of negotiation strategies with the Thai government and actively participated in political activism within their respective communities and via affiliations with the Communist Party in Hanoi, in defiance of the Thai government's stringent stance towards them. It needed to be more covert compared to public political movements before the Cold War period. The effectiveness of these *Việt Kiều* bargaining strategies lay in their ability to exert influence over the Thai state's behaviour in their interactions.

In examining transnational political activities among migrants or refugees, Østergaard-Nielsen (2003) highlights the dual engagement of migrants and refugees in both their home country's politics from abroad and the political institutions of their host country. Migrants have the potential to organise and take action in relation to politics in their home country, politics within their diaspora community, and politics that transcend local boundaries. The involvement of the *Việt Kiều* in Thailand in the domestic politics of their homeland spans support for and opposition to the political regime in their homeland. This engagement reflects the contestation of political power between North and South Vietnam. Furthermore, it is important to acknowledge that migrants' transnational endeavours are to acquire political clout in their nation of origin (Kaiser 1971). The scope of politics, particularly domestic political conflicts, extends beyond the inhabitants residing in a country to include non-resident citizens who can participate in the politics of their homeland from a distance (Boccagni et al. 2015). These domestic political challenges become transnational through the cultural circulation of people, which facilitates the dissemination of political ideas and behaviours.

One cultural strategy discussed is the utilisation of religious spaces to engage and seek assistance from local communities. This is particularly relevant, given the Buddhist and Christian heritage shared with the Thai and Vietnamese populations and rooted in a long historical relationship. Consequently, religious spaces, including temples and churches, serve as sanctuaries and strategic negotiation sites with the Thai state. However, the situation is complicated by the fact that these religious spaces have become battlegrounds for Vietnamese individuals supporting opposing political factions.

6.2 Transcending the Border of Death: Navigating Struggles in a New Land

Amid the French reestablishment of colonial rule in Indochina after the Second World War, the *Việt Kiều* were repressed and incited violence throughout Laos, including 1946 Thakhek incident, which compelled them to seek sanctuary in Thailand. The majority of *Việt Kiều* vividly recall that day, recounting how they determinedly leapt into the Mekong River and sought refuge in Thailand to evade imminent demise. A multitude of individuals perished on that fateful day, witnessed by those who managed to survive.

The Mekong River region serves not only as a border but also as a transnational zone that has long facilitated survival and fostered connections between populations on both banks. The Mekong is crucial in shaping the transnational existence of the overseas Vietnamese community. It became a transnational boundary, a place of escape and survival for Vietnamese in Laos who crossed into Thailand, where they were targeted and shot by colonial forces. Many Vietnamese were killed on that day (Khamwan 2019; Sripana and Trinh 2005).

Historically, the Mekong River was a natural channel connecting people on both sides before the emergence of nation-states. Even with modern border demarcation, these natural routes continued to connect communities. Thus, the Mekong River is considered a transnational border, contributing to the transnational life of the Vietnamese people. The influx of refugees, however, created challenges for both the *Việt Kiều* and the Thai authorities, who had to provide humanitarian assistance.

The fate of the Vietnamese in their new land was fraught with difficulties as they began their new transnational lives. This was particularly challenging as the global context shifted, and the Cold War prompted the Thai government to implement policies aimed at controlling the Vietnamese population. In 1946, the Thai government initiated surveillance of the *Việt Kiều* population, mostly as a reaction to the French colonial administration's efforts to quell the Việt Minh movement. More than 50,000 Vietnamese individuals relocated to Thailand from Laos by crossing the Mekong River. While a tiny faction of *Việt Kiều* expressed a desire to move to South Vietnam because of their limited family connections and affiliations with North Vietnam, the vast majority opted to establish themselves in Thailand (Poole 1970).

Although considered a security threat, the relationship between local Thais and the *Việt Kiều* has been perceived as robust due to the assistance offered since the 1946 incident. Both the Old *Việt Kiều* and local Thais provided assistance to the immigrants. Some individuals offered shelter assistance, while others offered employment opportunities. Local administrations in multiple provinces employed Vietnamese artisans, technicians, and physicians. Đàm (pseudonym), a 90-year-old *Việt Kiều* residing in Udonthani, recounted his experience as a survivor of the Thakhek tragedy. He stated that the politicians granted them permission to temporarily pass into Thailand for a duration of 15 days, but they never returned after fleeing Laos. According to him, the Vietnamese constructed a bamboo raft and successfully navigated to Tha Bo. Initially, they encountered difficulty in disembarking on the Thai side and were compelled to spend the night on the raft. Upon receiving the permission to disembark, they proceeded to the villages situated along the Mekong River and made their way to the Buddhist temple. The monks assisted in providing shelters, with some opting to reside in the school. According to Đàm's memory, "We were given food to eat every day by the very kind Thai people. Numerous Thais provided us with lodging, food, and clothing at that time, while a great number of Vietnamese crossed the border". For the *Việt Kiều*, temples and churches, while primarily places of religion and belief, also serve as sanctuaries from violence.

Bangkok considered certain *Việt Kiều*, especially those suspected to be affiliated with the Viet Minh Secret Society, as a distinct and autonomous entity within the state, commonly referred to as a "state within a state" (*rat-sorn-rat*). Consequently, stringent surveillance and administrative measures were enforced, organising villages into groups of ten families each led by a village head responsible for monitoring all movements. The government issued clear guidelines prohibiting political meetings or training, military training, the solicitation of funds to be sent abroad, the accumulation of excess weapons purportedly for the protection of life and property, and the possession of non-democratic political doctrines that contradicted Thailand's governance. Additionally, *Việt Kiều* residing in designated provinces were required to obtain a permit from the police chief 24 h before leaving the restricted area (Rueng Chon Suan Noi 1957). Deportation would be considered for individuals whose actions indicated membership in a clandestine group or who were deemed politically untrustworthy, prompting authorities to initiate investigations.

Thai authorities identified secret society directives containing several guidelines for expressing civil disobedience towards Thai authorities (Ansuchote 1960; Burutphat 1978). These instructions included shaving individuals' hair, donning black attire, lodging numerous complaints, fortifying houses with fences and bunkers, and fostering as many friendships with Thai people as possible (Khon Yuan Opphayop Nai Prathet Thai 1955). To garner support, *Việt Kiều* orchestrated merit-making ceremonies at temples and briefed monks on the persecution endured under the Thai government. They also depicted the suffering of both Thai civilians and government officials to elicit sympathy. Consequently, religious spaces emerged as strategic venues for the *Việt Kiều*, recognising temples as sanctuaries difficult for authorities to violate or intervene. Hence, the profound understanding of the *Việt Kiều* within Thai society, particularly in the Northeast region, renders it feasible to leverage cultural spaces across national borders as an effective strategy for negotiating with the Thai state. However, it is undeniable that instances of violence perpetrated by Thai officials against Vietnamese individuals have occurred frequently.

Although comprising various factions, including Pro-Hanoi, Pro-Saigon, or neutral groups, the *Việt Kiều* embodied precarious lives that left them vulnerable, particularly amid the complex political dynamics of Indochina and the stringent border policies enforced by Thailand. Many Vietnamese individuals who earned a livelihood through night-time fishing on the Mekong River found themselves unprotected and vulnerable to violence. On numerous occasions, they fell victim to lethal attacks by both Thai and Laotian officials. One such tragic incident occurred in 1956 in the Tha Uthen district of Nakhon Phanom, where Laotian security forces fatally shot five Vietnamese on a sandy beach in the middle of the Mekong River (Khon Yuan Nai Prathet Thai 1956). The Laotian authorities, fearing potential terrorism, were harsh towards Vietnamese sympathetic to Ho Chi Minh (Khon Yuan Nai Prathet Thai 1956). Meanwhile, despite appeals from the *Việt Kiều* for Thai authorities to take action, the Interior Ministry deemed the Vietnamese deaths a consequence of legal violations and refrained from seeking damages from the Lao side, instead noting the excessive use of force by Laotian soldiers. Furthermore, many *Việt Kiều* fell victim to robbery and suffered losses to their lives and property (Banthuk Kan-prachum 1961). For instance, some *Việt Kiều* were assaulted and robbed while purchasing chickens, others were attacked and robbed by aggressive customers at temple fairs, with police indifference exacerbating the situation. Additionally, those preparing to evacuate were targeted and killed by house robbers (Banthuk Kan-prachum 1961).

Despite the challenging circumstances faced by the *Việt Kiều*, the majority refrained from resorting to armed confrontation with authorities. Instead, they employed alternative strategies to exert pressure on Thai officials, with the establishment of Pro-Hanoi groups perceived as particularly threatening by the Thai authorities. The *Việt Kiều* were well-versed in the cultural traditions of the Isan (northeastern Thailand) people and utilised them as a means of forging connections with the locals, as part of a strategy devised by the Party Committee of the Overseas Vietnamese. As stated in this book's introduction, cultures develop inside societies and nations are influenced by social and geographical factors among racial and religious groupings in various nation-states. For the Việt Kiều, establishing transnational connections was crucial for their survival and served as a strategic element in their political movement, while garnering sympathy from the local community.

One such tradition, *Phuk Siew* (making friends), has been a long-standing practice among the Isan people. During house blessing ceremonies following marriages, participants would exchange vows, drink, and have their wrists bound with holy strings, symbolising a commitment to mutual assistance and solidarity during times of adversity (Chao Vietnam Phak Isan 1954). The *Việt Kiều* seized upon this tradition as an opportunity to engage in ceremonies with newly acquainted northeastern Thai individuals, with elders from the *Việt Kiều* community presiding over these gatherings. However, these activities aroused suspicion among the authorities, who feared that the *Việt Kiều* were plotting to indoctrinate Thais into communism or utilise them as future agents of communism. Consequently, numerous *Việt Kiều* were detained, and the Thai prime minister ordered the deportation of those involved in organising such ceremonies (The Vietnamese in the Northeastern Region 1954).

Another notable group that provided assistance to Hanoi was the Vietnamese women, who played a supportive role through their women's association. This demonstrates the influence of transnational politics on the *Việt Kiều*, who were organised by Hanoi's political movement organisation during the Cold War. They were organised into small groups, each consisting of no more than five individuals. These groups engaged in various activities such as rice cultivation, tending to other crops, animal husbandry, and food preservation, with the proceeds directed towards supporting the Association, particularly aiding soldiers in Vietnam. Additionally, they endeavoured to cultivate sympathy among the local Thai population in order to garner their support (Khamwan 2019).

Lan (pseudonym), an 89-year-old Vietnamese expatriate woman from Nakhon Phanom, recounted her experiences: "During the war, we, Vietnamese women living in northeastern Thailand, worked diligently to earn income by selling food, providing laundry services, and maintaining local Thai households". In addition, she described how the women's organisation mobilised Vietnamese women to various Thai villages, where they sold vegetables, food, and sweets to the locals. When confronted with mistreatment by the police or Thai officials, these Vietnamese women sought empathy from the Thai villagers. Lan stated, "During our journey to each village, we vocalised our distress, wept, and alerted bystanders about the oppression our fathers, spouses, and children were facing at the hands of the Thai authorities. The Thai people showed compassion towards us; some provided rice or food, while others paused to attentively hear about our difficult situation".

Their transnational lives were challenging and required them to navigate and devise strategic approaches to integrate into the local society. Believing that they were different from the local population, they felt the need to adjust in order to win the empathy of the local people. The ability of the *Việt Kiều* to elicit sympathy from the local population directly correlated with their capacity to safeguard themselves from actions taken by the Thai state.

Therefore, it is evident that the *Việt Kiều* employed cultural strategic methods to foster connections with the Thai populace. Demonstrating their allegiance to Hanoi, supporters invested both time and resources in numerous significant memorials. Upon the enforcement of repatriation to the DRV, monuments and enhancements to local infrastructure were erected in every district inhabited by *Việt Kiều*. Examples included the Nakhon Phanom Clock Tower, public parks, and various temple arches constructed by Vietnamese artisans. Furthermore, renovations were undertaken on dams belonging to the Royal Irrigation Department, highways, bridges and police stations. The depiction of the return to North Vietnam was characterised by a sense of delight. Vietnamese individuals with Thai acquaintances partook in the Thai blessing ceremony (*Su Khwan*) prior to their departure, exchanging blessings and keepsakes. The management of *Việt Kiều* organisations demonstrated effectiveness in political activism, even amid the heightened restrictions of the Cold War era. This contrasts with earlier periods such as World War II and the initial stages of the Cold War, during which they enjoyed greater autonomy in conducting political activities independently (Banthuk Kan-prachum 1961).

6.3 Religious Spaces as Arenas of Psychological Competition Between the Two Vietnams

The partition of Vietnam in 1954 had a profound impact on the *Việt Kiều* community, necessitating a detailed analysis of the causes behind their transnational existence and their efforts to establish lives beyond Vietnam. Understanding the concept of transnational politics among the Overseas Vietnamese is crucial, as the global Cold War influenced their political affiliations amid the conflicts between Hanoi and Saigon. The *Việt Kiều* in Thailand were deeply connected to the politics of their homeland while grappling with the challenges of navigating their transnational lives. Even religious institutions, which are typically perceived as apolitical, were undeniably involved in the transnational politics of the *Việt Kiều*. Hanoi inherited a robust organisational framework that effectively wielded influence over the community. In contrast, Saigon lacked strong connections and the capacity to assert authoritative control. A significant portion of the Vietnamese populace, comprising *Việt Kiều* had actively participated in the resistance efforts under the Viet Minh. A report from April 1955 in Saigon indicated continued support for the Viet Minh among the majority of this demographic (Báo cáo công tác tìm hiểu 1955).

The *Việt Kiều* community experienced increased fragmentation, discernible through the emergence of three distinct factions: those aligned with Hanoi, those supporting Saigon, and those adopting a neutral stance. The first two factions found themselves in direct competition, while the neutral group refrained from overt political affiliations, albeit relying on influential local entities for support. Owing to their geographical origins and familial connections in northern and central Vietnam, coupled with their ideological leanings, a significant number of *Việt Kiều* demonstrated allegiance to the DRV. They identified themselves as *Việt kiều cứu quốc* (Overseas Vietnamese for National Salvation), branding those supportive of the Republic of Vietnam (RVN) as *Việt Gian* (Vietnamese traitors), with the neutral faction labelled as *bọn lừng khừng* (fence-sitters) (Khamwan 2019). The pro-Saigon segment of *Việt Kiều Quốc Gia* categorised their adversaries as "*thân Cộng*" (pro-Communist). Chinh (pseudonym), a 93-year-old *Việt Kiều* residing in Nakhon Phanom, revealed that he was actively involved in a transnational movement. He explained that the pro-Saigon faction made efforts to establish connections with representatives from the RVN and worked together with Thai authorities. On the other hand, the pro-Hanoi group submitted to the authority of the Vietnam Workers Party. These conflicts led to the transnational existence and involvement of the *Việt Kiều*. The domestic political battles extended to include the participation of the *Việt Kiều* community from a distance, which demonstrates the fostering of a sense of brotherhood across long distances.

Thailand, therefore, became a transnational battleground for political ideology between North and South Vietnam during the Cold War, with Overseas Vietnamese playing a significant role. Both Hanoi and Saigon have implemented strategies to garner support for their respective causes. In the case of Hanoi, as mentioned, the substantial number and well-organised nature of Hanoi sympathisers, fostered through cultural policies predating those of South Vietnam, provided Hanoi with a significant advantage in mobilising support from the Vietnamese populace, particularly in the northeastern region. Following the establishment of the republic, Saigon initiated assertive measures beyond its borders to cultivate backing from Vietnamese expatriates. During the Ngô Đình Diệm era, the government utilised the Christian religious sphere to forge close ties with the aim of garnering support for its administration. In this context, Christianity functioned as a religion that transcended national boundaries. In Saigon, leveraging Christianity as a unifying element could cultivate a shared sense of national identity between the *Việt Kiều* in Thailand and South Vietnam. Christianity thus emerged as a crucial transnational belief system that was intricately connected to transnational political mobilisation. However, despite considerable efforts, achieving this goal proved challenging.

A "friendly delegation of émigré compatriots (*phái đoàn thân thiện các đại biểu đồng bào di cư*)" was dispatched to Thailand in early March 1955 under the command of Major Nguyễn Phước Đăng, Director of the Department of Psychological Warfare, Mental and Political Impacts at the Ministry of National Defense of South Vietnam. Their objective was twofold: to incentivise local *Việt Kiều* to consider returning to Saigon or to negotiate with Thai officials to ensure their safety should they opt to remain. Undoubtedly, this heightened transnational tension. The delegation comprised Buddhist representatives and a Buddhist youth commissioner, with Catholic delegates being priests stationed in Bangkok. This marked the inaugural psychological endeavour by the Saigon regime in Thailand. They showcased films and distributed literature highlighting Diệm's endeavours towards attaining independence and unity, the advancements of the Vietnamese national military, and the adversities endured by refugees from the DRV (Phái đoàn tác động tinh thần 1955).

Saigon endeavoured to establish connections with the *Việt Kiều* community, a task primarily undertaken by a small minority within this demographic, hailing from diverse backgrounds:

(1) Catholics and Buddhists who remained devout followers of their spiritual leaders.
(2) Intellectuals who had defected from the Viet Minh.
(3) Victims of the Viet Minh, whose relatives had suffered atrocities such as murder or abuse.
(4) Other individuals who had been expelled by the Viet Minh due to various transgressions or serious offences.

According to confidential reports from RVN envoys, the fourth group resorted to illicit or unethical means of livelihood, such as smuggling, prostitution, and the production of counterfeit political documents, with the aim of profiting from Thai authorities, either directly or through Vietnamese or international embassies (Kiều bào ở Ubol 1955).

Saigon concentrated on using religion to rally *Việt Kiều* to either support the South or remain neutral. They hoped to take advantage of Buddhism being Thailand's national religion. Aside from directing efforts towards the Vietnamese populace, Saigon also sought to engage with monks at a Vietnamese Buddhist Anam Nikaya temple located in Bangkok. It is indisputable that monks in these temples had been involved, to some extent, in the anti-colonial movement, aligning them closely with the nationalist movement in Hanoi. This presented yet another formidable challenge for the Saigon government, particularly given its strategic utilisation of Thailand as a component of its propaganda strategy throughout the Cold War.

However, the RVN believed that the *Việt Kiều* monks led the crowds and that the Viet Minh used them as a weapon. The Viet Minh exploited temples as communication and meeting venues, progressively proliferating them through the prestige of elderly monks. Saigon's goal was to counter communist propaganda that characterised Diệm's administration as an alien Catholic government. They hoped the Buddhist mission would show *Việt Kiều* the dangers posed by Communism to religion. Buddhists, like Catholics, were believed to be vehemently anti-Communist (Báo cáo công tác của phái đoàn đi Thái Lan 1955). This latter initiative had very mixed results, as a number of Vietnamese monks in Thailand were more sympathetic to the Viet Minh (Báo cáo công tác hoạt động tuyên truyền 1955).

Despite the understanding that Vietnamese monks played a crucial role in supporting the Vietnamese patriotic movement and maintained close ties with the Viet Minh, efforts were undertaken to establish contact with them at various temples, including Tu Te Pagoda and Canh Quoc Pagoda. According to confidential reports from Saigon, Monk Binh Luong expressed sympathy for the Viet Minh and endorsed the resistance war, despite his revolutionary inclinations. However, gaining trust during the delegation's brief mission proved challenging, although some monks indicated a willingness to provide assistance. Initially reserved, Monk Buu An's demeanour shifted to one of warmth, gentleness and consideration after engaging in conversation. Conversely, Monk Quang Thong appeared firmly aligned with the communist position, perceiving the Geneva agreement as a victory and heralding complete democracy for Vietnam (Báo cáo công tác hoạt động tuyên truyền 1955). Saigon held the belief that the Buddhist mission would effectively illustrate to the *Việt Kiều* the peril that Communism posed to religion. Their intention was to choose individuals from the Nationalist troops who had actively engaged in combat against the communists while serving among the Buddhist refugees (Báo cáo công tác của phái đoàn đi Thái Lan 1955).

Regarding Catholic *Việt Kiều*, numerous Vietnamese priests sought political asylum from Indochina in Thailand. Many individuals expanded their businesses and achieved stable social statuses. A majority opted to obtain Thai nationality, while many entered into marriages with Thai citizens. Additionally, some attained high-ranking positions within the Thai military and civil service (La situation des Vietnamiens 1955). Churches with substantial *Việt Kiều* congregations included Nong Saeng Church, Ban Chiang Yuen Church, Tha Uthen Church in Nakhon Phanom, and Thare in Sakol Nakhon. Similar to numerous other churches, Nong Saeng Church, primarily attended by Vietnamese Christians, was destroyed by French artillery from Thakhek during the Indochina conflict.

During the 1950s, American priests played a pivotal role in preaching within the Northeastern Missions. This became particularly significant in shaping perceptions within the *Việt Kiều* community, given the backdrop of the Cold War and the existence of anti-American sentiments among certain factions of the *Việt Kiều*. South Vietnam strategically utilised Christianity as a means of persuading Vietnamese expatriates to return to the southern region. In a noteworthy initiative, Ngô Đình Diệm dispatched Reverend Trần Hữu Thành and his delegation, led by Trần Quang Hoạt, to Thailand for a period of three to four months, recognising the potential to reach out to Catholics residing in Vietnam. (Ratthaban Vietnam Hai Batluang 1955; Ratthaban Vietnam Song Khon 1955) Despite being a mission sponsored by the South Vietnamese government, the Department of Civil Service organised a special police expedition to closely monitor their activities.

Trần Hữu Thành and Trần Quang Hoạt served within the Organization for Helping Migrants from North to South. The primary objective of their mission to Thailand was to provide clarification about the realities of the South Vietnamese government and to persuade expatriates to consider returning to the southern region instead of opting for North Vietnam. Ngô Đình Diệm held the belief that *Việt Kiều* were not inherently communists but rather that their patriotism had been exploited by the Viet Minh, who propagated the notion that Saigon remained under French rule. This task held significant importance, as voluntary repatriation to Hanoi posed a threat to the prestige and credibility of the Saigon government (Kan-opphayop Khon-yuan 1955). This demonstrates the government's competence in maintaining dignity and confidence in the international community while safeguarding its reputation and garnering support and acceptance from *Việt Kiều*, who were regarded as distant siblings. The perception that Vietnamese immigrants demonstrated greater loyalty to North Vietnam was regarded as a potential threat to the government's reputation in terms of international recognition.

According to the Ministry of Interior, both individuals embarked on their journey to the northeast via train, alongside American Patriarch Clarence J. Dulart. Upon arrival, they were assigned to various districts for their apostolic mission, spanning from Mak Khaeng Subdistrict in Udon Thani Province to Ban Wiang Khuk in Tha Bo District, Nong Khai Province. Each venue attracted a modest attendance of less than fifty individuals comprising both Thais and Vietnamese. The primary focus of their mission was to guide attendees in making informed decisions regarding their return to either Hanoi or Saigon (Ratthaban Vietnam Hai Batluang 1955).

In Ubon Ratchathani, four Vietnamese Catholics voluntarily opted to return to Saigon to face trials. Subsequently, in Nakhon Phanom, they visited Ban Nong Saeng Church, which boasted a sizable congregation of Vietnamese Catholics. During the preaching session, the priest emphasised religious teachings and urged the Vietnamese attendees to adhere to government directives while remaining vigilant against falling prey to Viet Minh propaganda (Ratthaban Vietnam Hai Batluang 1955).

Under the regime of Ngô Đình Diệm, Catholic *Việt Kiều* expressed optimism about aligning themselves with Saigon. Numerous letters were dispatched to Diệm, expressing admiration for the new leader and conveying hopes of reliance on the new government (Đại diện Công giáo 1955). Trần Đức Lịch, acting as the representative of Vietnamese Catholics Overseas in Bangkok, sought assistance in establishing contact with the national Catholic association (Đại diện Công giáo 1955).

In Nakhon Phanom, pro-RVN factions organised a celebration in honour of the president's patron saint (*lễ quan thầy thánh*). Following the announcement of election results in South Vietnam, they expressed pride in being citizens of the new republic and affirmed their commitment to genuine human dignity. They appealed to Ngô Đình Diệm to establish an embassy to provide direct spiritual and material assistance (Đại diện toàn thể *Việt Kiều* 1960). Likewise, *Việt Kiều* residing in Thare, Sakol Nakhon, sent a letter expressing their unwavering belief in the republican system spearheaded by the first president of South Vietnam. They referred to Diệm as "Ngô strong-willed scholar" (Ngô Chí sĩ). Their aspiration was for the RVN to uphold the principles aimed at safeguarding national security (Thư của một Việt Kiều 1963).

Even during the second republic, Catholic *Việt Kiều* maintained robust connections with Saigon. A handwritten letter penned by the parish priest at Nong Saeng Church in Nakhon Phanom was dispatched to Nguyễn Văn Thiệu, expressing gratitude for his support in funding the construction of a new Holy Temple. The priest viewed this financial assistance as tangible support for the construction of a church, representing another stride in the spiritual care of *Việt Kiều* who still identified as citizens of RVN. Additionally, he provided updates on the activities of Saigon's opponents, noting that supporters of Hanoi continued to amass significant funds and channel them to Hanoi. As a result, a considerable sum of money was amassed, with some individuals resorting to selling their homes and land to contribute to the resistance effort (Lá thư kính trọng 1971).

Therefore, these religious contestations created divisions within the *Việt Kiều* community that transcended national boundaries, directly influencing their transnational lives and leading to further fragmentation among the diaspora. The South Vietnamese administration believed that both Catholicism and Buddhism, as international religions, were staunchly opposed to Communism. In contrast, North Vietnam also utilised the religious sphere to incite a sense of nationalism. Transnational religion thus became a conduit for transnational politics, impacting the *Việt Kiều* community in Thailand and their connections to Vietnam.

6.4 Political Tensions Among the *Việt Kiều* Factions: Moving Beyond Religious Divides

As previously noted, tensions between Pro-Hanoi and Pro-Saigon *Việt Kiều* led to political conflict within the community. Nakhon Phanom emerged as a focal point for intense clashes between these two Vietnamese factions. In letters of grievance addressed to Pro-Saigon Vietnamese, who identified themselves as "Vietnamese in liberal democratic camps", complaints surfaced regarding the mistreatment inflicted by Pro-Hanoi Vietnamese. These individuals jointly signed a petition addressed to the Interior Minister, highlighting incidents of assault and gunfire targeting Nong Saeng Church. Within their grievances, they implored the Thai government to establish police security checkpoints at Nong Saeng Church to safeguard their participation in religious ceremonies, while also advocating for additional security measures (Jotmai Jak Chao-Vietnam 1957).

On 19 May 1957, violent clashes erupted between two Vietnamese factions in Nakhon Phanom, coinciding with Ho Chi Minh's birthday. Evidence from both sides corroborated that during a religious ceremony at Nong Saeng Church, supporters of Ho Chi Minh sang hymns dedicated to him and patriotic songs associated with the Democratic Republic of Vietnam (DRV). When pro-Saigon Vietnamese attempted to discourage this behaviour, citing the inappropriate nature of singing secular songs in a place of worship, tensions escalated. Subsequently, enraged Hanoi supporters launched an attack, wielding firearms and pelting the dissenter with stones and bricks. The victim managed to seek refuge in the home of a Thai individual, thus averting further harm. Despite sustaining severe injuries, the victim was fortunate to survive and was promptly transported to Nakhon Phanom Hospital for medical treatment (Jotmai Jak Chao-Vietnam 1957). In a letter of complaint addressed to Saigon supporters, it was reported that even after the assailants were apprehended by law enforcement officers, they continued to be threatened by their adversaries.

However, the diary of a Pro-DRV *Việt Kiều* who witnessed the incident provides a different account. According to this diary, there was a tradition of showing respect for both God and patriots. On 19 May, Ho Chi Minh's birthday, Vietnamese people of all ages, dressed in their finest clothes, would go to church to pray for Ho's long life. In 1957, the Catholic priest encouraged them to practise Ho's praises and revolutionary songs using the piano and organ, which resounded throughout the church.

The political harmony was disrupted when a pro-RVN Catholic man objected, shouting, "Why did the father allow singing Communist songs? I object!" This outburst incited the crowd, who began screaming, "Smash them to death!" The man's wife ran to retrieve a pistol and brought it to her husband. He fired three shots into the air, but a DRV supporter quickly kicked him in the back. The man stumbled, and a barrage of bricks was thrown at him from all directions. His white shirt became stained with blood, and he had to be taken to the hospital for treatment (Hoàng, n.d.).

Following the altercation, the police intervened and arrested nine people involved in the incident (Hoàng, n.d.). They were detained at Bangkhoang prison in Bangkok until their repatriation. Under the Rangoon Agreement, these nine individuals were deported to Vietnam as the first group to be sent back.

Pro-Saigon made up a small minority. They were frequently depicted as migrants who arrived in Thailand prior to the 1946 incident and as Catholics living in Catholic communities in the Northeast. Despite the fact that the pro-Saigon bloc did not consist completely of Catholics or Old *Việt Kiều*, as many Catholics backed the Hanoi government. Compared to Bangkok and other big cities, life was tougher in the northeastern regions, where the vast bulk of Hanoi sympathisers lived. Hanoi sympathisers rejoiced, danced and sang whenever a military coup occurred in South Vietnam. All of this insinuated sarcasm towards Saigon supporters (Báo cáo hàng tháng 1964).

It was difficult to determine whether *Việt Kiều* support was driven by political beliefs or other factors. Several individuals defected from the pro-Hanoi faction and joined the pro-Saigon party. Định (pseudonym), a former cadre (cán bộ) who worked for the Democratic Republic of Vietnam (DRV) in the northeastern region of Thailand for many years, stated, "Although many *Việt Kiều* had Thai citizenship, their hearts still loved the North and Uncle Ho. In the past, I joined them. Even my brother, a soldier, went to fight for the Viet Minh and died in Laos. My commitment was also voluntary because I believed it was the right course. They called it patriotism, and I followed them. As a cadre, I taught children at night and played mandolin, banjo, guitar, and violin at their wedding ceremonies".

The tense environment surrounding the transnational lives of the *Việt Kiều* stemmed from political tensions in Vietnam, compelling them to engage with the homeland's transnational politics. Their transnational existence intertwined with the contestation of patriotism, as the *Việt Kiều* remained connected to Vietnam's internal politics despite their physical absence from the country. They consistently demonstrated a strong transnational connection to their homeland.

However, Định later developed a strong devotion to Christianity. He recalled that whenever he travelled with Catholic priests, Hanoi supporters would curse at him. He said, "They were harsh. When I accompanied Western priests, I had to hide or cover my face. Similarly, women who wanted to meet the priest had to disguise themselves at night". This demonstrated that Christianity, a transnational religion, was deeply entangled in the transnational conflict, which had a significant impact on the lives of transnational individuals.

This account coincides with a narrative recounted by Alan (a pseudonym), a 96-year-old Western Catholic priest who resided in a church in northeastern Thailand during the Cold War. He noticed that there were two factions of Catholic followers: those supportive of Hanoi and those supportive of Saigon. These groups had a poor rapport. In addition, he noted that Christians had voiced apprehensions regarding congregations of Vietnamese supporters in Hanoi and had contributed funds to aid North Vietnamese soldiers.

Nam (pseudonym), a 94-year-old Catholic *Việt Kiều* residing in Udonthani, recounted that his friend, a doctor, encountered substantial obstacles from pro-Hanoi Vietnamese. The person who supported Hanoi deliberately refrained from seeking medical assistance from Nam's friend since that friend supported Saigon. Those who actively sought medical assistance from this doctor were stigmatised and branded as traitors (Việt gian). Despite this, people who lacked financial resources finally went to him out of necessity. Nam was accused of being a traitor because "he attended church and followed Western priests (nó đi theo Việt gian rồi, nó đi nhà thờ, nó đi theo Tây)". The Vietnamese who support Hanoi also asserted that "Vietnamese traitors, everything about them is negative (Việt gian, cái gì cũng xấu hết)". Someone went as far as labelling Nam a traitor (thằng Việt gian). He deliberately isolated himself from the Vietnamese community during his forties. Due to his inability to stay in the previous location, he made the decision to relocate to a different one.

One reason for this move, Nam claimed, was the belief that as free people, they must truly be free and not interfere with religion. He stated, "When you see a priest, you should not call them 'Westerners and American imperialists' (thằng tây thằng Mỹ đế quốc). Sometimes the priests traveled a long way, and I cooked for them. Some Vietnamese said, 'You do not cook for your parents, but you cook for them'". For this *Việt Kiều*, this community's ways of thinking did not match him. This demonstrates how transnational dynamics, particularly the conflicts in Vietnam, significantly impacted the transnational existence of the *Việt Kiều*. Despite relocating, they still faced expectations to engage in transitional politics. Adapting to a new existence proved neither effortless nor straightforward. They could not fully lead independent lives but must adjust and experience emotions in line with the norms of the *Việt Kiều* communities.

6.5 Conclusion

In Southeast Asia, the Cold War extended beyond high politics, impacting everyday people and often evolving into transnational movements. Throughout Vietnam's division into two nations, the *Việt Kiều* served as a crucial indicator of diplomatic relations between Thailand and the two Vietnams. Thailand was concerned that the *Việt Kiều* would complicate its efforts to contain Communist expansion.

Thailand became a battleground for competing narratives of the patriotic war, a contest in which the DRV successfully outmaneuvered the RVN. The Thai government constructed a public perception of the *Việt Kiều* as Communist rear guards and a security threat. To mitigate this perceived threat, they attempted to reduce the number of *Việt Kiều* by sending them to either Hanoi or Saigon. In dealing with the RVN, the Thai government was careful not to allow South Vietnamese officials to interfere with Thai authority over the *Việt Kiều*. Despite declaring an alliance with the RVN, Thai cooperation on the *Việt Kiều* issue remained limited due to fears that the *Việt Kiều* would exacerbate the spread of communist influence.

For Saigon, the older generation of *Việt Kiều* did not pose significant problems as they were primarily focused on making a living and held Thai citizenship. However, the newer generation of *Việt Kiều*, trained by the Viet Minh during the anti-French movement, exhibited communist tendencies and participated in the resistance against the US. The DRV succeeded in managing the *Việt Kiều* community through social pressure and economic sanctions, while Saigon faced greater challenges in winning their hearts and minds.

After the fall of Saigon, the troubles surrounding the *Việt Kiều* persisted. In 1976, tensions in northeastern Thailand between Thai authorities, some local Thais, and the *Việt Kiều* escalated into violence. Local Thais set fire to *Việt Kiều* businesses, and confrontations occurred in several northeastern provinces. These issues were eventually resolved following the establishment of diplomatic relations between Thailand and the DRV.

The Global Cold War significantly shaped transnational politics within Southeast Asia, particularly affecting the *Việt Kiều*. Political developments in Vietnam, especially the conflicts between Hanoi and Saigon, mobilised these communities. The *Việt Kiều* in Thailand, positioned amid this intrastate conflict, engaged in transnational political movements that often sought economic and political support.

The North–South Vietnam conflict extended to the *Việt Kiều* in Thailand, creating antagonism between different *Việt Kiều* factions and sparking a transnational contest for trust, support, and popularity between the Hanoi and Saigon administrations. This contest also permeated the religious sphere, previously seen as apolitical. As a result, Thailand became a crucial site for the transnational political activities of the Vietnamese expatriate community.

Transnational political engagement among the *Việt Kiều* in Thailand, therefore, involved complex dynamics of support and opposition, reflecting broader political contests that extended beyond national borders. This engagement encouraged long-distance membership and sustained political connections with their homeland.

References

Algie JG, Denis G, Nicholas G et al (2014) Americans in Thailand. Editions Didier Millet, Singapore

Ang CG (2010) Southeast Asia and the Vietnam War. Routledge, London

Ansuchote C (1960) The Vietnamese refugees in Thailand: a case study in decision-making. Dissertation, Thammasat University

Baker C, Phongpaichit P (2009) A history of Thailand. Cambridge University Press, Cambridge

Banthuk Kan-prachum Khong Phuthaen Chao Vietnam (Memorandum of meeting of Vietnamese representatives in Nakhon Phanom Province) (14 September 1961) (1) MorThor 3.1.7.4/7

Báo cáo công tác của phái đoàn đi Thái Lan (Report on the work of the delegation to Thailand) (30 April 1955) ĐICH21599

Báo cáo công tác hoạt động tuyên truyền Phật giáo (Report on Buddhist propaganda activities) (15 April 1955) ĐICH21599

Báo cáo công tác tìm hiểu Việt Kiều tại Thái Lan (Report on the study of overseas Vietnamese in Thailand) (15 April 1955) ĐICH21599

Báo cáo hàng tháng của Đại Sứ Quán Việt Nam Cộng Hoà tại Bangkok, địa phương Thatphnom (Monthly report of the Embassy of the Republic of Vietnam in Bangkok, Thatphnom locality) (14 September 1964) PTTG20332

Boccagni P, Lafleur JM, Levitt P (2015) Transnational politics as cultural circulation: toward conceptual understanding of migrant political participation on the move, mobilities. Mobilities:1–20. https://doi.org/10.1080/17450101.2014.1000023

Burutphat K (1978) Yuan obphayop (Vietnamese refugees). Duang Kamol, Bangkok

Chaloemtiarana T (2007) Thailand: the politics of despotic paternalism. Cornell Southeast Asia Program Publications, Ithaca

Chao Vietnam Phak Isan Tham Phithi Phukmit Kap Chao Phuenmuang (Vietnamese from the northeast hold a friendship ritual with the locals) (23 November 1954) (3) SorRor0201.24/1

Đại diện Công giáo Hải ngoại kính gửi Hội Công giáo Việt Nam tại Sài Gòn (Representative of Overseas Catholics to the Vietnamese Catholic Association in Saigon) (5 July 1955) PTTG29233

Đại diện toàn thể Việt Kiều Nakorn Phanom kính gửi Tổng Thống Việt Nam Cộng Hoà (Representative of all overseas Vietnamese in Nakhon Phanom to the President of the Republic of Vietnam) (6 May 1960) ĐICH21901

Fineman D (1997) A special relationship: the United States and military government in Thailand, 1947–1958. University of Hawaii Press, Honolulu

Goscha CE (1999) Thailand and the Southeast Asian networks of the Vietnamese revolution, 1885–1954. Curzon Press, Richmond

Goscha CE, Ostermann CF (eds) (2009) Connecting histories: decolonization and the Cold War in Southeast Asia, 1945–1962. Woodrow Wilson Center Press, Washington DC

Hoàng VP (n.d.) Bà con công giáo Noỏngxẻng 'Kính chúa yêu nước' (Nong Saeng Catholics 'respect god and patriot'). In: Nhật ký cá nhân của Hoàng Việt Phúc (Hoang Viet Phuc's personal diary), Nakhon Phanom, Thailand
Jayanama S (2017) Nayobaaj Khong Thai Tor Prathet Vietnam (Thailand's policy towards Vietnam). Siam Publishing House, Bangkok
Jotmai Jak Chao-Vietnam Nai Khet Thetsaban Muang Nakhon Phanom Thueng Ratthamontri Wakan Krasuang Mahatthai (Letter from Vietnamese people in Nakhon Phanom Municipality to the Minister of Interior) (30 September 1957) (2) KorTor16.3.7/81 (1957)
Kaiser K (1971) Transnational politics: toward a theory of multinational politics. International Organization 25(4):790–817
Kan-opphayop Khon-yuan Ork Jak Prathet Thai (Evacuation of the Vietnamese from Thailand) (1955) (3) SorRor0201.24/4 (1955)
Khamwan S (2019) The making of contracting nationalism: a case study of Viet Kieu in Nakhon Phanom. Dissertation, Naresuan University, Phitsanulok, Thailand
Khon Yuan Nai Prathet Thai Thuk Thahan Lao Ying Tai (The Vietnamese in Thailand were shot dead by the Lao soldiers) (22 February 1956) (1) MorThor3.1.2.8/124
Khon Yuan Opphayop Nai Prathet Thai Luang Homeronran (Migrant Vietnamese in Thailand by Pol.Maj.Luang Homronran). Prachathippatai (Democracy) (31 January 1955) Kor/Por7/2498/Bor1.1
Kiều bào ở Ubol gửi kiến nghị tỏ lòng hân hoan đối với Ông Tân Đặc sứ Mai Văn Hàm và triệt để tin tưởng Chính phủ Quốc gia do Ngô Tổng Thống lãnh đạo (Overseas Vietnamese in Ubol sent a petition expressing their joy for the New Special Envoy Mai Van Ham and thoroughly trusting the National Government led by President Ngo) (31 January 1955) ĐICH21599
La situation des Vietnamiens résidant en Thaïlande (The situation of Vietnamese residing in Thailand), Vietnam Presse (23 August 1955) (3) SorRor0201.44/11
Lá thư kính trọng tổng thống Việt Nam (Letter of respect to the President of Vietnam) (22 June 1971) ĐIICH4951
Naewkhit Thang Yutthakan Thi 114 Kandamnernkan Tor Chaoyuan-Khong Kongamnuoykan Pongkan Lae Prappram Communist (Operational Concept No. 114 Operations against Vietnamese immigrants of the Communist Suppression Operations Command) (1970) KorTor 90.3.3.1.1/1
Østergaard-Nielsen E (2003) The politics of migrants' transnational political practices. The International Migration Review 37(3):760–786
Panha Kiewkap Chao Yuan (Problems about the Vietnamese) (1967) (1) MorThor3.1.7.4/55
Phái đoàn tác động tinh thần sang Thái Lan (Spiritual delegations visited to Thailand) (31 March 1955) ĐICH21599
Phumplab M (2023) Making Cold War alliance: regional anti-communism for nation-building between Thailand and South Vietnam, 1955–1975. Dissertation, National University of Singapore
Poole PA (1970) The Vietnamese in Thailand: a historical perspective. Cornell University Press, Ithaca and London
Ratthaban Vietnam Hai Batluang Ma Kleaklom Yuan Opphayop Nai Prathet Thai (Vietnam Government has a pastor to persuade Vietnamese migrants in Thailand) (14 October 1955) (3) SorRor0201.24/6
Ratthaban Vietnam Song Khon Ma Klea-klom Yuan Opphayop Nai Prathet Thai (The Vietnamese government has sent people to persuade Vietnamese migrants in Thailand) (4 August 1955) (3) SorRor0201.24/6
Ricklefs MC (ed) (2010) A new history of Southeast Asia. Palgrave Macmillan, New York
Rueng Chon Suan Noi Jak Prathet Indochin (The story of minorities from Indochina) (1957) (3) SorRor0201.21.3/73
Sripana T, Trinh DT (2005) Viet Kieu Nai Prathet Thai Kap Khwamsamphan Thai-Vietnam (Vietnamese in Thailand and Thai-Vietnam relations). Institute of Asia Studies, Chulalongkorn University, Bangkok

The Vietnamese in the Northeastern Region conducted a ceremony to make friends with local people (9 December 1954) (2) KorTor16.3.7/15
Thư của một Việt Kiều tại Thare, Sakol, Thái Lan (Letter from an Overseas Vietnamese in Thare, Sakol, Thailand) (27 August 1963) ĐICH 9372
Westad OA (2005) The global cold war: third world interventions and the making of our times. Cambridge University Press, Cambridge
Wyatt DK (1984) Thailand a short history. Yale University Press, Michigan

Morragotwong Phumplab is Assistant Professor at the Faculty of Liberal Arts at Thammasat University, Thailand. Her interests encompass cultural, socio-political and diplomatic history, spanning from pre-modern times to the present day in Southeast Asia, with a particular emphasis on Vietnam and Thailand. Her current study focuses on Cold War studies, specifically the relationship between high-level political actors and ordinary people on the ground.

Chapter 7
Transnational Filipino-Bruneians: Language and Food as Conduits for Homing

Muhammad Arafat and Ailin Lai

Abstract This essay is about the transnational lives of mixed-heritage Filipino-Bruneians. Owing partly to Brunei Darussalam's proximity to the Philippines, the country sees a constant flow of people and goods from the latter. Consequently, the Philippines is ever-present in the lives of mixed-heritage Filipino-Bruneians who grew up in Brunei. Nevertheless, migration and diaspora studies have shown that identities and belonging are not given and must be actively reproduced. In-depth interviews with fourteen mixed-heritage Filipino-Bruneians were carried out to gather data. Despite displaying ambivalence about revealing their mixed heritage during their first encounters with others, our findings show our interlocutors' desire to acknowledge their connections to the Philippines. We argue that mixed-heritage Filipino-Bruneians negotiate and reproduce their Filipino-ness through learning and conversing in Filipino languages, despite their limited mastery, and acting as arbiters and ambassadors of Filipino food through consuming and selling it. Language and food, thus, serve as conduits for them to validate their claims of Filipino-ness.

Keywords Filipino-Bruneians · Language · Food · Transnational lives · Homing

7.1 Introduction

Transnational migration has contributed to the burgeoning of the world's population of mixed-heritage persons. As people migrate from one country to another for work, education, and so on, some find love and start families in the destination countries. Such families often navigate transnational social worlds as they interact and maintain relationships with family members, relatives, and friends across national borders (Childs et al. 2021).

M. Arafat (✉) · A. Lai
Faculty of Arts and Social Sciences, Universiti Brunei Darussalam, Gadong, Brunei Darussalam
e-mail: muhammad.arafat@outlook.com

H. M. Y. Ho and Y. Chan (eds.), *Transnational Southeast Asia*, Asia in Transition 29,
https://doi.org/10.1007/978-981-96-3608-2_7

This essay's point of departure is the identity conundrum of some mixed-heritage Filipino-Bruneians, all offspring of unions between Bruneians and immigrant Filipinos. We explore their relationships to their immigrant parents' home country—the Philippines and its culture—rather than Brunei Darussalam, where they were born and lived most of their lives.

On the surface, our interlocutors appear to distance themselves from Filipino cultural practices. Many do not readily reveal their mixed heritage upon their first meeting with others. However, our findings reveal a desire to connect to their Filipino heritage and community. They do so by taking steps to learn a Filipino language, usually Tagalog, and consuming, preparing, and selling Filipino food in Brunei. We argue that such acts constitute *homing* (Boccagni 2017a; Boccagni and Kusenbach 2020). Homing refers to actions performed to establish a sense of belonging to a group the actor wishes to associate with. They may do this by drawing a border between cultural insiders and outsiders while placing themselves among the former.

Being a mixed-heritage person can be a complex thing. Various contributors to a website about mixed-race people express exasperation at being asked to explain their identity backgrounds more frequently vis-à-vis monoracial persons (https://mixedracefaces.com/). They indicate that monoracial normativity persists despite the increased visibility and social acceptance of mixed-heritage people following centuries of human migration across nations and world regions. Mixed-heritage persons and families are still perceived as anomalies to a certain extent (Johnson 2024; Townsend et al. 2009). This leads some mixed-heritage people to embrace or distance themselves from aspects of their identities in response to their immediate social situations (Childs et al. 2021).

The second author, Ailin, a mixed-heritage Filipino-Bruneian, was disparaged by her Filipino teacher at a Chinese-stream elementary school in Brunei for identifying herself as Bruneian-Chinese. Still only eight years old at that time, she was left confused. Before family visits to the Philippines, her paternal Chinese relatives in Brunei would insinuate that she was 'returning' to the Philippines briefly. To her relatives in the Philippines, Ailin was a 'visitor'. Thirty-year-old Sofia endured stares in the Philippines for wearing the hijab. Twenty-four-year-old Nani felt like an outsider when experiencing a language barrier among her cousins in the Philippines. Hilmi, twenty-five years old, was bullied in his elementary school in Brunei for 'looking different'. Intended or not, the actions of others generated feelings of alienation among mixed-heritage Filipino-Bruneians, including Ailin.

This essay does not focus on the factors contributing to whether they feel a sense of belonging to the Philippines or Brunei. Instead, we spotlight our interlocutors' attempts to forge connections to the Philippines through language and food. These cultural elements transcend their utility in communication and sustenance, respectively. Research in migration and diaspora studies has highlighted their function as potent symbols of identity (De Guzman 2021; Halloran 2016; Jo 2017; Kudo 2015; Nititham 2017; Parasecoli 2014; Seiger 2019; Song and Gutierrez 2015). Language and food thus become conduits for our interlocutors to perform and claim Filipinoness. One might assert that anyone can learn Filipino languages and eat Filipino food. Nevertheless, our interlocutors are motivated to do these because some Filipino cultural practices and objects have always been part of their consciousness and lived experiences.

Aguilar Jr. asserts that the Philippines have experienced a 'migration revolution' since the mid-1960s as part of the state's strategy of capitalising on the 'structural needs of different parts of the global capitalist system' (2014, p. 1). While there are Filipinos living in various countries today, Brunei's proximity to the Philippines figures prominently in shaping the diasporic experiences of Filipinos in the country. According to Keasberry et al. (2022), Filipinos constitute the largest expatriate community in Brunei. They work in various sectors and levels of society and the economy, including civil service institutions, commercial enterprises, and private homes. Consequently, Keasberry et al. remark: 'It is impossible not to encounter a Filipino in Brunei' (2022, p. 68). Additionally, there are flights shuttling passengers between the two countries daily. With these in mind—the ubiquitous presence of Filipinos living in Brunei and the constant flow of Filipino persons and goods into the country—the Philippines is present in the daily lives of mixed-heritage Filipino-Bruneians vis-à-vis members of the Filipino diaspora in other parts of the world. In short, their social worlds are unequivocally transnational. For this reason, we employ the term 'Transnational Filipino-Bruneians' (TFB) to refer to our interlocutors in the rest of this essay.

Ailin gathered data for this essay through in-depth semi-structured interviews with fourteen TFBs aged between twenty and thirty-three years old. Three of these TFBs have lived in the Philippines. Two are now based in London and Sydney, respectively, while the others reside in Brunei. Seven volunteered to be interviewed after seeing Ailin's invitation on social media. Ailin recruited the other TFBs through her network of fellow TFBs. All interviews were conducted online from November 2021 to February 2022 during the COVID-19 pandemic. The shortest interview was one hour, while the longest lasted slightly over two and a half hours. Throughout the data collection and analysis, Ailin had to exercise care not to project her lived experiences as a TFB onto the interlocutors'.

Generally, the interviews started formally. However, the awkwardness dissipated once Ailin revealed herself as a TFB. Their shared identity as *kababayan* (fellow Filipinos) eased the atmosphere, leading the interviewees to discuss their lived experiences more freely. Some interviewees initiated speaking in Tagalog, making it easy for Ailin to ease into her mother tongue for part of the interview.

Ailin and the TFBs occasionally laughed whenever they realised they shared similar childhood experiences, even though Ailin and some interviewees had never met before the session. Nevertheless, there were divergences in their lived experiences as TFBs in Brunei. This discovery led to awkward silences during some of the interviews. To overcome this situation, Ailin or the interviewee came up with random recollections of more childhood experiences to pick up the conversation.

It is useful to note that all fourteen TFBs are first-generation offspring of interracial marriages between Filipinos and Bruneians. Additionally, the narrow range of the TFBs' ages means that all are in a similar life cycle stage. Consequently, our research findings might not reflect the views and life situations of TFBs who have transitioned to more advanced life stages.

7.2 Homing, Identity, and Transnationalism from Below

The rapid increase in global connectivity in recent decades has generated much scholarly interest in cross-border activities. On the one hand, the initial enthusiasm surrounding globalisation led scholars to decentre the study of social life away from specific national territories into an imagined global space (Kearney 1995). On the other hand, many scholars recognise that 'transnational processes are anchored in and transcend one or more nation-states' (Kearney 1995, p. 548). Nation-states and national borders continue to impact transnational activities and lives.

Transnationalism refers to various cross-border activities, including economic, political, and social initiatives (Portes et al. 1999; Tedeschi et al. 2022). While cross-border interactions are not new, Portes et al. (1999) argue that the proliferation of increasingly cost-efficient communication technologies has contributed to the massive growth of transnational activities of governments and states, corporations, social groups, and individuals.

Considering the breadth of activities that might be deemed transnational, scholars have asserted the need for researchers to 'delimit' the types of activities in their respective investigations. Tedeschi et al. (2022) offer a typology of transnationalism in their survey of the field of transnational studies, of which *transnationalism from below* is particularly relevant. It 'concerns cross-border (sociocultural, political, and/or economic) activities, practices, and behaviours that are meaningful, affect the identity and sense of belonging of people, and are carried out on a regular (not exceptional) basis in the everyday lives of individuals' (Tedeschi et al. 2022, p. 615). Taking the individual as our point of departure, we observe how the TFBs' feelings of connection to or disconnection from the Philippines figure in their daily lives and actions.

Steven Vertovec (2001) suggests that transnationalism and identity are conceptual bedfellows because members of transnational networks often converge based on an imagined shared place of origin and the cultural beliefs and practices associated with that place. Nevertheless, they negotiate their identities with other places in mind, not least the current society of residence (Anthias 2001; Vertovec 2001). Take foodways, for example. Investigating the maintenance and transformation of foodways among migrant communities, Parasecoli asserts that 'an analysis of food-related practices of migrant communities and their connection to specific environments, cultural needs, and embodied desires may counterbalance the detachment from physicality that we can at times detect in globalisation theories' (2014, p. 426).

Much of the literature about transnationalism and identity has focused on migrants, leading Vertovec (2001) to implore more research into the lived experiences of the offspring of immigrants, especially those born and raised in destination countries. To take this further: How do children of mixed-heritage families navigate their cultural identities, particularly in relation to their immigrant parents' heritage? To this end, Seiger (2019) and Kudo (2015) have discussed the identity dilemmas that the persisting myth of ethnic homogeneity in Japan engenders among the offspring of multiethnic marriages, known derogatorily as '*haafu*'.

Hoon and Fermin (2022) examine the strategic situational identity manoeuvres of the Filipino-Malay Bruneians in a Bruneian society where the Malay-Muslim ethnoreligious identity is dominant. Specifically, they assert that the presence of many Filipinas working as domestic helpers (*amah*) in Brunei has led TFBs to be stereotyped as 'anak amah' (child of a domestic helper). In this context, the authors reveal that many of their TFB interlocutors face an identity conundrum: 'Their half-Filipino background and the accompanying racial stereotypes continue to subject them to cultural, linguistic, and religious marginalisation and deny them full acceptance as a Malay in spite of their Malay proficiency and Muslim familial orientations' (2022, p. 231). Consequently, the TFBs distance themselves from their Filipino identity in some situations and emphasise it in others when it benefits them. While their study is highly relevant to ours, we take the Philippines, the home country of the TFB's immigrant parent, as our interlocutors' point of reference for identification rather than Brunei and its *Melayu Islam Beraja* (MIB or Malay-Muslim Monarchy) national ideology. With this shift in context, we observe how TFBs validate their claims to identify as Filipinos.

Home is a compelling concept for discussing a person's social location vis-à-vis places, groups of persons, country, and so on. Boccagni asserts that *home* is often used as a metaphor 'to trace a boundary between *the inside* and what lies outside, the latter being whatever is unfamiliar, stranger, unusual, or perceived/felt as remote in space, time, culture, or morality' (2017a, p. 4). Nevertheless, its processual potential is overlooked when perceived as a product, a 'particular spatial setting' (Boccagni 2017b, p. 23). Indirectly, Nititham highlights the importance of the processual approach when studying belonging: 'Being Filipino or of Filipino descent does not give one an automatic ticket to belonging. Belonging must be actively cultivated' (2017, p. 111).

Boccagni, thus, suggests a shift towards *homing*, which he defines as 'the aggregate of micro-practices whereby people try to make themselves at home in their life circumstances, and to the moral and emotional economies underpinning them' (2017a, p. 10). Emphasising the importance of praxis, he directs attention to the homing practices that people carry out to achieve security, familiarity, and control over their day-to-day life situations (Boccagni 2017a). This seeking of 'insiderness' often remains an aspiration rather than fully achieved (Boccagni 2017a, b; Boccagni and Kusenbach 2020). Similarly, as we shall in the rest of this essay, our TFB interlocutors' claim to Filipino-ness is constantly negotiated and reproduced rather than given.

7.3 Language

Brunei does not appear to be a favourable social environment for acquiring Filipino languages despite the substantial presence of Filipinos in the country (Hoon and Fermin 2022). The TFBs in this research picked up a Filipino language while growing up in Brunei, albeit attaining varying degrees of fluency. They identify with the language but are quick to acknowledge their limited mastery. Nevertheless, they assert that language fluency should not be the sole criterion of one's Filipino-ness.

English and Malay enjoy the status of being the most widely spoken languages in Brunei (Deterding 2020; Haji-Othman 2012; McLellan et al. 2016). According to Haji-Othman (2012), official language policies have contributed to this trend. On the one hand, Brunei's educational policies have promoted the acquisition of the English language for its transnational utility since the nation regained independence in 1984. On the other hand, Malay's status as a central aspect of national consciousness and identity began when Brunei was a British protectorate. This status is maintained to this day. Consequently, official policies shaped a linguistic landscape favourable to English and Malay language acquisition at the expense of other languages, including indigenous (Deterding 2020; Haji-Othman 2012) and possibly foreign ones.

Language acquisition, however, occurs through various means and activities outside official domains and institutions. Family may play a key role in encouraging language learning through early cultural and language exposure, although other societal factors may influence a child's language acquisition (Seiger 2019; Stephan 1991).

Some TFBs learn a Filipino language, usually Tagalog, through popular media such as television drama series and music. Their forays into Filipino popular culture begin when they are introduced to films by a family member, usually a parent, at an early age. Some would explore more of Filipino popular culture of their own volition as they progressed in life. Their self-directed consumption of Filipino popular culture when the consumer is spoilt for choice—entertainment products from around the world are now accessible at one's fingertips—reveals some gravitation towards their Filipino heritage and identity.

Five of our TFB interlocutors claim to be fluent in Tagalog. Some benefitted from their early introduction to the language during childhood, while others developed proficiency from spending extended time in the Philippines later in life. The remaining nine TFBs have limited comprehension of the language, of which only three can speak it with minimal proficiency.

Christian, Daniel, and Kimberly are three TFBs who claim to be fluent in Tagalog. Not coincidentally, they have also lived in the Philippines. Christian moved to the Philippines after his parents separated. Daniel and Kimberly are siblings who spent five years pursuing tertiary education in Manila. All three were minimally proficient in Tagalog before moving to the Philippines. They grew up in predominantly English-speaking households in Brunei.

Christian recalls feeling isolated for not being able to communicate with other Filipinos and the difficulties he faced in trying to engage with his peers and even his relatives in the Philippines. It took him five years to become comfortable conversing in Tagalog, which he learned from watching television and interacting with friends. Daniel and Kimberly, too, attribute their fluency to their friends. Thirty-one-year-old Daniel, who lived in Metro Manila for four years, shared his experience:

> Daily exposure to the language. It is like, one day, you wake up suddenly, and then you understand everything. It is because you hear the language everywhere. On the radio, the

> TV, and people talking around you, your friends, also talking. And slowly, you pick up the words. And you know, say a few words here and there and eventually become fluent.

Exposure to homeland television plays a factor in maintaining and improving the home language among the children of immigrants (Al-Jumaily 2015; De Guzman 2021). In Hong Kong, where immigrant children are unlikely to encounter their parents' native languages in public spaces, some Filipino parents subscribe to Filipino television channels to expose their children to the language (De Guzman 2021). Similarly, most of our TFB interlocutors reveal that much of their contact with Filipino language and culture occurs through watching television shows.

Kimberly began watching Filipino television shows before becoming fluent in Tagalog. Syasya and Nur, two other TFBs fluent in Tagalog but have not lived in the Philippines, attribute their Tagalog fluency to their exposure to and consumption of Filipino popular culture. Nur, twenty-four years old, recalls resorting to watching Filipino movies on the Internet during her teenage years:

> I did this a lot when I was young, like in my preteen or teenage years, but not in adulthood. I think it is because Bruneian TV rarely shows those movies in Tagalog. So, I usually watch most of them on YouTube.

Ailin, too, remembers watching television shows such as *Pangako Sa 'Yo*, *Marimar*, and the *Ang Tanging Ina* during her childhood despite not fully understanding what the actors were saying. It had never occurred to her until now how peculiar it was to watch a Filipino TV series on a Malaysian channel with Malay subtitles. Nevertheless, she is grateful that she could bond with her mother and aunt as they watched the drama unfold, gasping and laughing together at the right moments. Some of our TFB interlocutors share Ailin's experiences of watching Filipino drama series with their families:

> Sofia: It depends on the movies, but usually, it is a family thing. I think, especially ones like *Sana'y Wala Nang Wakas* or *Pangako Sa 'Yo*, where everyone just rushed home to watch it together. Yeah, those were the memories.
>
> Miza: My mum loves watching Filipino movies on DVD. We have a few of them. And mostly, they're comedies. Again, I just read the subtitles. But I think that's one of the influences on how I came to understand Tagalog.

Consuming popular culture contributes to the TFBs learning the language of their Filipino parent. Nevertheless, this activity helps them achieve minimal proficiency only. Consequently, many of them feel inadequate when conversing with other Filipinos. Hoon and Fermin show that the TFBs' lack of confidence is not unfounded: 'Although the ability to speak Tagalog has given them access to the Filipino-speaking community, their non-native competency in the language sometimes leads to questions about their authenticity as a 'real' Filipino' (2022, p. 235).

Ten of the fourteen TFBs do not reveal their mixed identity upon meeting someone for the first time, preferring to let it surface organically. Twenty-four-year-old Miza, who was an exchange student at Ateneo de Manila University, said:

> I never told them because they would expect me to speak in Tagalog. I do understand them, but imagine if it's in a lecture—educational institutions talk and, like having a class, speak

> Tagalog the whole time—I wouldn't understand that right, especially the complex words. I only know daily words in Tagalog.

Despite their self-perceived inadequacy and the fear of being called out as inauthentically Filipino, some TFBs admit to speaking in a Filipino language when the situation benefits them. Typically, this involves the TFB saying something in Tagalog, which leads the other Filipino to identify the TFB as *kababayan*—a cultural insider—and then provides the TFB with special treatment like a discount.

Nur finds it easy to talk to other Filipinos in Brunei and even consciously attempts to interact with Filipino servers and salespeople. She suggests that doing this makes her feel at 'home' even though she has yet to step foot in the Philippines. She claims that she feels an automatic bond in these interactions:

> If I spoke Tagalog, they would instantly ask:' Oh, *Filipino ka pala? bat di mo sinabe na Pinoy ka*?' (Oh, you're Filipino? Why didn't you say so?) And then we just instantly, you know, are in a conversation. And then, at the end of it, I get a discount, so a discount is a win-win situation. Sometimes, I would over-exploit that privilege and then get those discounts. But yeah, it's that feeling of home I'm yearning for from those interactions; they make me feel like I'm home with my other relatives.

Similarly, Faiq uses his identity to his advantage, particularly when shopping. However, unlike Nur, Faiq reveals his identity only in situations of extended interactions, like when he shops for clothes. He claims that he gets a more comfortable shopping experience when the staff knows of his Filipino identity, and it's a win–win situation because the Filipino sales assistant 'gets the help, I get the friendliness'.

In some cases, the *kababayan* notices the TFB first. While on a one-year twinning program in Sydney, Nani said a Filipino paintball vendor quoted a 'special price' after her identity surfaced during their encounter:

> He kept looking at me and asked: 'Are you like any chance mixed or anything?' I was like, oh yeah, yeah, I am. Then I said I am half Filipino, and that was when he said, 'Oh *kababayan*.' ... and then because of that I found out ... that guy told my other group saying that 'Oh, I gave your friend a special price because, like Pinoy, we're *kababayans*', and I'm like oh wow, okay, that's nice of him.

Nani did not have to speak Tagalog with the vendor to prove her identity, yet she received a discount. She also capitalises on her identity at a hair salon she frequents in Brunei 'just to be besties, you know', implying that she receives better treatment from the Filipino staff when she makes her identity known.

The TFBs who are not proficient in Tagalog negotiate their relationships and interactions with their Filipino relatives through various means, including bonding over food and relying on social cues. Hilmi shared that the language barrier between himself and his Tagalog-and Bisaya-speaking relatives in the Philippines does make him feel out of the loop, to a certain extent, but he also said:

> But it comes to a point where if you spend enough time, even though you don't understand it, you kinda get what they say because, for Filipinos, when we speak, there are a lot of hand gestures, body language, and social cues because they're very expressive. Even if I can't understand the language, I pick up on those notions. So, it somewhat mitigates that to a certain extent.

On the other hand, Sofia recalled the awkwardness between herself and her cousins in the Philippines because of language barriers. While visiting the Philippines as a fourteen-year-old, she chose not to talk much. Sofia knew some Tagalog, while her cousins could understand English. Nevertheless, she felt that her speaking in English would intimidate them.

Although most of the TFBs agree that knowing how to speak or understand Tagalog is significant to (re)claiming their Filipino identity, they all disagreed with the notion that it is the sole factor one should focus on:

> Nani: Other than languages, other factors make us Pinoy. One example is our culture. Second is the way we live and how it influences our everyday lives.
>
> Hafiz: I'd say it's more like how you live. It's more of a lifestyle thing, more so than language.
>
> Syasya: It's a just language, you know. Everyone can learn it; everyone can learn any language. But for me, it's about if you know and by heart, the culture of the Philippines and the way of living. If you really know what it feels like to be a Filipino in the Philippines, it should be enough to prove that you're Filipino for me, like knowing the culture, experiencing it, and living there. Language is not really that important.

7.4 Food

> I've been learning Mandarin for over three years now but there are still days when language fails me, when food feels like the only thing I have to tie me to this other home my family brought to me from far away. (Powles 2020, chap. 'Hungry Girls')

Nina Powles encapsulates the identity struggles of many people of multiethnic heritage. Where language identification wavers, some gravitate towards food around which to moor their identities. More than fulfilling a primal need for survival, food is fundamental to constructing and reproducing individual and collective identities (Fischler 1988). Parasecoli suggests: 'Participants in the same culinary culture acknowledge each other by the way they eat, by what they eat, and by what their diets exclude' (2014, pp. 424–425). Consuming and identifying certain foods as part of one's heritage constitutes an act of homing, whereby the consuming subject draws a line separating cultural insiders from outsiders while simultaneously locating herself among the former.

Food can mitigate the sense of distance from the homeland and help people feel connected to their families and friends in their homeland and other places (Parasecoli 2014). De Guzman finds that 'food was never far away from any conversation or experience that I had during my fieldwork' (2021, p. 4) while researching Filipino migrants in Hong Kong. For Nititham, whose interactions with Filipinos in Ireland often began and ended around the kitchen table, food became a social adhesive: 'I began to understand food as a pathway which can ascribe and inscribe a shared sense of belonging, creating a sense of familiarity for ourselves—however fleeting' (2017, p. 105).

Conversations with the TFBs in this research perked once we shifted focus towards food. The consternation that some displayed when discussing Filipino language fluency dissipated, replaced by confidence. Food became a solid platform for the TFBs to perform Filipino-ness by consuming, sharing, and selling Filipino food.

The TFBs frequently mention *adobo*, *sinigang*, *tapsilog*, *bangus*, and *leche flan*. For dishes traditionally made with pork, some restaurants in Brunei, such as Diners!, Suria Azlan, and Pinoy Taste, make them accessible to Brunei's predominantly Muslim society, including some TFBs, using halal alternatives. Of course, Jollibee, the iconic Filipino fast-food chain, is also available!

The home is where those born in the diaspora, including the TFBs, are often introduced to Filipino food as part of their heritage (De Guzman 2021). Most TFBs interviewed grew up with Filipino food in Brunei, either cooked by their Filipino parents or bought from *tita*s (aunties) in their Filipino parents' network of *kababayans*. Usually, they prefer home-cooked food to those from Filipino restaurants.

None of the TFBs questioned the authenticity of the Filipino foods prepared by their parents. Ashraf reminisces about his mother's cooking while highlighting the importance of knowing the right ingredients—and which brands—to produce authentic flavours:

> My mum's chicken adobo. That is one of my favourites. I wait for the Mama Sita brand to come to Brunei. I usually ask my mum to get them if they are in stock because that is the closest authentic Filipino chicken dish I can eat. That is one of my favourites, apart from eating spaghetti. These are the two Filipino foods that I grew up eating.

On the other hand, the TFBs view Filipino food served in commercial establishments with some scepticism. Maria opts for imported Filipino ice cream and avoids the popular dessert *halo-halo* altogether simply because the latter, though readily available in Brunei, tastes inauthentic:

> Every *halo-halo* that I've tried here in Brunei is so-so. I would rather eat plain ube ice cream from Hua Ho (a supermarket chain) because there's a brand of ice cream that only comes from the Philippines, but now you can get it at Hua Ho. Back when I was a kid, you could not even find it in Brunei. We had to buy it from the airport in the Philippines. We no longer have to do that because it's available at Hua Ho. So, I would rather eat that plain than seek out *halo-halo* in Brunei because, at least, I know that one is authentic.

Despite implying that *halo-halo* in Brunei tastes inauthentic, Maria goes on dates at Filipino restaurants with her fellow TFB boyfriend 'because he has embraced his Filipino side so much'. He is often the instigator of going to Filipino eateries on their dates. These include the food courts in malls like Times Square and Yayasan Shopping Complexes or Jollibee whenever they could not choose a place to eat. Maria and her boyfriend become agents of homing for one another. Their frequent consumption of Filipino food and calling Jollibee 'the mothership' improved their bond as a romantic couple and their connections with Filipino food and culture.

Nur, however, remains reluctant even to try the food at Filipino restaurants in Brunei:

> I was a bit hesitant to try. I already saw the menus. Mainly, people mention *tapsilog*. It looks decent in the pictures. The taste, though. Will the authenticity still be there? Or will they alter some ingredients to suit the Bruneian palate?

Aside from acting as arbiters of authentic flavours as consumers of Filipino food, the TFBs also assume the role of ambassadors and promoters of Filipino cuisine as producers and sellers or simply sharing food with Bruneian friends. Aisha recalls giving food items as *pasalubong* (souvenirs) to her friends. Identification with cultural products and objects embodies the diasporan's negotiation of her social location within the society of residence and the 'homeland' (Nititham 2017). For Aisha, gifting friends with a piece of her other home—the Philippines—is a pleasant way to solidify her identity as a TFB among her Bruneian friends and, to a certain extent, herself:

> Either my mum or her friend just came back, and you know how they like to bring *pasalubong*, and it's usually like Goldilocks and whatnot. I remember my mum saying: 'We have a lot of packs; you can give one to each of your friends.' And, I remember just being so proud of myself: 'Oh my god. Yeah, I have Filipino products!'

Like migrants who 'employ their culinary knowhow to occupy social and economic positions in the territory of the Other by transforming the members of the host community into consumers of their products' (Parasecoli 2014, p. 430), some TFBs in Brunei have started Filipino food businesses to share a piece of their heritage with others.

Using the recipes passed down in his mother's family since the eighteenth century, Hilmi and his family have started a home-based business selling Filipino food, primarily from the Mindanao region. When asked about the reception towards their business, Hilmi shared that their customers comprise Muslim Filipinos and local Bruneians and that the latter group was driven mainly by their curiosity about Filipino foods beyond *adobo* and *sinigang*. To him, the business allows him to 'give back' to the community while upholding his Filipino roots, making him feel wholesome: 'I am helping these Filipino Muslims not to miss out on their food. It brings the 'family' together because we're doing this to help each other.'

Hilmi and his family have built a favourable reputation in the local scene through their participation in pop-up events and collaborations with community organisations, and the Philippine Embassy has also recognised their efforts. Hilmi and his family consciously try to retain the authentic Filipino taste, sourcing ingredients from a local Filipino product supplier and substituting the traditionally non-halal ingredients with their halal counterparts.

Like Hilmi, Sofia also runs a small online business selling *buko pandan*, a Filipino dessert made from coconut, pandan leaves, and sago. She has watered the dessert down to suit the Bruneian palate. This move garnered criticism from her mother, who argued that the dessert should be creamier to retain its authentic taste. Sofia laughed while recalling her counterargument: 'Mum, the name is (redacted) Malay Filipina. A mixture of Malay and Filipino tastes'.

The centrality of food to identities and culture lends it to sentimentalisation, especially among migrant communities. Parasecoli, who was entertained with a hearty spread while visiting his grandaunt and cousins upon arriving in the United States from Italy, pondered: 'I soon realised some of the dishes served had the same names as those I used to eat back home, but they looked and tasted different' (2014, p. 415).

He even remarked that how the food was served—with no sequencing from appetisers to desserts—was new to him. De Guzman muses about the migrant foodways in a similar vein:

> This notion of eating food to connect to the Philippines can be seen as largely imagined, perhaps even romantic. The idea that one can share a table with family members back in the Philippines because one is eating *adobo* or *sinigang* can be comforting, but in reality, those in the Philippines do not always eat traditional Filipino food day in and day out. (2021, p. 9)

Like Parasecoli's Italian American cousins and De Guzman's migrant Filipino interlocutors in Hong Kong, we have seen that the TFBs gravitate towards food to reproduce their connections to the Philippines and its culture and society. It may well be that the Filipino food they prepare, consume, or sell in Brunei is 'inauthentic', as seen from the example of Sofia's *buko pandan* above. Authentic or not, food serves as a homing medium for the TFBs. It reminds them, and others, of their connections to the Philippines. Through consuming and sharing food, the TFB constructs a sense of belonging to the Philippines and performs their TFB identity to others at opportune moments.

7.5 Conclusion

In this essay, we have discussed the TFBs' negotiation of their relationships with the Philippines, the home country of their immigrant parent. Specifically, we examine the TFBs' use of language and food as conduits of homing. According to Tedeschi et al. (2022), many scholars agree that much of transnationalism is 'from below'. Our investigation of the TFB's maintenance of their connections to the Philippines through language and food contributes to the study of transnationalism by showing how transnationalism from below manifests in the everyday lives of transnational persons.

The transnational paradigm has been equally instructive to our study. Sociological investigations of mixed-heritage people have tended to use their mixed identities to measure their assimilation into society despite the growing recognition that the relatively higher occurrence of intermarriage involving immigrants has significantly contributed to the increasing population of such persons (Newman 2021). Alyssa Newman asserts that 'very little is known about how these immigrant roots affect the multiracial experience' (2021, pp. 26–27). Put differently, the transnational aspect of the mixed-heritage people's lives has been neglected. We contribute to trimming this blind spot by focusing on the TFB's negotiation and management of their immigrant roots rather than their status within Brunei, their society of residence.

Through learning and conversing in Filipino languages as well as consuming, preparing, and selling Filipino food, the TFBs in this research negotiate and reproduce their position as Filipino cultural insiders while living outside the geographical borders of the country. Despite the TFBs' ambivalence and self-doubt about their abilities to communicate in Filipino languages, all have committed effort, albeit in varying degrees, to learn the language of their Filipino parent while living in Brunei. Some have successfully used the language to establish rapport with other Filipino *kababayans*, especially in situations that benefit them. Nevertheless, they were quick to point out that there are alternative ways to be Filipino. We have discussed how the TFBs perform Filipino-ness confidently through food. As consumers, they assume the role of arbiters of authentic Filipino flavours when they rate the food served at eateries in Brunei. Some TFBs go further to take on the role of ambassadors of Filipino cuisine when they set up food businesses and 'share' this aspect of their heritage with others. These examples show that the TFBs live in transnational social worlds while growing up and living in Brunei.

Acknowledgements We wish to thank Lian Kwen Fee, Maria Carinnes Alejandria, and the editors for comments on earlier drafts of this chapter. More importantly, our heartfelt gratitude goes out to our Transnational Filipino-Bruneian interlocutors who graciously shared with us stories about a fascinating, and yet deeply personal, aspect of their lives.

Data Availability Ailin Lai collected data for this essay while preparing her undergraduate thesis under the guidance of Muhammad Arafat at Universiti Brunei Darussalam. Subsequently, Muhammad Arafat wrote the first draft of this essay while Ailin Lai contributed comments and suggestions for the essay's improvement.

References

Aguilar FV Jr (2014) Migration revolution: Philippine nationhood & class relations in a globalized age. Kyoto University Press, Singapore; Kyoto, NUS Press

Al-Jumaily S (2015) Parents' first language is their children's second language. Language loss: Arabic is a case. International Journal of English Language Teaching 2(2):19–31. https://doi.org/10.5430/ijelt.v2n2p19

Anthias F (2001) New hybridities, old concepts: The limits of 'culture'. Ethical and Racial Studies 24(4):619–641. https://doi.org/10.1080/01419870120049815

Boccagni P (2017a) Saving home from the pitfalls of the home, through homing: towards a "positive deconstruction" of the social meanings, functions and moralities of home. HOMInG working paper no. 1/2017, 7 December 2017

Boccagni P (2017b) Migration and the search for home. Palgrave Macmillan US, New York

Boccagni P, Kusenbach M (2020) For a comparative sociology of home: relationships, cultures, structures. Current Sociology 68:595–606. https://doi.org/10.1177/0011392120927776

Childs EC, Lyons A, Jones SL (2021) Migrating mixedness: exploring mixed identity development in New York City. Journal of Ethnic and Migration Studies 47(4):782–801. https://doi.org/10.1080/1369183X.2019.1654153

De Guzman JA (2021) Home cooking and identity formation in migrant Filipino families in Hong Kong. Stud Ethn Natl 21:158–171. https://doi.org/10.1111/sena.12348

Deterding D (2020) Indigenous languages and English in the globalised modern era in Brunei Darussalam. KEMANUSIAAN The Asian Journal of Humanities 27:165–178. https://doi.org/10.21315/kajh2020.27.2.9

Fischler C (1988) Food, self and identity. Social Science Information 27:275–292. https://doi.org/10.1177/053901888027002005

Haji-Othman NA (2012) It's not always English: 'Duelling Aunties' in Brunei Darussalam. In: Rapatahana V, Bunce P (eds) English language as hydra. Blue Ridge Summit: Multilingual Matters, Bristol, pp 175–190

Halloran VN (2016) The immigrant kitchen: food, ethnicity, and diaspora. The Ohio State University Press, Columbus

Hoon CY, Fermin M (2022) "I Don't Want to be Called Anak Amah": the identity conundrum of biracial Filipino-Malay Bruneians. International Journal of Asia Pacific Studies 18:227–254. https://doi.org/10.21315/ijaps2022.18.2.10

https://mixedracefaces.com/Mixedracefaces: storytelling platform & diversity hub for the mixed-race community. https://mixedracefaces.com/. Accessed 13 Jun 2024

Jo J-YO (2017) Homing: an affective topography of ethnic Korean return migration. University of Hawai'i Press, Honolulu, Hawai'i

Johnson K (2024) The multiracial family as an anomaly: experiences of monoracism. Journal of Marriage and Family. https://doi.org/10.1111/jomf.12995

Kearney M (1995) The local and the global: the anthropology of globalization and transnationalism. Annu Rev Anthropol 24:547–565. https://doi.org/10.1146/annurev.an.24.100195.002555

Keasberry C, Le Ha P, Alas Y (2022) Tales of Filipinos in Brunei Darussalam: identities, cultural flows, histories, and personal narratives. In: Lee JW (ed) The Sociolinguistics of Global Asias. Routledge, New York, pp 68–83

Kudo M (2015) Transnational families in a global circulation context: the case of cross-border marriages between Japanese women and Pakistani migrants. Bulletin of the National Museum of Ethnology 40:71–84. https://doi.org/10.15021/00003795

McLellan J, Noor Azam H-O, Deterding D (2016) The language situation in Brunei Darussalam. In: Noor Azam H-O, McLellan J, Deterding D (eds) The use and status of language in Brunei Darussalam. Springer, Singapore, pp 9–16

Newman AM (2021) Revisiting the marginal man: bridging immigration scholarship and mixed-race studies. Sociology of Race and Ethnicity 7:26–40. https://doi.org/10.1177/2332649220933302

Nititham DS (2017) Making home in diasporic communities: transnational belonging amongst Filipina migrants. Routledge, Oxon and New York

Parasecoli F (2014) Food, identity, and cultural reproduction in immigrant communities. Social Research: An International Quarterly 81:415–439. https://doi.org/10.1353/sor.2014.0015

Portes A, Guarnizo LE, Landolt P (1999) The study of transnationalism: pitfalls and promise of an emergent research field. Ethn Racial Stud 22:217–237. https://doi.org/10.1080/014198799329468

Powles NM (2020) Tiny moons: a year of eating in Shanghai. The Emma Press, Birmingham

Seiger FK (2019) 'Mixed' Japanese-Filipino identities under Japanese multiculturalism. Social Identities 25:392–407. https://doi.org/10.1080/13504630.2018.1499225

Song M, Gutierrez CO (2015) 'Keeping the story alive': is ethnic and racial dilution inevitable for multiracial people and their children? Sociol Rev 63:680–698. https://doi.org/10.1111/1467-954X.12308

Stephan CW (1991) Ethnic identity among mixed-heritage people in Hawaii. Symb Interact 14:261–277

Tedeschi M, Vorobeva E, Jauhiainen JS (2022) Transnationalism: current debates and new perspectives. GeoJournal 87:603–619. https://doi.org/10.1007/s10708-020-10271-8

Townsend SSM, Markus HR, Bergsieker HB (2009) My choice, your categories: the denial of multiracial identities. Journal of Social Issues 65:185–204. https://doi.org/10.1111/j.1540-4560.2008.01594.x

Vertovec S (2001) Transnationalism and identity. J Ethn Migr Stud 27:573–582. https://doi.org/10.1080/13691830120090386

Muhammad Arafat is Assistant Professor of Sociology and Anthropology at Universiti Brunei Darussalam. His research interests include the relationship between transculturality, homing, place, and belonging, Arabia-Southeast Asia connections, the study of Malay-Muslim societies, and the anthropology of modern and contemporary art of Southeast Asia.

Ailin Lai double majored in English Studies and Sociology & Anthropology at Universiti Brunei Darussalam. She completed her undergraduate studies in 2022 and earned a Bachelor of Arts [1st class]. Her lived experiences as a person of mixed parentage—Bruneian-Chinese and Filipino—inspired her interest in research about the Bruneian-Filipino community. She aspires to pursue a master's degree in Anthropology.

Chapter 8
Language Shifts and Transnational Migration in the Belait Community: A History of Resilience and Adaptation

Ashrol Rahimy Damit and Zulfadzlee Zulkiflee

Abstract This study examines language shift among Bahasa Belait speakers in Brunei as a means of adapting to the social challenges of migrating to Brunei. Regarded as one of the minority languages in Brunei (Martin 1995), Bahasa Belait is a non-Malay isolect (Nothofer 1991) of the Baram–Tinjar subgroups of the North Sarawak language (Blust 1972). Martin (1995) reported that Brunei Malay is the first language for Belait people under 35, which indicates that a language shift from Bahasa Belait to Brunei Malay occurred a few decades earlier. This shift, resulting from the speakers' migration from North Sarawak to Kuala Balai in Belait District, can be traced through the phonological, morphosyntactic, and lexical aspects of the language, which features assimilation of aspects from local languages and dialects. This study explores how the Belait people adapted to their 'new home' through assimilation into Brunei culture and engaging in transnationalism with Belait communities back in Sarawak. It uses historical linguistics, interviews with Belait speakers and examination of factors that influence language shift. Culminating in tracing the community's journey towards being legally recognised as one of the seven *Puak Jati* (indigenous groups) in Brunei Darussalam.

Keywords Language shifts · Transnational migration · Bahasa Belait · Assimilation · Indigenous language

---

A. R. Damit (✉) · Z. Zulkiflee
Faculty of Arts and Social Sciences, Universiti Brunei Darussalam, Gadong, Brunei Darussalam
e-mail: rahimy.damit@ubd.edu.bn

Z. Zulkiflee
e-mail: 24h8045@ubd.edu.bn

H. M. Y. Ho and Y. Chan (eds.), *Transnational Southeast Asia*, Asia in Transition 29,
https://doi.org/10.1007/978-981-96-3608-2_8

8.1 Introduction

Language holds immense significance within cultural contexts and plays a crucial role in shaping community identity (Gelman and Roberts 2017). Community languages are essential for the development and preservation of group identity, individual identity and multigenerational communication. It plays a fundamental role in the manifestation of culture, influencing how people view themselves and the groups they associate with. Through language, we can also represent the world and highlight various identities, groups, and categories that make up social life (Silverstein 2022). Furthermore, it promotes a sense of collective identity and unity by acting as a medium for promoting values, beliefs, and customs. In particular, languages serve as a medium for transmitting a community's tradition and folklore from one generation to another (Bakar et al. 2014). Languages also carry markers that represent various aspects of a speaker's identity. Accents, dialects, and vocabulary are examples of linguistic features that represent an identity's social, historical, cultural, and geographic influences. Communities also often develop shared linguistic patterns that differentiate them from others, encouraging a sense of belonging and shaping their identity. As language is closely tied to culture, language preservation is therefore essential for preserving cultural legacy as losing a language can result in the loss of a person's or a community's culture and identity.

A language can be lost through a process of what is called language shift, a process that is influenced by various social, political, and economic factors such as transnationalism or the geographical movement of minority languages. A group of speakers can be defined as experiencing a language shift, sometimes referred to as language transfer or language replacement, when the group progressively adopts a language that has been recently introduced over an extended period. Brown (2022) also identifies that language shifts can be defined by a process in which a community transitions away from their minority language to a majority or more dominant language. This process of language shift from a minority to majority language is often the case due to the majority language frequently being considered to have a higher status, with the associated privileges that come with speaking it. Language shift and switching due to differences in the status languages is linked to what Phillipson (2018) calls linguistic imperialism, which involves the imposition of one language onto a group of people due to said language possessing a dominant status, often times through its association to elites that possess supremacy status. Such imposition can be attributed to the effects of colonialism, which enables select groups to attain supremacy status and subsequently the capacity to impose language. The widespread adoption of this dominant language in turn can negatively impact linguistic diversity, potentially leading to language shift and switching (Phillipson 2018).

This process of language shift and switching, and the factors that drive them, is an instance that can be observed in the Belait community. In the late 1800s, the Belait community is alleged to have experienced migration from Baram, North of Sarawak, with Hughes–Hallet (1980) asserting that the Belait tribe arrived in Brunei between 1830 and 1890. The Belait communities' migration to Brunei has shaped much of their way of life and status, as they have come to be defined as a *puak jati* Brunei under the *Perlembagaan Negeri Brunei 1959* (Government of Brunei 1961). As prescribed in the *Perlembagaan Negeri Brunei 1959,* Brunei Darussalam consists of 7 tribes, namely Belait, Bisaya, Brunei, Dusun, Kedayan, Murut, and Tutong (Government of Brunei 1961, pp. 118–120). Each tribe has its own distinctive language and culture that serves as identity markers. As the Belait group have experienced international migration, they have come to possess their own distinctive culture and language, primarily because of daily environmental influences on their language and culture. Transnational migration has also significantly impacted the language dynamics of communities, including the Belait. When communities migrate across borders to new countries, the languages and linguistic practices of the migrant's home country influences the linguistic practices and linguistic landscapes of both their new and old homes (Gu and Tong 2019). Thus, language use and language practices are impacted by the complex effects of transnational migration.

Reversing language shifts and bringing minority languages back to life are often difficult but valuable endeavours, as success is achievable with patience, commitment, and access to necessary tools. Minority language preservation may depend heavily on appreciating the advantages of multilingualism and accepting linguistic variety in cultures where language shift is common. Despite the difficulties of preserving a language, valuing and recognising minority languages can help create a linguistic environment that is more diverse and inclusive (Bedi and Lafrance 2023). Hence, it is vital to study and continue investigating the Belait language to revitalise it and document it as a minority language in Brunei. Therefore, this study will explore how the Belait people adapted to their 'new home' through assimilation and transnationalism by looking at various factors motivating the language shift process in the Belait group and how this language shift reflects the community's journey towards being recognised as of the seven *Puak Jati* Brunei.

The Belait community, amid transnational migration, faces the challenges of language shift and adaptation. Transnational migration brings about linguistic changes as migrants navigate multilingual environments, leading to language contact, borrowing, and code-switching which in turn shapes the linguistic repertoire of migrants and host communities (Bermingham 2021). Hence, understanding the impact of transnational migration on language dynamics and the role of language in shaping international migration is essential for addressing the linguistic implications of migration for the Belait community. By recognising the significance of language in cultural contexts and its role in shaping community identity, further efforts can also be made to support the resilience and adaptation of the Belait community amid language shifts brought about by transnational migration. Through tracing language shift in the Belait community, this study can also fulfill its aim of investigating how the Belait people's migration from North Sarawak to Brunei resulted the tribe being recognised as one of the seven indigenous groups or *puak jati* under the *Perlembagaan Negeri Brunei 1959* (1962).

8.2 Background on Belait People in Brunei

The term Belait, within a Bruneian context, can refer to three items; the Belait river itself, one of Brunei's 4 districts, and a group of people considered as one of Brunei's seven ethnic groups. The Belait people are mainly situated in the Belait District as well in some parts of Tutong, such as in the villages of Kiudang, Lamunin, and Bukit Sawat (Noor Alifah 2004). According to Harrison (1958), the Belait people are believed to be a mixture of Kelabit speakers and Lemeting or Meting speakers. The Lemeting people were originally native to the Tinjar River, a tributary of the Baram River in Sarawak, Malaysia. The Lemeting people then later migrated to the area of Belait and eventually integrated with the natives in the Belait district and Brunei itself. In terms of language, according to Martin (1995) on the transition of languages within the Belait people, the Belait language spoken in Brunei today is the Lemeting language as this was the language from the Baram region where the speakers lived prior to migration to Belait. Additionally, Hudden (1949) stated that the Belait people have origins based on the Lelak community, which had assimilated with the Lemeting and subsequently settled in the Belait district of Brunei.

The Belait language also belongs to the Malayo–Polynesian sub-group of Austronesian languages. Blust (1972) argues that the Belait language is classified as one of the subgroups within the Rejang–Baram linguistic cluster, specifically located in the northern region of Sarawak. The language is also classified as critically endangered, as it exhibits an inadequate number of younger individuals proficient in its use. With its limited use, Martin (1990) therefore posits that the Belait people in Brunei represent a minority group primarily due to their limited number of speakers.

The Belait have also assimilated the cultures and lifestyles of communities native to the place that they have settled, particularly to the Brunei Malay culture. According to Martin (2005), an overwhelming majority of individuals who speak the Belait language also identify as Muslim, with only a small minority professing a non–Muslim faith. As a result of settling near coastal regions and living alongside Brunei Malays, the cultural practices of the Belait community have been impacted by the coastal (Malay) culture, resulting in the assimilation of these customs into Belait culture (Martin 1996b). Nevertheless, the Belait community exhibits various distinct forms of cultural practises and traditions. For example, the *'Beras Rujut'* practice is considered a tradition among the Belait generation and is not associated with animism or religious rituals (Nur Fa'iz 2015). However, it is worth noting that conventional traditions also encompass ceremonial elements, exemplified by the ritual known as *'Mengalai Padi'*.

With their origins based in migration, particularly into Brunei Malay communities, the Belait community has undergone a language shift, which can be attributed to modernisation, globalisation, and socio–political transformations. Martin's (1995) survey revealed that in the early 1990s, there was a noticeable decline in the usage of *puak jati* minority languages, such as Belait, among the younger population. The study conducted by Martin (1995) provides compelling evidence that demonstrates the decline in minority languages in Brunei. Martin's (1995) research also revealed that the proportion of individuals in the younger generation who speak the Belait language as their native tongue is only 9%, in contrast to preceding generations of grandparents and parents, in which reported rates were at 96% and 60% respectively.

Various possible factors can be attributed to this language shift of minority languages including the Belait language. The shift can be attributed to the impact of culturally dominant languages, namely Brunei Malay and English which are adopted and utilised more extensively over that of minority languages. The subordinate status of minority languages within a diglossia context involving Standard Malay and English has also played a role in the language shift. In this case, both Standard Malay and English are adopted widely at the cost of minority language usage.

Language is strongly connected to cultural adaptation, as language can function as an instrument for cultural expression or for the protection of cultural identity when encountering new and unfamiliar cultures (Kumaravadivelu 2008). Therefore, how language shift and language acquisition occur within the Belait people can be connected and influenced dynamically by the ways in which they assimilate, adapt, or integrate into unfamiliar and new social structures. In the case of the Belait people, language shifting from Belait to Brunei Malay can also be an instance of Belait speakers adapting to new social structures and communities by adopting the language of the natives within the area they moved into, which in this case is the Brunei Malay community. This adaptation however, can come at the consequence of cultural preservation, as the Belait people face the task of balancing between preserving their cultural uniqueness through their traditional language while also adjusting to the cultural conventions of their new surroundings that they have migrated into, achieved by the adoption of Brunei Malay. As the Belait migrate and adapt to different environments, the dynamic relationship between language and cultural adaptation plays a vital part in this process, as language functions as an instrument for the manifestation of cultural expression and the protection of identity (Kumaravadivelu 2008).

8.3 Literature Review

In the present era, where global mobility and interconnectedness have reached unprecedented levels, language plays a crucial role in shaping the identity and unity of communities. The study of language shift, particularly in minority language communities, has become a complex and multifaceted field. The occurrence of transnationalism has led to significant transformations in the linguistic landscapes of minority communities, affecting not only language usage but also the way cultural identities are negotiated in our increasingly interconnected world. As language shift within a community can also occur gradually or abruptly, the linguistic and thus cultural landscapes of minority communities can change rapidly (Batubara and Shafira 2023). In minority communities, where a smaller linguistic group exists within a larger dominant linguistic environment, various social, economic, and political factors can accelerate the process of language shift (Cenoz and Gorter 2017). Economic opportunities, educational policies, and exposure to media can also play a significant role in influencing the direction of this accelerated language shift, which poses challenges to the sustainability of minority languages (Granadillo and Orcutt–Gachiri 2011). The dynamics of language shift also involve intergenerational transmission, language proficiency, and daily interactions within the community. As younger generations adopt dominant languages for practical reasons, the minority language's vitality becomes at risk, potentially leading to its endangerment or extinction. This complex interaction between languages in the languages of minority communities sets the stage for the influence of transnationalism.

Numerous research studies have been carried out exploring the phenomena of language shifts and transnationalism, particularly in regions such as the Borneo Island which is the third largest island globally and spans approximately 744,360 sq. km. These studies have adopted various concepts and approaches to understand the intricate relationship between language and culture in diverse and multilingual contexts. Borneo, known for its rich linguistic and cultural tapestry, serves as a fascinating case study for examining the complexities and dynamics of language shifts within the context of transnationalism. Within Borneo Island, it is the Austronesian language family that encompasses the Island (McLellan 2014). The language can be identified as transnational in nature as the language family exists divided among three countries. The language can be found in Kalimantan, the largest portion of Borneo Island which belongs to Indonesia, Sabah and Sarawak, the smaller portions which are federal states of Malaysia and the independent nation of Brunei Darussalam which holds the smallest share of the Borneo Island. According to Blust (2013), Borneo's languages can be classified into ten major languages, such as Iban, Brunei Malay, Kedayan Dayak, Dusun, Kadazan, and Lawangan, and ten minor languages, including Punan Aput, Ukit, Tanjong, and Lengilu.

Studies on dialects and languages in Brunei are currently limited and not extensively conducted. Previous research has focused on Austronesian languages in Brunei, including Penan, Mukah, and Iban which were conducted by Nothofer (1991). However, there is still a lack of comprehensive studies specifically on the Belait language. Non-linguistic aspects, such as cultural and societal perspectives, have been explored by researchers such as Hughes–Hallett (1938), Shariffuddin (1980), Ak Johari (1983), Noor Ehsan (1983), Dzulkefli (1984), and Martin (1990, 1995, 1996a, 1996b, 2005). From a linguistic standpoint, there have been limited studies in the Belait language. Some research on this area includes Agustinus' research (1995) which explored passive construction in Belait and Noor Alifah's research (2004) on the structure of Belait.

Martin's (1995) study on the Belait ethnic group demonstrates examples of language shift among the Belait ethnic and the decline in the use of Belait in Brunei Darussalam. The study aimed to assess language maintenance and shift among different generations of the Belait ethnic group. The findings revealed a significant decrease in the percentage of Belait speakers from the grandparents' generation to the children's generation, with only 9% of the children retaining, practicing, and using Belait as their mother tongue. In addition, Martin's study also explored various factors contributing to language shift among the Belait ethnic group. One key factor identified was the migration of Belait speakers from Kampung Kuala Balai to other areas, particularly the coastal regions. This migration resulted in the adoption of cultural characteristics from other ethnic groups that reside in coastal regions, such as the Brunei Malays, and the influence of the Brunei Malay dialect (DMB) on the language shift. The study also highlighted intermarriage and education as factors influencing language shift among Belait speakers. Interethnic marriages were also found to potentially overshadow the identity of the Belait ethnic group, as stated by Hughes–Hallet (1938) in his study on the Orang Belait of Brunei. Education was also identified as a contributing factor, although further details were not provided in Martin's (1995) text. Overall, Martin's study emphasised the need for language researchers to direct attention towards the Belait community due to the decreasing number of Belait speakers and the decline of the Belait language. These findings shed light on the importance of addressing factors contributing to language shift and implementing measures to preserve and promote the Belait language within the community.

8.3.1 Methodology

Language development is a multifaceted field that encompasses numerous aspects such as migration, language contact, and language change. Migration plays a crucial role in shaping the evolution of languages, as migrants frequently strive to establish a sense of belonging to native communities with their new homes, while simultaneously preserving their native language (Zou 2020). This area of study delves into the ways in which languages evolve over time, and how this evolution gives rise to distinct communities with their unique languages and cultures.

The research approach employed in the study of language development is also largely qualitative in nature. Qualitative research aims to scientifically describe data and engage in discussions based on established terms within the field of language study. Qualitative research, as defined by Pathak et al. (2013), is characterised as a humanistic or idealistic approach. This method is utilised to gain insights into beliefs, experiences, attitudes, behaviours, and interactions among individuals. It seeks to understand the nuances of human language and the impact of various sociocultural factors on language development. In the context of migration and language contact, qualitative research plays a crucial role in unravelling the complex dynamics of linguistic adaptation and cultural change within migrant communities.

One specific area of focus within the study of language development is the study of language shifts among migrant communities. This phenomenon is particularly prevalent in transnational contexts, where migrants bring their native languages into contact with the dominant languages of their host countries. Such language shifts often lead to the formation of hybrid linguistic forms and the emergence of distinct language varieties within migrant communities. Within the context of Brunei, the Belait people, who did not originally hail from the Belait district in Brunei, have experienced significant linguistic and cultural changes because of migration and contact with the dominant Brunei Malay language in Brunei. Through qualitative research methods, it is possible to gain a deeper understanding of how the Belait language has evolved and adapted within the transnational context of Brunei.

Within this research, methods in studying the Belait language among the migrant community in Brunei involved various methods including participant observation, interviews, and analysis of linguistic artifacts. These methods allow close examination of the language practices, beliefs, and experiences of the Belait community members, shedding light on the ways in which their language has been influenced by migration and contact with Brunei Malayas well as the complexities of language shift and adaptation within the Belait community. This approach also allows for a comprehensive exploration of the sociocultural, historical, and linguistic factors that have shaped the development of the Belait language within the transnational context. By engaging with the beliefs, attitudes, and behaviours of community members, researchers can construct a rich and nuanced understanding of the language dynamics at play within the Belait migrant community.

The study also employs lexicostatistics and glottochronology techniques to investigate the kinship percentage of Belait language variants and the grouping of these languages. Lexicostatistics and glottochronology are two closely linked techniques used to explore the historical connections between languages (McMahon and McMahon 2012). Lexicostatistics involves comparing basic vocabulary lists to determine the percentage of cognate words, providing an estimate of linguistic distance (Zulkiflee et al. 2022). Glottochronology, on the other hand, calculates the time of language divergence or duration of separation between related languages based on the percentage of cognate words (Crowley and Bowern 2010). The glottochronology method involves the use of a specific formula to compute the time of language divergence, denoted by the variable t. The formula can be expressed as follows:

$$t = \log C/2\log r$$

In this formula, the variable t represents the number of thousands of years that have passed since the two languages diverged. The variable C denotes the percentage of cognate words when comparing the basic vocabulary of the two languages. R is a fixed change factor, with a determined value of 0.805 (Crowley and Bowern 2010). The utilisation of lexicostatistics and glottochronology techniques in analysing the Belait language has yielded significant findings regarding its familial relationships and categorisation. Through the comparison of fundamental vocabulary lists from various Belait dialects, researchers have been able to determine the classification and kinship of Belait language variants, consequently enhancing comprehension of language migration, evolution, and development.

8.4 Research Data

This study aims to gather empirical data on the Belait language and culture through interviews with 7 informants who are of Belait *jati*; conditionally, the parents of informants must be both from the Belait group and the informant themselves must still reside in the Belait District, and actively use the Belait language in their daily lives. The informants were selected based on a specific criterion in which they are limited to an age range of fifty to seventy–five years old. They are also screened for their linguistic and cultural background and gender. A total of four males and three females were interviewed for this research. Interviews were conducted in a familiar setting, mainly the informants' homes, to ensure their comfort and to obtain authentic and comprehensive data. The interview sessions, which lasted approximately one and a half-hour to two hours, were scheduled at a time convenient for the informant's preferences. The interview was also carried out in a semi-structured manner which allowed the interviewer the flexibility in probing the informants on various topics related to Belait language, culture, and traditions.

The main purpose of the interviews was to gain insights into the Belait language, its usage in different contexts, the cultural significance of the language, and its role in shaping the identity of the Belait community. The informants were also asked about their experiences growing up with the Belait language, how it is used in daily interactions, its influence on cultural practices and customs, and any changes they have observed in the language over time. Furthermore, the interviews also explored informants' perspectives on the preservation and promotion of the Belait language and culture, as well as their thoughts on the impact of external influences and modernisation.

The informants' personal narratives, memories, and reflections provided valuable qualitative data that shed light on the intricacies of the Belait language and its cultural significance. Given the rich insights obtained from the interviews, the study aims to contribute to a deeper understanding of the Belait language and culture, and to document the perspectives of older generations who have experienced the language in its traditional context and explore how the Belait people adapted to their new home through assimilation and transnationalism. The findings are expected to offer valuable insights for linguistic and cultural preservation initiatives, as well as for researchers and scholars interested in the diversity of languages and cultures within Brunei Darussalam.

8.5 Discussion

8.5.1 Tracing Migration of Belait People in Brunei

The evolution of the Belait language has been observed in numerous linguistic studies, leading to various theories about the language's origin. Scholars have observed that in the northern region of Sarawak (Blust 1972), the Belait ethnic group speaks one of the Rejang Baram group's minority languages (Hudson 1978). Some scholars such as Hudden (1949), Harrison (1958) and Martin (1995) argue that the Belait language, especially in Brunei, consists of speakers of the Lemeting language to the extent that the Belait language at the present time is considered to be defined as the Lemeting language. According to Blust (1972), Bahasa Belait is also a language under the 'Lower Baram' category in the northern part of Sarawak. Blust also has the view that Bahasa Belait is not only spoken in Brunei, but also covers some areas of Sarawak, such as North Sarawak, specifically in Baram. This view corresponds with a narrative provided by one of the informants on the origins of the Belait people:

> *...urang Belait ani berasal dari Baram yang namanya Sungai Meting. Sungai Meting atu di dalam Sungai Tinjar, indung sungai atu Sg Baram. Jadi kami dari Sg Baram basar masuk ke Tinjar ada lagi di ujung sana, Sg Meting...* (The belait people originally came from the Meting river in the Baram (river). The Meting river was inside of the Tinjar river, which was a tributary of the Baram river. We were from the Baram river, and entered into the Tinjar river, and further down was the Meting river...)[1]

Based on the response above, Belait informants mentioned that the origin of the Belait tribe in Brunei is from Baram, north of Sarawak, specifically in the Meting River area. The informant explains that the Meting River is in the big Baram River and then goes back into Tinjar, and at the end of the region lies the Meting River. When these locations are carefully examined, much of the Belait people were originally from remote and small areas, as this was due to the tribe avoiding capture and attack by the Kayan[2] people, who at the time were well–known for their skills in headhunting and attacking other tribes including the Belait people. The informant also highlighted that the Belait tribe, when they previously lived in Meting, it was when the terms 'Meting people', 'Meting speakers', or 'Meting tribe' arose. From interviews with informants, they also believed that the spoken language called Meting was the Belait language. Overall, clearly identifying the origins of the Belait language remains challenging, however scholars in linguistic studies of Bahasa Belait, such as those of Hudden (1949), Harrison (1958), and Martin (1995) suggests or asserts that the origin of the Belait tribe and their language consisted of speakers of the Meting language.

For instance, Martin (1995) compares both Belait and Meting languages in order study the differences and similarities between both languages. In his study, he identified similarities between the Belait language and Lemeting or Meting. For example, he compared and identified similarities between Belait words with Lemeting or Meting ones such as; [ŋan] and [ŋana] which means 'hand', [putai] and [putay] which means 'banana', [səmaq] and [səma] which means 'blood', and [mətam] and [mitam] in Belait and Lemeting which means ‘black’. From a linguistic point of view, there can be observed small differences such as changes of the phoneme sound /ə/ to /i/ and the addition of the phoneme /a/ in the word [ŋana]. Both phoneme changes occur sporadically at the final syllable of the word in the Meting language.

The terms Meting ‘people’, ‘tribe’, and ‘language’ also came to be when the Belait tribe first made foot and temporary settlements in the Meting River. According to Ray (1913), the term ‘people’ or the ‘tribe’ of Meting was formed based on the name of ‘Sungai Meting’ in Ulu Baram Sarawak. It also suggests that the Belait tribe settled in the Meting River for a somewhat long time, long enough that they came to rename themselves and be referred to or identified as the Meting people or tribe. Indirectly, this also meant that the Belait tribe’s origins as the Meting people came to be because of the Kayan tribe’s threats which caused them to move to Meting river and rename themselves geographically, which in turn led them to be identified as Meting people or speakers.

> *Turun dari Sg Meting atu, melalui Sg Tinjar, keluar Tinjar mengikut sungai anu lagi aliran sungai ke laut ke Merudi. Tinjar atu di ulu lagi, kalau pakai perau 2 jam dari Merudi baru sampai sana tu bah. Sampai di Merudi singgahlah, kiranya di sana bekampung lah kan.*

After the Belait settled on the Meting River, they then moved to a second stage of migration in which they moved to the Merudi area through the Tinjar River, towards the sea and eventually to Merudi. When they arrived at Merudi, the Belait temporarily settled there.

> *Durang banar2 kan mengapuskan urang Belait ani bah. Inda lama sama, bepakat lagi, baiktah kitani keluar lagi dari sini. Turun, kira turun lah ikut sungai. Nah dari Merudi atu, jalan kaki ni inda pakai anu lagi. Sebabnya tanah ada bukit apaindah kan sungai inda tau ke mana lagi nada tempat lagi. Pakai jalan kaki, masuk sampai satu tampat yang dikatakan Tanajur namanya di Labi* (Leaving the Meting river, through the Tinjar River, we exited the Tinjar river following a river that flowed into the sea to Merudi. The Tinjar river was far in the Ulu, using a boat, it would take 2 hours to arrive to Merudi. At Merudi, we stayed temporarily and built villages there.).

According to an informant’s response above, at one time, the Kayan people discovered their presence at Merudi. This had led the Belait tribe to carry out a third stage of migration to another area called Tanajor in Labi, which is in the Belait district of Brunei. This is where the Belait tribe’s migrations to Brunei first began. Subsequently, the Belait tribe stayed in Labi and unintentionally introduced Belait culture, such as planting crops such as rice (*padi*), to live in rural areas similar to how the the Dusun tribe did.[3] However, the Belait continued migrating and did not stop in Labi.

> *Jalan sungai pakai perau pelahan2 lah sampai tempat di satu sungai, ada simpangnya sungai basar. Sg damit dari Labi sg damit, sampai tempat keluar sg basar macam Merudi atulah namanya sg Belait. Di sini baiktah kali, inggan ani tah kitani dulu, kitani inda mampu lagi jauh ni. Ok apabila sampai di sana, urang Belait masih bepikir bahawa kitani harus bertahan di sini dan sini buat rumah pertahanan yang digelarkan rumah skada, amin skada. Amin skada atu dua tingkat, satu tempat urang tinggal. Di sini atu durang pasang anu badil2 mariam menyubarang sungai cematu. Jadi bila sampai masanya datang, idup urang Belait sana atu di Kuala Balai...* (Following the river route, we used a boat to slowly reach a river, which lead to a bigger river. From a small river in Labi until we emerged into a bigger river

like the one in Merudi, the name of that river was Belait river. We decided it was better to stay here, as we could no longer travel further. Once we were there, the Belait people were still thinking that we should build our defenses here and to make defensive houses that were called rumah skada, or amin skada. The Amin Skada had two floors, one floor was for living. Here, they installed cannons facing the river. Until the time came, the people of Belait lived in Kuala Balai…).

The Belait continued their journey after being attacked again by the Kayan in Labi. They moved to an area in the Belait district from the Labi River to the great river called the Kuala Balai. Based on interviews carried out, the informant shares that the Belait tribe completed their migration and settled in the river area, by building houses called 'skada' or 'amin skada'.[4] Kuala Balai is the fourth stage of migration where the Belait tribe also evolved and developed their community and culture. In Kuala Balai, the Belait tribe had managed to build all the necessities for living. In the twentieth century, after the migration of Belait tribe to Kuala Balai, Brunei also started to develop as oil and gas was discovered in Seria, Belait. The Belait tribe also slowly developed economically as work and economic opportunities were provided.

Some of the Belait also began to move to the district of Tutong. According to an interviewee, the existence and movement of some of the Belait tribe into Kiudang occurred because they were posted to Kiudang to help higher authorities regarding an issue that occurred there. Figure 8.1 below represents the migration patterns of Belait tribe from North Sarawak to Brunei:

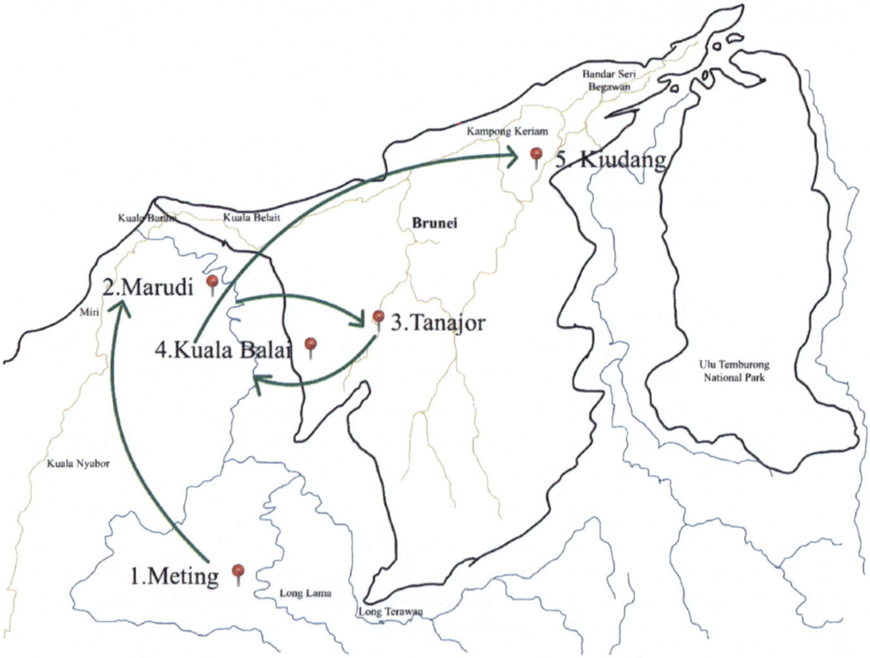

Fig. 8.1 Authors' Visualisation of Migration of Belait people from North Sarawak to Brunei

Using historical linguistic research, specifically through the comparison of Swadesh data collected from study informants, the migration of the Belait tribe to Brunei can be estimated to have occurred around the year 1857, which is 167 years ago. Evidence supporting this time frame can be identified by using lexicostatistics and glottochronology techniques (Crowley and Bowern 2010).

By analysing using the lexicostatistic method, the percentage of cognates between Belait language variants is determined and identified by dividing the number of cognate words by the total number of words compared. The formula for the percentage of cognates for the Belait language variants is as follows:

Number of cognate words	: 187 words
Total number of words compared	: 201 words
Percentage of cognate words	: 93%

The percentage of cognates between Belait language variants using the lexicostatistic technique is 93%. Subsequently, the separation time between Belait language variants is analysed using the glottochronology technique. The formula for the separation time between Belait language variants is as follows:

$$
\begin{aligned}
t &= \log C/2 \log r \\
t &= \log 0.93/2 \log 0.805 \\
t &= 0.1673 \\
t &= 0.1673 \times 1000 \\
t &= 167 \text{ years ago}
\end{aligned}
$$

As per the formula above, the estimated separation time between Belait language variants is approximately 167 years ago. It can then be concluded that the languages diverged around the year 1857. In addition to this, these calculations which demonstrate the separation time between Belait language variants aligns with the grouping provided by Crowley and Bowern (2010):

Based on the level of language relationship groups as shown in Table 8.1, the Belait language variants belong to the dialect of a language, considering their relationship and cognate percentage is between 81 and 100%, specifically 93%, which is approximately 167 years ago. This result also aligns with research conducted by Hughes–Hallett (1980) who stated that the arrival of the Belait tribe was around the nineteenth century, between 1830 and 1890.

Migration is a complex social phenomenon that holds great significance in human history. Various factors drive tribes or social groups to relocate, with one common reason being the pursuit of improved economic opportunities. Migration determinants are multifaceted and encompass economic, political, social, and demographic factors (Arif 2020). These factors may involve seeking better job prospects or gaining access to more abundant natural resources. In instances where economic conditions are unfavourable, individuals or social groups often move to areas offering more

Table 8.1 Level of language classification (Crowley and Bowern 2010)

Level of subgrouping	Cognate percentage (%)
Dialects of a language	81–100
Languages of a family	36–81
Families of a stock	12–36
Stocks of microphylum	4–122
Microphyla of a mesophylum	1–4
Mesophyla of a macrophylum	0–1

promising economic prospects. Research indicates that the level of economic development in the area of a group's origin significantly influences the propensity for migration (Wajdi et al. 2017). For instance, the migration of the Iban tribe to Brunei, particularly the Belait district in the early twentieth century, was spurred by an ongoing rebellion at the time (Haji Buyong 1981). This rebellion led to disruptions in job and economic opportunities, leading some to seek job opportunities in Sabah, Kalimantan, Sumatra, the Philippines, and the Belait district (Austin 1976).

Additionally, social and political factors play a significant role in driving social group migration. Political conflicts or wars, for example, can compel social groups to seek safer living environments to safeguard their well-being and community stability. Throughout history, conflict has been a major catalyst for substantial migration movements (Simpson 2022). For instance, the migration of the Belait tribe to Brunei can be attributed to the 'ayauan' activities, also known as bandit and headhunting activities, of the Kayan people against the Balait tribes in Baram Tinjar, North Sarawak (Wan et al. 2018). Conflict or interactions with the Kayan people have prompted the Belait tribe to find new lands and embrace new identities shaped by their new environment (Omar and Norahim 2020). Aside from conflicts, similar to the Iban, the migration of the Belait into Brunei could have also been motivated by the search for better opportunities.

Overall, migration encompasses the act of moving from one location to another and navigating unfamiliar pathways, adapting to new daily patterns, and ultimately building a sense of belonging in a previously unknown environment (Román–Velázquez and Retis 2021). This journey to finding a new sense of belonging in a new unknown environment can also be used to describe the migration experiences of the Belait. Based on this study's research, the emigrating Belait tribe has performed with outstanding results in adapting to the process of migrating from Baram, Sarawak to Brunei. They have also maintained their transnationality, maintaining connections with the Belait people in Sarawak.

In the *Perlembagaan Negeri Brunei 1959* (1962), it was stated that Brunei Darussalam consisted of seven recognised tribes or in Malay, referred to as *puak jati*. These groups namely are; Belait, Bisaya, Brunei, Dusun, Kedayan, Murut and Tutong (Government of Brunei 1961, pp. 118–120). Under the *Perlembagaan Negeri Brunei 1959,* the Belait community was officially acknowledged as a *puak jati* or indigenous group. This status both indirectly and directly ensures the safeguarding of their identity, language, and culture. This constitutional acknowledgment also reflects the Bruneian government's dedication to safeguarding the rights and welfare of its people. Consequently, the Belait community is constitutionally assured the protection of their cultural legacy, enabling them to uphold their distinct identity within Bruneian society. Moreover, the Belait community benefits from Bruneian citizenship, which grants them access to various social services, free education, free health care, and economic advantages (Peters et al. 2017). These benefits include a monthly old-age pension (*gaji orang tua*) of 250 Bruneian Dollars for elderly citizens, educational opportunities for all citizens (Chua et al. 2019), free dental and medical services (Maxwell 2001), subsidised housing (Gunn 1997), and an annual 'draw' for Muslims employed in government ministries to undertake a subsidised pilgrimage (*naik haji*) to Mecca and Medina. This citizenship status allows them to seamlessly integrate into Bruneian society, enjoying equal rights and opportunities alongside their fellow citizens.

The *Undang–Undang Taraf Kebangsaan Brunei 1961* (Government of Brunei 1961) further solidified the citizenship and legal status of individuals, promoting their integration while preserving their cultural and ethnic identity. The Belait community has established cross-border cultural networks, fostering unity and shared identity among the Belait people on both sides of the Sarawak–Brunei border. These networks have allowed the community to maintain their cultural traditions, language, and customs despite being geographically displaced. This transnational cultural connection has created a sense of belonging and shared identity among the Belait people, surpassing national boundaries.

The *Undang–Undang Taraf Kebangsaan Brunei 1961* also outlines a pathway for Belait people to acquire Bruneian nationality. For instance, Belait people who are born in Brunei to both Belait parents or those who have Belait bloodlines are automatically eligible for Bruneian nationality. This provision guarantees members of the Belait community the pathway to secure their status as Bruneian citizens, granting them access to the rights and privileges associated with citizenship. At the same time, they can also practice and take part in social, economic, and cultural activities across the border, thereby facilitating the maintenance of a cross-border ethnic identity. This transnational circumstance strengthens their transnational connections and affiliation between the two regions.

By providing a legal framework under the *Undang–Undang Taraf Kebangsaan Brunei 1961* for the citizenship of the Belait community, the enactment ensures the preservation and promotion of their cultural heritage within Brunei, including the protection of their language, customs, and traditional practices. The legal integration facilitated by the nationality enactment complements the constitutional recognition of the Belait community, reinforcing their status as an integral part of Bruneian society while acknowledging their transnational ties.

8.5.2 The Adaptation and Resilience of the Belait People in Brunei

The process of negotiating identity in lands near borders is intricate, as within borders, national, ethnic, and religious identities frequently overlap, leading to the emergence of a 'post-colonial stage' characterised by the redefinition of traditional boundaries (Horstmann 2002). According to Martin (2005), the Belait language is predominantly spoken in the upstream village of Kuala Balai, where a majority of its speakers are Muslims. This is significant as it underscores the influence of Islam on the cultural heritage of the Belait community. The acceptance of Islam has brought about the integration of certain cultural traditions, subsequently impacting the Belait language and traditions. Notably, the utilisation of Malay, the official language of Brunei, has become more prevalent among the Belait populace, especially in their interactions with other ethnic groups. Moreover, the Belait community has relocated to different regions such as Labi and Kiudang village in the Tutong district, due to environmental circumstances and intermarriage with other ethnicities. This migration has resulted in a cultural shift, where the Belait people have embraced certain customs from other ethnic groups such as from the Dusun and Chinese communities (Martin 2005). For instance, intermarriage with Muslims has led to the adoption of Malay customs, particularly among individuals who have embraced Islam. This cultural interchange has led to the fusion of traditions, with the Belait community incorporating elements of Malay culture into their own traditional practices.

The adaptation of the Belait people in Brunei is also reflected in their language, which has undergone significant changes over the years. The Belait language, which belongs to the Austronesian language family, has undergone notable changes due to the influence of Malay and other languages spoken in Brunei. Other languages have an undeniable impact on one another through language contact, including the process of cultural assimilation when viewed through the lens of sociolinguistics as a form of language change (Zulkiflee 2024). This language contact has resulted in the emergence of a distinct dialect that differs from the original Belait language which is Meting. Notably, the use of Malay loanwords has become more prevalent in informal conversations within the Belait language.

Additionally, the Belait people have also adjusted to the changing economic landscape of Brunei. Traditionally, they were primarily engaged in subsistence farming. However, with the discovery of oil in the 1920s, Brunei's economy underwent a significant transformation which in turn led to new economic opportunities for the Belait. Martin (1995) mentions that the migration of Belait people was caused by the foundation of the Oil and Gas industry in Seria in 1929. As stated before, employment prospects play a crucial role in driving migration, particularly in developing nations where individuals are compelled to seek places with more abundant job opportunities due to reasons such as limited agricultural income, unemployment, and underemployment (Thet 2014). In response to this change, many Belait individuals have sought employment in the oil and gas industry, which has become a major contributor to the country's economy. Consequently, their lifestyle has also shifted as they have migrated from rural areas to urban centres in search of employment opportunities.

With the privileges received as a Bruneian citizen, and enjoying the rapid development in terms of infrastructure, education, job opportunities with both government and non-government agencies, the Belait people are now experiencing a further language shift, where Brunei Malay and English are replacing Belait language at home. As Martin (1995) highlights, much the younger generation of the Belait are brought up in a language that is not their parent's mother tongue. This has implications on how the Belait shape their identity as they have to negotiate between being Belait, being Malay and also being Bruneian. Most might claim they are Belait people even when some no longer speak the Belait language.

However, it is important to note that Martin's (1995) research was conducted two decades ago and as such is not a complete reflection of the status of the Belait language. Today we still see the younger generation of the Belait people have the ability to speak Belait language. Language transmission from their parents and grandparents of younger Belait are therefore still happening, and as we have observed, there are still Belait families in which conversations in the Belait language still occur. These conversations even occur across three generations; from the grandparents, to the parents and lastly to the children. This maintenance of the use and the delay of the extinction of the Belait language could also be attributed to the sense of pride and the positive attitude Belait speakers have to the Belait language as a mother tongue.

Additionally, the preservation of the Belait language can be fostered by language revival and preservation initiatives, made possible by government support which has never faded in ensuring the languages of the ethnic groups to be maintained, actively used, and supported in whatever means. Various government agencies also play a role in contributing to this endeavour. For example, *Dewan Bahasa dan Pustaka Brunei* actively publishes language research on Brunei's seven ethnic groups. *Jabatan Muzium Brunei* have published research on culture of indigenous groups such as the Belait people; these include *Warisan Daerah Belait* (2016) and *Adat Perkahwinan Etnik–Etnik di Negara Brunei Darussalam* (Dayangku Norazah and Dayang Masnah 2004). The Ministry of Education through Universiti Brunei Darussalam's Language Centre also offers Belait language as part of its curriculum. The Faculty of Arts and Social Sciences under Universiti Brunei Darussalam also advocates for research to be carried out on minority languages across its undergraduates and post-graduate studies. *Radio Televisyen Brunei* under the Prime Minister's Office also plays a vital role in supporting minority groups by broadcasting radio and television programmes that are carried out using minority languages such as in the '*Bahasa Mengenali Bangsa*' segment.

8.6 Conclusion

In this chapter, we have explored the language situation of the Belait community in Brunei, through a lens of transnationalism which offers valuable insights into the intricate dynamics of language shift among Belait speakers, highlighting their remarkable stability despite significant changes brought by modernity, globalisation, and sociopolitical reforms. Through their transnational relationships, the Belait community has preserved cultural connections to their roots in Northern Sarawak, affecting their linguistic practices in Brunei. The Belait people have managed to preserve aspects of their culture and pass them down to future generations. Their adaptability to changing circumstances is truly commendable. Whether adopting Islam, assimilating Malay culture, or incorporating new practices, the Belait community has consistently shown resilience. They have also embraced new economic opportunities, particularly in the oil and gas industry, leading many to move from rural to urban areas. Government support, family gatherings, and a strong sense of pride in their language and culture have been crucial in maintaining the Belait language across generations. Institutions in Brunei Darussalam, such as the *Dewan Bahasa dan Pustaka Brunei*, the *Muzium Brunei*, and Universiti Brunei Darussalam, have also played significant roles in cultural preservation.

The Belait people's language, customs, and traditions have evolved, reflecting their interactions with various ethnic groups in Brunei. As Brunei continues to progress, the Belait people are poised to thrive, maintaining their cultural heritage while embracing modernisation. In conclusion, despite undergoing significant language shifts due to internal and external influences, the Belait community's resilience and adaptive strategies have safeguarded their cultural identity. As the country advances towards a more developed society, the Belait people will continue to enjoy a high standard of living, preserving their culture while meeting the demands of the modern world.

Notes

1. This quote and all interview quotes in this chapter are obtained from an interview with Hj Busu bin Ahmad, conducted on 15 November 2022. Hj Busu is an 80-year-old male of Belait ethnicity from Kuala Balai. He resides in Pandan, Kuala Belait. The interview was conducted to gather insights on the origins of the Belait people and the history of their development, migration and Belait lexical data. The interview was conducted by the co-author of this chapter Zulfadzlee Zulkiflee.
2. The Kayan indigenous peoples residing in Sarawak have a historical connection to the Kayan population located in Kalimantan, Indonesia. According to historical accounts, it is widely believed that their origins may be traced back to a specific region in Kalimantan called the Apo Kayan, situated along the Kayan River in the central part of the island of Borneo (Wan et al. 2015).
3. Yabit (1994) asserts that the Dusun tribe in Brunei identifies themselves as 'Sang Jati' (our people), a term believed to reflect their strong connection to rice fields and fruit farming.

4. Amin Skada is a tall house made by the Belait tribe with a long wall where they live and shelter in.

References

Ahli Jawatankuasa Penyediaan Pameran Galeri Muzium Daerah Belait (Belait District Museum Gallery Exhibition Preparation Committee Member) (2016) Warisan Daerah Belait (Belait District Heritage). Jabatan Muzium-Muzium, Bandar Seri Begawan

Agustinus DK (1995) Konstruksi pasif dalam Bahasa Belait (Passive construction in Bahasa Belait). In: Dialek memperkaya bahasa (Dialect enriches language). Dewan Bahasa dan Pustaka (Language and Literature Bureau), Brunei Darussalam

Ak Johari PP (1983) Adat beranak puak asli Belait (Traditions of the Belait indigenous group). Berita Muzium (Museum News) 7(2):12–17

Arif I (2020) The determinants of international migration: unbundling the role of economic, political and social institutions. The World Economy 43(6):1699–1729. https://doi.org/10.1111/twec.12889

Austin RF (1976) Some demographic characteristics of the Iban population of Brunei – Part 1: 1947–1960. Brunei Museum Journal 3(4):64–69

Bakar AA, Osman MM, Bachok S et al (2014) Analysis on community involvement in cultural activities: transmission of ethnic language. Planning Malaysia 12(4):105–132. https://doi.org/10.21837/pmjournal.v12.i4.127

Batubara MH, Shafira P (2023) Language shift: in Gayo society. Journal of Linguistics, Literature, and Language Teaching (JLLLT) 2(2):98–103

Bedi S, Lafrance S (2023) The linguistic diversity of pluralist cultures: status of linguistic minorities in India and Canada. Inclusion and diversity. Routledge, India, pp 11–28

Bermingham N (2021) Language ideologies and transnational migration: a study of Cape Verdeans in Galicia. Languages 6(2):99

Blust R (1972) New subgrouping of the languages of West Borneo. Borneo Research Bulletin 2(1):4–5

Blust R (2013) The Austronesian languages, revised. The Australian National University, Australia

Brown JR (ed) (2022) The verticalization model of language shift: the great change in American communities. Oxford University Press

Brunei Nationality Enactment, 1961 (1962) Retrieved from https://www.agc.gov.bn/AGC%20Images/LAWS/ENACTMENT/1961/E4.pdf. Accessed 13 July 2024

Crowley T, Bowern C (2010) An introduction to historical linguistics. Oxford University Press

Cenoz J, Gorter D (2017) Minority languages and sustainable translanguaging: threat or opportunity?. Journal of Multilingual and Multicultural Development 38(10):901–912. https://doi.org/10.1080/01434632.2017.1284855

Chua RY, Kadirvelu A, Yasin S et al (2019) The cultural, family and community factors for resilience in Southeast Asian indigenous communities: a systematic review. Journal of Community Psychology 47(7):1750–1771. https://doi.org/10.1002/jcop.22224

Dayangku Norazah PHM, Dayang Masnah L (2004) Adat perkahwinan etnik–etnik di negara Brunei Darussalam (Marriage traditions of ethnic groups in Brunei Darussalam). Ezy Printing Services & Trading Co, Brunei Darussalam

Dzulkefli HY (1984) Adat kematian puak asli Belait (Funeral traditions of the Belait indigenous group). Berita Muzium (Museum News) 6(12):30–34

Gelman SA, Roberts SO (2017) How language shapes the cultural inheritance of categories. Proceedings of the National Academy of Sciences 114(30):7900–7907. https://doi.org/10.1073/pnas.1621073114

Government of Brunei (1961) Undang–undang taraf kebangsaan Brunei (Brunei Nationality Laws). In: Surat–surat Perlembagaan Negeri Brunei (Constitution of Brunei). Brunei Darussalam, pp 115–135
Granadillo T, Orcutt–Gachiri HA (eds) (2011) Ethnographic contributions to the study of endangered languages. University of Arizona Press
Gu M, Tong HK (2019) Constructing classed linguistic practices across borders: family language policy in South (east) Asian families in Hong Kong. Journal of Multilingual and Multicultural Development 41(7):581–599
Gunn GC (1997) Language, power and ideology in Negara Brunei Darussalam. Ohio University Press, Ohio
Haji Buyong A (1981) Sejarah Sarawak (Sarawak History). Dewan Bahasa dan Pustaka (Language and Literature Bureau), Kuala Lumpur
Harrison T (1958) Origins and attitudes of Brunei Tutong–Belait–Bukit–Dusun, North Borneo "Dusun" and Sarawak "Bisayan", Meting and the other Peoples. The Sarawak Museum Journal 8(11):293–321
Horstmann A (2002) Incorporation and resistance: borderlands, transnational communities and social change in Southeast Asia. Tokyo University of Foreign Studies, Japan
Hudden DC (1949) The Baram district [Cited from report in 1939]. Sarawak Gazette 1093–78–79
Hudson AB (1978). Linguistic relations among Bornean peoples with special reference to Sarawak: an interim report. Studies in Third World Societies 3:1–44
Hughes–Hallett H (1938) An account of a Berhantu ceremony called "perakong" by the orang Bêlait (Belait people) of Brunei. Journal of the Malayan Branch of the Royal Asiatic Society 16(1):102–108
Hughes–Hallett H (1980) An account of a Berhantu ceremony called "perakong" by the orang Bêlait (Belait people) of Brunei. Brunei Museum Journal 5:41–48
Kumaravadivelu B (2008) Cultural globalization and language education. Yale University Press
Martin PW (1990) The orang Belait (Belait people) of Brunei: linguistic affinities with Lemeting (Meting). Borneo Research Bulletin 22(2):130–138
Martin PW (1995) Peralihan bahasa di antara puak Belait (Language shift among the Belait tribe). In: Dialek memperkaya Bahasa (Dialect enriches language). Dewan Bahasa dan Pustaka (Language and Literature Bureau), Brunei Darussalam, pp 193–217
Martin PW (1996a) Brunei Malay and BM: a sociolinguistic perspective. In: Martin PW, Ozog C, Poedjosoedarmo G (eds) Language use & language change in Brunei Darussalam. Center of International Studies, Athens, Ohio, pp 27–36
Martin PW (1996b) Sociohistorical determinants of language shift among the Belait community in the sultanate of Brunei. Anthropos, 91(H. 1./3.):199–207
Martin PW (2005) Language shift and code–mixing: a case study from Northern Borneo. Australian Journal of Linguistics 25(1):109–125. https://doi.org/10.1080/07268600500113682
Maxwell AR (2001) Malay polysemy and political power: census categories, ethnicity, and citizenship in Brunei Darussalam. South East Asia Research 9(2):173–212
McLellan J (2014) Strategies for revitalizing endangered Borneo languages: a comparison between Negara Brunei Darussalam and Sarawak, Malaysia. Southeast Asia: A Multidisciplinary Journal 14:14–22
McMahon A, McMahon R (2012) Lexicostatistics and glottochronology. In: The encyclopedia of applied linguistics. Wiley. https://doi.org/10.1002/9781405198431.wbeal0701
Noor Alifah A (2004) Struktur Bahasa Belait (Bahasa Belait structure). Dewan Bahasa dan Pustaka (Language and Literature Bureau), Bandar Seri Begawan
Noor Ehsan HNK (1983) Adat perkahwinan puak asli Belait (Marriage traditions of the Belait indigenous group). Berita Muzium 7(1–2):32–43
Nothofer B (1991) The languages of Brunei Darussalam. In: Steinhauer H (ed) Pacific linguistics series A-84: papers in Austronesian lingustics, vol 81(1). The Australian National University, pp 151–176

Nur Fa'iz AR (2015) A study of the Belait ethnic group of Brunei Darussalam: an emic perspective. Dissertation, Sociology and Anthropology, Universiti Brunei Darussalam, Brunei Darussalam
Omar AH, Norahim N (2020) Lower and upper baram sub–groups. In: The GLOCAL proceedings Asia 2020, Bintulu, Sarawak, 5–8 February 2020, pp 92–100
Pathak V, Jena B, Kalra S (2013) Qualitative research. Perspectives in Clinical Research 4(3):192.
Peters F, Vink M, Schmeets H (2017) Anticipating the citizenship premium: before and after effects of immigrant naturalisation on employment. Journal of Ethnic and Migration Studies 44(7):1051–1080. https://doi.org/10.1080/1369183X.2017.1367650
Phillipson R (2018) Language challenges in global and regional integration. Darnioji daugiakalbystė (Sustainable Multilingualism) 12(1):14–35.
Ray SH (1913) The languages of borneo. Sarawak Museum Journal 1(4):196
Román–Velázquez P, Retis J (2021) Narratives of migration, relocation and belonging: Latin Americans in London. Springer International Publishing
Shariffuddin PD (1980) Undang–undang dan adat istiadat orang–orang Belait asli (Laws and traditions of the Belait indigenous group). Brunei Museum J 1–10
Silverstein M (2022) Language in culture: lectures on the social semiotics of language. Cambridge University Press, Cambridge
Simpson NB (2022) Demographic and economic determinants of migration. IZA World of Labor 2022. https://doi.org/10.15185/izawol.373.v2
Thet KK (2014) Pull and push factors of migration: a case study in the urban area of Monywa township, Myanmar. News from the World of Statistics 1(24):1–14
Wajdi N, Adioetomo SM, Mulder CH (2017) Gravity models of interregional migration in Indonesia. Bulletin of Indonesian Economic Studies 53(3):309–332
Wan R, Renganathan S, Kral I (2018) Tekná–a vanishing oral tradition among the Kayan people of Sarawak, Malaysian Borneo. Indonesia and the Malay World 46(135):218–234
Wan R, Renganathan S, Philip B (2015) What is the point of us talking? ethnic language and ethnic identity in Northern Borneo, Malaysia. Humanities and Social Sciences Review 4(1):109–120
Yabit A (1994) Reconstruction of pre–Dusun and the classification of its descendant. MA dissertation, University of Hawai'i Manoa, Manoa, Hawai'i
Zou C (2020) Inter–generational language shift and maintenance: language practice observed in Guangzhou Hakka families. Asian Ethnicity 23(2):362–376
Zulkiflee Z (2024) Rekonstruksi varian–varian Bahasa Belait (Reconstruction of Bahasa Belait variations). Unpublished MA dissertation, Malay Language and Linguistics, University Brunei Darussalam, Brunei Darussalam
Zulkiflee Z, Damit AR, Alas Y (2022) Leksikostatistik dan grotokronologi Bahasa Tutong dan Dusun (of Lexicostatistics and glottochronology of Bahasa Tutong and Dusun). Brunei Museum J 92–108

Ashrol Rahimy Damit is Lecturer at Universiti Brunei Darussalam where he teaches sociolinguistics, language and communication, and communication and industrial revolution in the Malay Language and Linguistics Programme. His research interests include language maintenance and language shift, endangered languages of Brunei Darussalam, Malay language studies, language use in cyberspace, and natural language processing.

Zulfadzlee Zulkiflee is a postgraduate student at Universiti Brunei Darussalam, within the Malay Language and Linguistics, in Malay Studies programme. Currently, his research focuses on minority languages, particularly the Belait language, using comparative linguistics. His expertise lies in language and applied linguistics, with a specialisation in historical linguistics and sociolinguistics, particularly in relation to the indigenous languages of Brunei.

Part II
Contestations

Chapter 9
"Slave" Stories: Historical Fiction and the Voice of Young Female Bondservants (*Mui Tsais*) in Singapore

Hema Kiruppalini

Abstract In the nineteenth and early twentieth centuries, there were young girls from China and those born locally who were indentured as bondservants in British colonies in Asia such as Hong Kong and British Malaya. Often due to poverty, female children and teenagers—better known as *mui tsais* ("little sisters" in Cantonese)—were either sold or transferred to affluent families for domestic labour and this practice was widely perceived as a cultural tradition unique to the Chinese. In perusing the rich body of literature on the *mui tsais*, there remains a gap in scholarly work on the subject that explores historical fiction in Singapore within the broad context of transnational Asian connections. By presenting two novels—Catherine Lim's *The Bondmaid* (1995) and Eva Nava Wong's *The House of Little Sisters* (2022)—that deal with the *mui tsai* experience in colonial Singapore and by way of an interdisciplinary and comparative approach, this chapter considers the interrelated themes of gender, domestic servitude, transnationality, and underclass voices. The texts, written approximately three decades apart, capture narratives that blur the boundaries between domestic servants and adopted daughters; ethnic identity and cultural cosmopolitanism; and, reality and imagination. In so doing, these literary works bring into question historical memory about the past and how *mui tsai*s lived experiences are remembered and retold. By situating these novels in relation to a larger corpus of primary sources—colonial office records, oral history interviews, and newspaper articles—drawn from the National Archives in Britain and Singapore, this chapter seeks to uncover the significance of historical fiction in foregrounding socio-political, cultural, and transnational context(s) that are unique to Singapore.

Keywords *Mui tsai* · Domestic servitude · Historical fiction and social memory · Adopted daughters · Gender · Voice and agency · *The Bondmaid* · *The House of Little Sisters* · Singapore

H. Kiruppalini (✉)
Asia Research Institute, National University of Singapore, Singapore, Singapore
e-mail: hema@nus.edu.sg

H. M. Y. Ho and Y. Chan (eds.), *Transnational Southeast Asia*, Asia in Transition 29,
https://doi.org/10.1007/978-981-96-3608-2_9

9.1 Introduction

Since the nineteenth century, the *mui tsai* phenomenon has inspired historical fiction about them, both within China and in diasporic communities as far as America. Among the earlier works include books by Reverend J.A. Davis which bring to focus a young girl's conversion to Christianity and her dedicated efforts within the evangelical movement: *The Chinese slave girl: A story of a women's life in China*, which was published in 1880, and followed by *Leng Tsao: Sequel to Chinese slave girl*. In San Francisco, Michelle Lettieri sheds light on nineteenth-century historical fiction such as *A Chinese Slave Girl: Highbinder and Tong Wars* and *Little Ah Yee of the Opium Dens* (1886) by Emma R. Cable who wrote her book based on her experiences as a house-to-house visitor (Lettieri 2006). In the context of Southeast Asia, Janet Lim's bestselling memoir—*Sold for silver: an autobiography of a girl sold into slavery in Southeast Asia*—published in 1958 offers insights into the life of a *mui tsai* in British Malaya. Janet Lim details her life history as a *mui tsai* in 1930s Singapore and traces her remarkable journey through the inter-war years up till the 1950s when she was appointed as Singapore's first Asian hospital matron of St Andrew's Mission Hospital (Lim 1958, reprint 2004). Janet Lim's book has been explored from literary and sociological perspectives (Holden 2010; Miers 1994). Notwithstanding Lim's autobiography, first-hand accounts of the *mui tsais* as a social underclass in British Malaya are few and far between.

At the National Archives of Singapore, there is a modest collection of oral history records from the 1980s and 1990s that shed light on the historical memory of the *mui tsai* system during the colonial era. However, these archival transcripts, which will be discussed in this chapter, are articulated from the perspective of Chinese families that had "slave girls" (*mui tsais*). These personal recollections in an institutional archive privilege individual accounts of families that purchased them. Nonetheless, this gap in historical inquiry which neglects the voices of an underclass can be primarily attributed to several factors. In part, the absence of written records is a consequence of the socio-economic profile of the vast majority of female bondservants who were uneducated and unwaged. And, even if among them some went on to pursue an education, the social stigma and shame associated with being a *mui tsai* have also given shape to women's—and archival—silence. Arguably, then, historical fiction gives expression to the voices of female children and young girls who experienced domestic servitude, and in so doing, it allows us to break through the silence of untold stories albeit in an imaginative fashion. Inside the voice of the character, their innermost thoughts and feelings, fears and hopes surface.

A critical examination of historical fiction presents an opportunity to explore the themes of gender, culture, and transnationality vis-à-vis a female underclass in multiracial Singapore and to make sense of the enduring social memory of "slavery" in the contemporary period. In her analysis of historical imagination, Jane Wong highlights that:

> … at the heart of this unique genre [historical fiction] is its sensitivity to local conditions and its ability to recast and reimagine those sensibilities in narratives that demand a different kind of historical inquiry. It is a kind of inquiry that dares to imagine, to speak, and to revolt,

> and within that locality that is Asia, where the diversity of languages, cultures, and political ideologies converges (Wong 2018, pp. 13-14).

At the core of historical fiction lies its sensitivity to local contexts and adeptness in shaping particular sensibilities into narratives that prompt a distinctive mode of historical investigation. For example, in his analysis of Amitav Ghosh's *The Glass Palace*, Maitrii Aung-Thwin argues that historical fiction offers a pedagogical conduit to study and teach transnational history in the context of South and Southeast Asia and further asserts that novels not only challenge conventional historical narratives but also promote an understanding of regional intersections (Aung-Thwin 2018, pp. 155–179). Through historical fiction, a sense of interconnected transnational dynamics and how historical processes, ideologies, and power dynamics have unfolded across borders and shaped the Asian region is revealed.

In Singapore, there are a handful of historical fiction that give life to the *mui tsai* phenomenon during the colonial era. These novels include: *The Bondmaid* (1995) published by acclaimed Singapore writer Catherine Lim and two contemporary books; one by Anjana Rai Chaudhuri, *The Scent of Frangipani: A tale of love, rivalry and murder in colonial Singapore* (2019), and another recent publication by Eva Nava Wong, *The House of Little Sisters* (2022). Through historical fiction published in the post-colonial and contemporary periods, we delve into narrative representations of Singapore's colonial past; encounter countries, characters, and cultures that unsettle regional boundaries; and explore the intersections between race, gender, and domestic servitude in a cosmopolitan context. The aforementioned novels have yet to be examined comparatively and this chapter considers the literary texts by Lim and Wong (Lim 1995; Wong 2022).

Set in the 1950s, *The Bondmaid* tells a tragic love story of Han—who at age of four was sold as a bondservant to a family—and Wu, the young master of the household. The book is divided into three parts—girl, woman, and goddess. The story begins with Han as a playmate to Wu who then goes overseas for his studies. When he returns from abroad, Wu's family matches him with the daughter of another wealthy family. However, Han and Wu love each other secretly and he tries though unsuccessfully to make Han his secondary wife. Both women give birth at the same time: Wu's wife delivers a baby girl while Han, the bondmaid welcomes a much-desired baby boy who is stolen from her. Subsequently, Han commits suicide and transforms into a ghost/goddess and Wu is portrayed as an aged man tending to a shrine in her honour. In contrast, *The House of Little Sisters* is set in the early 1930s at a time when a bill promulgating the registration of *mui tsais* was being passed in Singapore—the 1932 Mui Tsai Ordinance. Wong narrates a coming-of-age story of sixteen-year-old *Lim Mei Mei* (Ah Mei) who in embracing her life as a *mui tsai* in a Peranakan family, Mister Lee's household, discovers a terrible secret lurking in the shadows of their past and subsequently, navigates her path to independence. The novel interweaves the story of *Mei Mei* alongside a cast of other female characters who as servants—*ahmas*, *majies*, etc.—also worked within the intimate realm of a domestic household. The inter-racial love story that develops between Mei Mei and Hassan, an Indian Muslim boy, is part of a larger plot about girls and women negotiating their freedom and emancipation.

Against the backcloth of domestic servitude, both texts deal similarly and differently with the following themes: culture and social class; romance and forbidden love; and women's silence and agency. By juxtaposing both these novels, readers can appreciate how the position of a *mui tsai* within Chinese social structures and kinship is (re)cast and at certain junctures, re-visited by Catherine Lim and Eva Nava Wong. Both novels challenge the long-established narrative of *mui tsai* as voiceless and hapless victims. While Lim presents Han as a *mui tsai* who defiantly seeks ingenious ways to protect her body from the sexual advances of predatory men, Wong portrays Mei Mei's agency within the parameters of both legislative measures and supernatural elements. In both texts, characters are privy to the ill-treatment, sexual abuse, and acts of cruelty towards *mui tsais*. Yet, in doing so, they refuse to articulate the kind of subservience and victimhood that one might expect from the book, and instead employ for example, magical realism as a literary device and metaphorical sub-text to reinterpret our preconceived assumptions, compelling us to recognise the agency exercised by young *mui tsais*. Critically examining both novels by way of an interdisciplinary and comparative approach will allow us to explore the interrelated themes of gender, domestic servitude, transnationality, and underclass voices. In what follows, the chapter will consider how both literary texts capture narratives that blur the boundaries between domestic servants and adopted daughters; ethnic identity and cultural cosmopolitanism; and reality and imagination.

9.2 2 "Adopting Daughters" as Domestic Servants: Gender, Class Hierarchies, and Sexual Abuse

Through the lens of historical fiction, insights into the reasons underpinning the *mui tsai* system and its characteristic features are fleshed out. Both novels, set in colonial Singapore, reflect aspects of the practice such as patriarchal attitudes. Girls and women are perceived as a lesser sex and the preference for sons over daughters in the past showcases socio-cultural attitudes in a transnational context. In *The House of Little Sisters*, Ah Mei's father views her as a "useless girl" and laments about "what can she do for me and the family?" (Wong 2022, p. 40). Furthermore, in the Lee household, prayers are done to "carry on the Lee name" and they turn to a *sinseh* "to buy boy producing herbs" (Wong 2022, p. 153). In both novels, also seen is how superstitions such as girls born in the Chinese Year of the Tiger are considered unfortunate and discarded. In addition, socio-economic factors, notably poverty and social ills such as gambling are also underscored as reasons that drove the practice of selling daughters into domestic servitude. In the opening chapter of *The Bondmaid*, Catherine Lim describes the overcrowded living conditions of Han's family to emphasise poverty. Conversely, Wong portrays Ah Leong (Ah Mei's father) as a compulsive gambler who relinquishes "his own flesh and blood" and how Eminent Mister Lee "paid a lot of good money for Ah Mei" (Wong 2022, p. 38). In terms of the origins of the *mui tsais*, Lim's narrative lacks explicit elucidation. Historically, they predominantly hailed from China but Eva Nava Wong's book sheds light on locally born *mui tsais* (see Appendix for statistics on *mui tsais* born in Malaya, 1934). For example, Lim Mei Mei, the central figure in *The House of Little Sisters*, grew up in a *kampong* in Singapore inhabited by mainly Chinese families; a new *mui tsai* that replaced another was from Malaysia; and another character adopts a daughter who previously indentured as a *mui tsai* and hailed from a tin mining town in Malaysia.

Domestic service encompassed a range of roles and demographics with each bearing its unique characteristics. In particular, *The House of Sisters* draws attention to this multifaceted nature. Apart from *mui tsais*, the book features Chinese women (or Black and White Amah) (Ooi 1992) who served generations of the family and also *majies*. Furthermore, through Ah Ping Yee's character, readers gain insights into female servants who distinguished themselves by maintaining celibacy, a commitment evident in their distinctive hairstyling practices. These women were initiated into spinsterhood through a ceremonial hair-combing ritual. Taken together, these girls and women were entrusted with the care of children and the management of household affairs. Through the lens of differently situated domestic servants, Wong's novel sheds light on the intersection between cultural traditions and social roles.

Both texts highlight how the *mui tsai* system resembled a form of trade where young girls were being treated as goods and reflected entrenched class hierarchies within the Chinese diasporic community in Singapore. In one scenario in Wong's book, a little girl was "recused from a market where fish, vegetables and little girls were traded" (Wong 2022, p. 171) and this paints a vivid picture of their commodification. During an oral history interview conducted by the National Archives of Singapore in the 1980s, Betty Seow recounted her family's experience with what is termed as "slave girls". Speaking about the diversity within the Straits Chinese community, Betty of Peranakan background, notes that there were "actually very poor Peranakan families around" (Oral history interview with Betty Guat Beng Seow, NAS 1989) in Singapore. She mentions that her own family purchased some girls from these families and explains that:

> In those days poor people sell their daughters. Then you give them an "ang pao" or you give some rice, sugar for sweetness, always with meaning, and they come and stay with you (Oral history interview with Betty Guat Beng Seow, NAS 1989).

Indeed, the transactional nature of this system surfaces in both literary texts. In *The Bondmaid*, the matriarch of the House of Wu received villagers presenting themselves and offering to inspect their female children before an adoption or sale. Catherine Lim describes how:

> From the chair of authority, she [the matriarch] delivered judgments on the poor pathetic creatures, inspecting the offspring to see if they had healthy scalps and teeth, and completed the transaction with a sum of money in a sealed red envelope. She now owned them body and soul, for they had become her bondmaids (Lim 2011).

The passage above offers a poignant portrayal of the power dynamics, uneven class hierarchies, and financial transactions inherent in the process of acquiring bondmaids. The exchange of an *ang pao* (red envelope) symbolises the formalisation of ownership and control. The entire inspection process not only establishes the matriarch's position of power and authority but it underlines the marginalised status of *mui tsais*. Similarly, in *The House of Little Sisters*, Ah Mei had to undress, and her body was inspected to check if she was a virgin; this act, in effect, illustrates the dehumanising nature of the process and how *mui tsais* were stripped of agency and autonomy.

Contrary to newspaper reports that highlight that the *mui tsai* system in British Malaya "does not ordinarily expose a girl to moral dangers" and that instances of "cruelty or sexual crime are rare" (*Daily Telegraph* 1937), the depiction of sexual abuse in both novels draws attention to the harsh realities that many young girls encountered and foregrounds the nebulous nature of their status as "adopted daughters". Their exploitation by both male and female members of the household brings to the fore the various forms of abuse. The narratives unveil the physical abuse endured by *mui tsais*, inflicted by the wives within the household, who moreover play a role in orchestrating their sale into prostitution upon pregnancy or skin diseases among other factors. In *The House of Little Sisters*, the abuse primarily stems from within the family with the patriarch and matriarch exploiting victims. However, in *The Bondmaid*, the abuse extends beyond the family as *mui tsais* are also sexually exploited by outsiders such as a Reverend of the temple who frequently visits the House of Wu. In one notable scenario in *The House of Little Sisters*, during an exchange between the Chinese Protector Mister Pickering and Eminent Mister Lee, the latter stroked Mei Mei's left thigh and pats her buttock while conveying to Pickering that "*mui tsai* means little sister, Mister Pickering, or if you will, little daughters" and exclaims that "I have in my household a number of these little girls whom I've adopted, providing their families with a source of income and saving their daughters from a life of disgrace, destitute and deprivation". Eminent Mister Lee then proceeds to grab Mei Mei's face and exclaims: "See this *mui tsai*? She is my daughter and I do what the hell I please with her" (Wong 2022, p.137). The dialogue illustrates a disturbing contradiction between Lee's self-professed benevolence and the exploitation he enacts. Significantly, it highlights the liminal positionality of *mui tsais* who ambiguously existed neither fully as daughters nor domestic servants but as a vulnerable space between the two which subjected them to ill-treatment by those who "adopted" them.

The ambivalent positionality of *mui tsais* subtly masked under the euphemism of "adopted daughters" has, historically, been a contentious issue. In a report by colonial administrator George Maxwell, he explains that although the system did not entail legal processes of adoption, "there was an element of 'quasi-adoption' in the transaction" (CO 323/1257/11 1934). This ambiguity is elucidated in an oral history interview with Betty Seow. When queried about the status of *mui tsais* in her household in the past, she responded that:

> Betty: One of the things my ancestors did was to bring girls to become slaves...My grandmother had three or four slaves, sort of adopted daughter. Then after a while, they get married, but they worked in the house. We usually engaged a cook. But these girls will help do things around the house.
>
> Interviewer: Do you remember if they were actually a part of the family?
>
> Betty: No
>
> Interviewer: Distant family?
>
> Betty: No... (Oral history interview with Betty Guat Beng Seow, NAS 1989)

In this excerpt from an oral history interview, Betty recalls her family's practice of having *mui tsais* and uses the terms "slaves" and "adopted daughters" somewhat interchangeably. Her response suggests that although these girls assisted in domestic duties, they were not treated as daughters per se thus, indicating a clear distinction in their status as domestic labourers. On the contrary, another oral history account with Yau Chung Chii who also uses the term "slaves" to describe *mui tsais*, suggests instead that her grandmother treated them like a family member. Chung Chii adds that after their *mui tsai* was married off, her grandmother maintained a close relationship with her and insisted that she be referred to as an aunt when visiting the family (Oral history interview with Yau Chung Chii, NAS 1984).

The afterlives of *mui tsais* like those depicted in the novels, are often deeply entwined with the institution of marriage and related notions about female virtue. Historically, when these girls reach the age of eighteen, discussions about arranging marriages for them become commonplace. In *The House of Little Sisters*, "Mei Mei thinks about the responsibility of Number One Madam to find her a suitable marriage when she turns eighteen years old" (Wong 2022, p. 142) and similarly, in *The Bondmaid*, the matriarch remarks how it is time to "decide her [Han's] fate" and since she was good looking, she could potentially "end up as concubine or secondary wife to some rich man" (Lim 2011, pp. 162–163). Herein, there is much emphasis either on the importance of female virginity or on becoming of "marriageable quality again" (Wong 2022, p. 171) in an event of sexual abuse. For example, in Wong's novel, a sexually abused *mui tsai* who was sent to the Poh Leung Kok—a factual depiction of a shelter home that existed in the past for girls and women in Singapore—was spoken about in terms of being "mended" before being married off to a suitable family. At that juncture, Mei Mei thinks to herself how the language used, "mended", suggests that these girls were not seen as a person but "a piece of furniture or a broken bowl, something that can be fixed when broken" (Wong 2022, p. 171). Lim's novel too sheds light on the "bodies of young girls destroyed" (Lim 2011, p. 202) by the House of Wu over the years. The double standard regarding sexual behaviour as Wong emphasises in her novel—"a boy can sow his wild oats but a girl must remain virginal" (Wong 2022, p. 162)—underscore deeply ingrained gender norms and socio-cultural inequalities. Although the notion that young women's value lies primarily in their purity and sexual innocence is expressed, the realities faced by them within the domestic sphere reveal stark contradictions. Against the cultural bounds of Chinese tradition, both literary texts navigate how *mui tsais* found themselves in vulnerable situations. Readers are confronted with the contradictions inherent in the construction of female virtue and their lived experiences within patriarchal societies.

9.3 Local Identities and Cosmopolitanism: Situating Transnational Chinese-Ness

The diversity within the Chinese diasporic community in Singapore, shaped by interactions with various ethnic groups is a notable feature in *The House of Little Sisters*. *The Bondmaid* is more of a Chinese-centric narrative with a limited focus on their interactions with other ethnic groups. However, what emerges in *The House of Little Sisters* is the development of localised Chinese identities, and in particular, Wong foregrounds the complexity of Chinese identity vis-à-vis Peranakan culture. Their distinctiveness and adaptations in language, cuisine, and customs are a defining feature of the cosmopolitan landscape that shapes the narrative of the book. The Lee family is Peranakan and in one scenario the Straits Chinese are described as follows:

> They're Chinese but not Chinese. Speak Malay but are not Malay. Eat pork because they are not Muslim and pray to Chinese gods, but act like Europeans (Wong 2022, p. 69).

The Straits Chinese are a unique cultural group with a blend of Chinese, Malay, and indigenous influences, resulting in a distinct identity that is neither fully Chinese nor fully Malay. In their analysis of how being Peranakan is related to (not) being Chinese in Singapore—*Orang Cina bukan Cina* (Chinese but not Chinese)—Rocha and Yeoh contend that the "social construction of race has created a hybrid yet ethnically Chinese identity" (Rocha and Yeoh 2023, p. 568) and foreground the "contradictions and messiness inherent in living this dichotomy" (Rocha and Yeoh 2023, p. 573). The Peranakan community, as they further highlight, positioned itself as "the King's Chinese": an in-between population, which was English-educated (Rocha and Yeoh 2023, p. 570). Through the Lee family, novelist Wong explores the blurring of national boundaries and how Singapore is a transnational space that is home to multiple identities and cultures.

Both within and beyond the context of a Chinese domestic household, the multicultural and multiethnic nature of Singapore society emerges. For example, in the Lee household, apart from domestic servants from a cross-section of Chinese society, there is also Nur, a Malay washerwomen, and Chandra, an Indian gardener. Another minority character is the shopkeeper, Encik Abdullah, who can communicate with his customers in Hokkien. The local market in Singapore is described by the protagonist as a sensory space where:

> A melange of languages and the woody, earthy scents of herbs and spices find their way into my ear and up my nose. We are a melting pot of lively chatter churning with sounds of Tamil, Malay, Kristang, Cantonese, and Fukienese all being stirred together with the smells of turmeric, ginger, garlic, fresh coriander, and spicy chilies (Wong 2022, p. 53).

The passage above captures the essence of Singapore's cosmopolitanism where diverse languages, cuisines, and cultural influences coexist and intersect, creating a unique sensory landscape that reflects the rich social fabric of the nation's heritage. Against the narrative backdrop of domestic servitude, the inclusion of characters from different racial and religious backgrounds not only serves as a vehicle for social commentary on issues such as class and labour but also sheds light on shared experiences and cultural integration.

At the heart of both novels lies a tale of romance and forbidden love. In one it's an intra-racial love story between Wu and Han fraught with class divisions, while in the other, it's an inter-racial love story that develops between Hassan and Mei Mei. During his childhood years, Wu, the young master in *The Bondmaid* is discouraged from playing with Han let alone marrying her. Although they both love each other and Wu wants to marry her, he is instead married off to a woman chosen by the two rich families. Yet, the couple persists in clandestine meetings and the story ends tragically with the death of Han. Class separatism in Singapore, thus, acts as a formidable societal barrier that impacted the lives of individuals. In contrast, in *The House of Little Sisters*, Hassan—an Indian Muslim boy with Bengali heritage stirred by Rabindranath Tagore poetry—and Mei Mei develop a close friendship and deep affection for each other. Mei Mei remarks that:

> Nobody must even know that I even speak to a Muslim boy. I would bring shame to the Lee family and I can't afford to do that. And, nobody must know that I even like Hassan. I mean, I have never seen anyone like me married to anyone like him. It is just not done (Wong 2022, p. 60).

Both Hassan and Mei Mei speak of freedom that awaits them and their relationship is depicted as defying societal norms. She exclaims how their "love for each other crosses the different oceans in this world, and is the bridge that will bring our two cultures together" (Wong 2022, p. 188). The book ends with a sense of hope and optimism, "the bridge" symbolises how their love has the transformative power to transcend social, religious, and cultural barriers. As a melting pot of diverse ethnicities and cultures, colonial Singapore as a diasporic home is characterised by the co-existence of myriad identities and the intermingling of traditions from different parts of the world.

9.4 Between Real and Imaginary: Literary Devices that Foreground Voice and Agency

Both novels employ magical realism as a literary technique to give voice and agency to a social underclass. The texts are replete with folklore and tales; ghosts and spirits; and gods and goddesses. Magical realism as a literary device blurs the boundaries between the real and the imaginary and through dreams, spirits, and supernatural elements, the authors illuminate the inner thoughts of characters and explore their hopes and fears as well as struggle for agency. For example, in both novels' material items like traditional Chinese jewellery feature as a symbol of protection. In *The House of Little Sisters*, as Mei Mei packs her things to leave her family, her grandmother gives her a jade amulet to wear every day and indeed in times of adversity, she rubs her Guanyin pendant and seeks the protection of the goddess. Similarly, in *The Bondmaid*, Han is given a jade bracelet by her biological family, and the jewellery which features the Sky God was meant to be "a protective jewellery for wearing on chests and wrists" (Lim 2011, p. 31).

Everyday conversations with gods and/or ghosts are normalised in both literary works. Crucially, as Gina Wisker highlights, "ghosts and ghost seers reveal secrets, seek revenge for inequalities, silencing, brutality" (Wisker 2022, p. 189). In the preface of *The House of Little Sisters*, the author establishes that British Malaya is a place "where legends and folklore abound" and underlines that "ghosts and spirits live among the people of this land" and therefore, "when spirits start to talk, you must listen". In the chapters that follow, there is mention of the Pontianak—the ancestral spirit of the indigenous Malays; the Lee household prepares for the Hungry Ghost festival; and most significantly, Mei Mei hears a voice in her head—that of Ah Lian, a former *mui tsai* who died in the Lee household—that urges her to "rage against the injustice" (Wong 2022, p. 18) meted out to them. In one scenario, as Eminent Mister Lee approaches Mei Mei's room one fateful night, the rattling door triggers an outburst from her. In her distress, she hears an unfamiliar voice cry out invoking the judgement of the Guanyin Goddess upon him. Then a haunting vision unfolds before her and in the spectral shadow, Mei Mei witnesses the harrowing tale of Ah Lian. She sees Ah Lian's anguish over her pregnancy; Eminent Lee's failed promise to her to not worry about being "hungry, poor and destitute"; her brutal punishment at the hands of Number One Madam; and, the heart-wrenching cries of her newborn torn from her arms. Through the ghostly apparition—"Look at me. I am a ghost, hungry to take revenge" (Wong 2022, pp. 149–150)—the hidden secrets of the Lee household emerge and readers encounter a chilling reminder of the injustices endured by *mui tsais* like Ah Lian. In a similar vein, in *The Bondmaid*, as Han's child is forcibly taken away from her embrace, her impassionate prayer to the Sky Goddess marks a pivotal moment as ultimately, although the story ends with her tragic death, Han transforms into a goddess/ghost. In the aftermath, Wu, as an old man holds vigil over the pond—where they used to meet and where she died—and mourns the rest of his life while tending to the goddess's shrine. In both texts, the crimes committed on *mui tsais* and the recurring motif of "everyday haunting" (Wisker 2022, p. 193) suggest that while the trauma of historical events such as these may have faded from the collective social memory of a nation, they have a lingering impact with unresolved anguish that transcends generations.

Women's ghost stories, Wisker writes, expose the "links between the political and the personal, global genocides and local domestic tyrannies" and through the ghostly body, "hidden histories of gendered and racial injustice" surface (Wisker 2022, p. 2). By inviting spirits to speak or simply exist, both Lim and Wong bring to light the "strangely wavering space between the world of the imaginary and of the tangible, revealing each to be equally real (or unreal)" (Wisker 2022, p. 15). Ghosts and spirits, therefore, serve as a powerful instrument for critiquing gender dynamics especially those marked by abuse and oppression. As Wisker further contends:

> Their [Ghosts] insistence that they be seen and heard, and listened to, make them perfect vehicles for gendered critique of the contemporary fast-growing cities of Southeast Asia, particularly Singapore, much is erased, re-written, re-represented, and among that the abused and silenced lives of the women, Chinese, Malay, and Tamil (Wisker 2022, p. 187).

Ghosts as entities demanding visibility underscore the erasure of voices and experiences and serve as a potent symbol for recognising and addressing gendered injustices and inequalities.

Both texts are rooted in historical contexts thereby further obfuscating the demarcations between reality and fiction. In *The Bondmaid*, historical events are more implicit. However, *The House of Little Sisters* is deeply entrenched within historical frameworks, replete with references to factual laws and real personalities. Percy Lee, the well-educated scion of the esteemed Mister Lee, is portrayed as a progressive and compassionate character dedicated to advocating for the social welfare of *mui tsais*. During an interaction with Mei Mei, he demonstrates his moral convictions by discussing his study of the Slavery Abolition Act of 1833 during his tenure at law school in London and articulates that "indentured servitude is a type of slavery" (Wong 2022, p. 90) that warrants cessation. He delves into an ordinance being passed in Singapore that would make it illegal to keep *mui tsais* without registering them—his narrative intricately parallels actual legal developments pertaining to them. In addition, Percy Lee pays Lim Mei Mei a monthly wage and she gradually saves up and starts her own small business sewing and selling handkerchiefs. Furthermore, the novel features characters such as Pickering and Shirin Fozder, respectively, representing a Chinese Protector and a Baha'i women's rights activist in Singapore who were both, in reality and fiction, actively involved in the social welfare of vulnerable populations. These characters and the legislative measures depicted have a significant influence on the trajectory of *mui tsais* and resultantly, the empowerment of Lim Mei Mei through legal rights, wages, and economic autonomy is seen.

In the archives, there are implicit references to the impact of legal measures in colonial Singapore such as the Mui Tsai Ordinance of 1932. In an oral history account, Yau Chung Chii candidly shares in Cantonese that her mother would threaten to sell her off as a *mui tsai* if she did not study hard. Recounting her childhood around 1933 or 1934, when she was about 7 years old, she alludes to the legal changes happening in Singapore at the time:

> We had slaves, did you know that? I remember the year, I don't know the date actually, the government clamp down on slavery. I remember we were quite excited...From what I gather, they wanted to marry of our *mui tsai* as soon as possible otherwise the government will fine us you know, for keeping a slave (Oral history interview with Yau Chung Chii, NAS 1984).

As mentioned earlier, arranging a marriage for *mui tsais* was not uncommon. However, in the novel, the significance of marriage is recast through the voices of the social underclass. In a pivotal exchange between Wan Jie (an older domestic servant) and Mei Mei, she imparts a profound piece of advice and asserts that "a woman's freedom lies not in the man she marries but in the purse strings that she holds her own" (Wong 2022, p. 200). Mei Mei's symbolic act of sewing a phoenix as embroidery on the handkerchief—"*mui tsai* will rise from the fire a powerful and liberated phoenix" (Wong 2022, p. 194)—becomes emblematic of her journey towards empowerment and liberation.

In a succinct analogy, Dale Porter articulates that "historical fiction is an elusive child of mixed parentage" as it claims the right of invention reserved for fiction while maintaining claims to be based on historical reality (Porter 1993, pp. 315–335). Authors often would include a disclaimer to their work and highlight historical events and figures that inspired it. For example, *The House of Little Sisters* is considered an #OwnVoices book, and Wong who identifies herself as Peranakan Chinese writes that the novel is based on "snippets of family lore and history". She writes that Ah Ping Yee, "although a fictional character, was inspired by a real person. In my family, we were privileged to have a black and white servant who was my nanny for some years". In a similar vein, *The Bondmaid* is based on Catherine Lim's family's past, and during an interview, she candidly shares that she was inspired by her grandparents to write this story. She adds that "my grandmother had bondmaids" and she "virtually bought them off the doorstep, not bought them, she would give an ang pao, a little gift of money, you know as a sign of acquisition" (Catherine Lim on BBC Peschardt's People ½, YouTube 2010). The anecdotes by both authors provide insights into the historical context of domestic servitude within the parameters of family history and cultural practices.

9.5 Conclusion

In contemporary Singapore, arguably the social memory of *mui tsais* as a category of gendered labour is overshadowed by say, the *samsui* women or black and white *amahs*. However, through the medium of historical fiction, the lives of *mui tsais* unfold and readers gain valuable insights into various aspects of this phenomenon. Against the backdrop of tradition and tyranny, the literary texts by Catherine Lim and Eva Nava Wong throw light on the interplay between domestic servants and adopted daughters; ethnic identity and cultural diversity, and reality and imagination. A pivotal scene in Wong's novel poignantly portrays one *mui tsai* who was unable to stop what was happening to her as she "did not have a voice" (Wong 2022, p. 162). Nevertheless, both authors depict the resilience and agency of these young girls despite their status as a social underclass. By employing magical realism as a literary device, they blend supernatural elements with the real world, and in so doing, render visible *mui tsai* voices in a transnational context.

Appendix

(See Table 9.1).

Table 9.1 Mui Tsai registered in British Malaya, 30 June 1934

Analysis of Mui-Tsai registered on 30.6.34.

	Straits Settlements			Federated Malay States				Unfederated Malay States.		Malaya
	S'pore	Penang	Malacca	Perak	Selangor	Negri Sembi-lan.	Pahang	Johore	Kedah	Total
No. of registered Mui-Tsai	589	492	249	448	457	203	64	150	97	2749
No. of Mui-Tsai who are without parents whose present addresses are known to Govt.	373	268	134	445	338	164	40	132	75	1969
No. of Mui-Tsai whose parents are in Malaya and whose addresses are known to Govt.	58	59	50	3	42	37	13	16	21	299
No. of Mui-Tsai whose parents are not in Malaya and whose addresses are known to Govt.	158	165	65	-	77	2	11	2	1	481
No. of Mui-Tsai born in Malaya	128	183	91	166	110	82	19	53	60	892
No. of Mui-Tsai born in Fukien	109	-	90	25	16	29	10	38	6	323
No. of Mui-Tsai born in Kwangtung	259	-	43	67	30	13	12	42	15	481
No. of Mui-Tsai born in Kwangsi	-	-	-	-	-	1	-	-	-	1
No. of Mui-Tsai born elsewhere in China	2	-	-	-	-	-	-	-	-	2
No. of Mui-Tsai born in China but province unknown	9	156	21	141	248	73	19	11	3	681
No.of Mui-Tsai born in Hongkong	2	2	-	-	1	-	-	-	-	5
No.of Mui-Tsai born elsewhere than in China, H.K. or Malaya	21	40	-	7	3	3	-	2	9	85
No.of M.T. whose birthplace is unknown	59	111	4	42	49	2	4	4	4	279
No.of M.T. under 10 years of age.	120	89	81	96	94	59	13	35	43	630
No.of M.T. of 10 years & over but under 15 years of age	368	283	139	269	291	87	36	91	50	1614
No.of M.T. of or over 15 yrs. of age	101	120	29	83	72	57	15	24	4	505

Source: The National Archives, Kew, London

References

National Archives of Singapore

BMA 255–45: "Poh Leung Kok - Women and Girls Welfare", Oct 1945 - Mar 1946

From the oral history interview of CHII, Yau Chung (Mrs) (Accession No. 000427, Track No. 1), Oral History Centre, National Archives of Singapore

From the oral history interview of SEOW, Betty Guat Beng @ Lim Koon Teck (Mrs) (Accession No. 001048, Track No. 2), Oral History Centre, National Archives of Singapore

From the oral history of WONG, Hing (SK Wong) (Mrs) (Accession No. 000093, Track No. 4), Oral History Centre, National Archives of Singapore

The National Archives, Kew, London

CO 323/1257/11: Mui-tsai system, 1934

CO 323/1257/12: Mui-tsai system, 1934

CO 882/16: "The Mui Tsai system in Hong Kong and Malaya", Reports, correspondence and legislation regarding the administration of the system of supply and treatment of domestic servants, 1934–1939

"Mui tsai in Straits Settlements: A Registration Bill", The Manchester Guardian, 11 March 1932

'Chinese child slavery: "Mui tsai evils disappearing", Daily Telegraph, 1 March 1937

Aung-Thwin M (2018) Connections, contact, and community in the Southeast Asian past: teaching transnational history through Amitav Ghosh's The Glass Palace. In: Wong JYC (ed) Asia and the historical imagination. Palgrave Macmillan, Singapore, pp 155–179

catherinelimsg. "Catherine Lim on BBC Peschardt's People Part 2/2". Youtube Video, 13:04. 7 August 2010. https://www.youtube.com/watch?v=r2SySms_j9g
Chaudhuri AR (2019) The scent of Frangipani: a tale of love, rivalry and murder in colonial Singapore. Monsoon Books
Holden P (2010) Interrogating multiculturalism and cosmopolitanism in the city-state: some recent Singapore fiction in English. Mobilities 5(2):277–290
Lettieri MA (2006) Maximize our missions: Mui Tsai, missionaries and material representation. Dissertation, University of Nevada
Lim C (2011) The bondmaid. @Catherine Lim 1995, Marshall Cavendish
Lim J (2004) Sold for silver: an autobiography of a girl sold into slavery in Southeast Asia. Singapore, Monsoon
Miers S (1994) Mui Tsai through the eyes of the victim: Janet Lim's story of bondage and escape. In: Jaschok M, Miers S (eds) Women and Chinese patriarchy: submission, servitude, and escape. Zed books, pp 108–121
Ooi KG (1992) Domestic servants par excellence: the black and white amahs of Malaya and Singapore with special reference to Penang. Journal of the Malaysian Branch of the Royal Asiatic Society 65(2):69–84
Porter DH (1993) The Gold in Fort Knox: historical fiction in the context of historiography. Soundings: An Interdisciplinary Journal 76(2/3):315–335
Rocha ZL, Yeoh BSA (2023) Orang Cina bukan Cina: being Peranakan, (not) being Chinese and the social construction of race in Singapore. Identities 30(4):568–587
Wisker G (2022) Contemporary women's ghost stories: spectres, revenants, ghostly returns. Palgrave Macmillan
Wong EN (2022) The house of little sisters. Penguin Random House Southeast Asia
Wong JYC (2018) Asia's Other History. In: Wong JYC (ed) Asia and the historical imagination. Palgrave Macmillan, Singapore, pp 1–14

Hema Kiruppalini is a Research Fellow at the Asia Research Institute (ARI), National University of Singapore (NUS). She is a historian specialising in migration and transnationalism in inter-Asian contexts. Her research delves into themes such as race, colonialism, gender, soldiering, and social histories. She is particularly interested in exploring the interconnections between the Himalayas and Southeast Asia. Her doctoral dissertation and publications, which reconstruct the social worlds and transnational history of Nepali Gurkha families in post-war Asia, have received several scholarly accolades such as the BNAC PhD Dissertation Prize and the Craig A. Lockard Prize among others. At ARI, she is currently working collaboratively on a research project titled "Archiving the Underclasses: Knowledge, Law, and Everyday Agency in Modern Southeast Asia".

Chapter 10
Remaking Identities: Thai Yunnanese Transnational Entrepreneurs in Everyday Life Practice and Performance

Kian Cheng Lee

Abstract This chapter argues that Thai Yunnanese transnational entrepreneurs have multiple identities in interstitial and relational contexts through everyday practice and performance. Set within a marginalised stigmatised background, these Yunnanese descendants with Thai citizenship have attempted various strategies to reconstruct their identity: business aestheticisation, professionalism, and socialisation; identifying with a Taiwanese identity; accentuating a Thai identity; distinguishing Thai Yunnanese identity; translocalising a People's Republic of China (PRC) Yunnanese identity; and transnationalising a Chinese identity. Through these aspects, this chapter demonstrates how the Thai Yunnanese transnational entrepreneurs have ingeniously adopted polychotomisation and multi-complementarisation in their assertion of multiple identities to gain diverse positive outcomes. Critically, this chapter, based on qualitative research, has contributed to the understanding of ethnic identities through the lens of an oft-neglected group—Thai Yunnanese—in the nexus of transnationalism, entrepreneurship, and other cultural markers. Finally, this chapter departs from the debilitating nature of assimilation explanation and dualistic perspective in articulating the remaking of a new wave of entrepreneurs that has gone beyond the narrow definition of utilising common background and migrant experience in identity construction.

Keywords Thailand · China · Thai Yunnanese · Transnational entrepreneurship

K. C. Lee (✉)
Faculty of Political Science and Public Administration, Chiang Mai University, Chiang Mai, Thailand
e-mail: kiancheng.lee@cmu.ac.th

H. M. Y. Ho and Y. Chan (eds.), *Transnational Southeast Asia*, Asia in Transition 29,
https://doi.org/10.1007/978-981-96-3608-2_10

10.1 Introduction

This chapter argues that Thai Yunnanese transnational entrepreneurs define a new image characterised by multiple interstitial identities with polychotomisation and multi-complementarisation in relationality through everyday life practice and performance. This chapter begins by showing a brief historical context of Thai Yunnanese transnational entrepreneurs while explicating the prevailing marginalised conditions behind their identity reconstructions. Subsequently, this qualitative research-based chapter explores the negotiation of their identity as a practice of skilful performativity. This is accomplished through the unravelling of the multiplicity of dynamics as follows: business aestheticisation, professionalism and socialisation; identifying with Taiwanese; accentuating Thai identity; distinguishing Thai Yunnanese identity; translocalising People's Republic of China (PRC) Yunnanese identity; and transnationalising Chinese identity. Finally, this chapter ends with a critical discussion and conclusion that surfaces salient perspectives. This includes a departure from the assimilationist and dualistic limitations in extant literature while showing the complexities involved in the new wave of entrepreneurship ushered by Thai Yunnanese transnational entrepreneurs.

10.2 Brief Historical Context

The selection of Thai Yunnanese transnational entrepreneurs in this study was not made randomly. From a historical perspective, Thai Yunnanese have experienced many political, social, and cultural challenges. Set against their historical marginality, their engagement of multiple interstitial identities in transforming their social status is of great interest.

According to Chang (2001), the Yunnanese Chinese in Thailand can be understood in three historical phases: as war refugees (1943–1961), resettlement in Northern Thailand since the 1960s, and the new changes and migrations since the 1980s. The Yunnanese Chinese normally do not reveal their Yunnanese origin in public. This is attributed to a prevalent negative Thai perception of the Kuomintang (KMT) Yunnanese who are associated with their former military and drug-trafficking activities (Chang 2001). They are also referred to as Jin-ho, which contains pejorative meanings, ascribing the KMT Yunnanese as violent, backward, and engaging in illegal trade (Hill 1998, p. 18). In fact, their migratory history in the latter half of the twentieth century has largely been viewed from the perspective of KMT activities (Wang 2006, p. 357). This perspective inevitably influenced the Thai government to create policies related to national security and border control that further contributed to the marginalisation of the Yunnanese (Wang 2006, p. 357).

This chapter is interested in unravelling how Thai Yunnanese—Yunnanese Chinese who have acquired Thai citizenship—attempt to gain upward class mobility through multiple identifications to escape from village life and resistance to refugee status (Boyle et al. 1998, pp. 208–216). This chapter is situated within the transnational political dynamism, which takes the significance of this research to a whole new level. It is set against the developments of the intricate political orientations of Thai Yunnanese which situates in-between Thailand, Taiwan, and PRC. Politically, Thai Yunnanese can be said to be historically aligned with Taiwan instead of PRC. Regardless of whether they are Han or Hui Yunnanese origin, they are united ideologically with Taiwan in an anti-communist stance (Wang 2006, p. 348). Among the aid projects given to the Yunnanese Chinese, Taiwan has, with particular importance, long-valued Chinese education (Chang 2004). Through the learning received by the Yunnanese Chinese, Taiwan hopes to inject the political idea that Taiwan is their ancestral home, especially prior to the 1990s (Chang 2004). Coupled with the attractive economic development that Taiwan offers, Taiwan has promised many Yunnanese youth full Chinese citizenship when they study or work in Taiwan (Chang 2004). Thus, it seems logical to assume that Thai Yunnanese would be aligned politically with Taiwan. Yet, this chapter is poised to question and explore the dynamism involved with regard to patriotism. In fact, the question of political re-alignment in this research is sharpened considering the meteoric rise of the PRC. An apparent paradox has already been built into the political scenario with the progression of time and the economic growth of the PRC in the region. Not long after the Cold War, foreign media has already labelled the PRC as a "potential superpower" (Yan 2001, p. 33). PRC has begun to witness a long-term economy boom that has been brought about with the Open Door Policy adopted by Deng Xiaoping (Yan 2001, p. 35). With real annual gross domestic product averaging 10% in 2011, the PRC has become one of the fastest-growing economies in the world (Tong 2014, p. 14). With reference to Thailand, since 2012, both countries have signed a series of deals, which cover agriculture, transport, and elevated relations to a comprehensive strategic partnership (Guangjin and Wa 2012). During Thailand's flood in 2011, the PRC offered emergency humanitarian aid to Thailand, which included US$1 million in cash remittances and US$1.57 million worth of relief materials (Guangjin and Wa 2012). By 2011, PRC had become Thailand's largest export market and second-largest source of imports with trade figures hitting US$64.7 billion (Guangjin and Wa 2012). Consequently, the issue of patriotism is set within the juxtaposition of a historical political affiliation with Taiwan on one hand, as well as, on the other, the increasing economic dominance of the PRC in the global arena. This issue of patriotism is further accentuated as the Thai Yunnanese transnational entrepreneurs are based in Chiang Mai, Thailand. Hence, this chapter explores the dynamism of Thai Yunnanese transnational entrepreneurs in their negotiation of patriotism vis-à-vis Thailand, Taiwan, and the PRC through their multiple interstitial identities.

10.3 Negotiating Identities as Practice of Skilful Performativity

This section aims to unravel the negotiation of Thai Yunnanese identities as a practice of skilful performativity. According to Biernacki (2007, p. 3607), a focus on practice emphasises the impact of "taken-for-granted, pre-theoretical assumptions on human conduct." Practice is, therefore, a routinised way of acting where the "pretheoretical assumptions and routines affect how we act, especially how we manage our bodies, handle objects, treat subjects, describe things, and understand the world" (Ritzer 2008, p. 650). Further, practice can be understood as "the regular, skillful 'performance' of human bodies" (Reckwitz 2002, p. 251). As skilful performativity, practice involves the engagement of the mind in mental activities, the routinised training of bodily performances, and the use of objects (Ritzer 2008, p. 650). The analysis of practice therefore must consider the interaction of body, mind, and object (Ritzer 2008, p. 650). In this section, Thai Yunnanese multiple interstitial identities are examined as a form of practice, which involves skilful performances with the interaction of body, mind, and object. In these expressions, different peculiar examples are highlighted in a non-exhaustive manner, which reflect saliency and diversity. All interviewees' names have been anonymised to maintain confidentiality, and individual consent for the interviews was sought and granted.

10.4 Deconstructing Stigma and Reconstructing a New Identity

10.4.1 Desire to Remove Deep-Seated Stigma as Drug Traffickers

YSQ, the founder and managing director of a company dealing with bee-related products in Chiang Mai, is candid to share about his identity transformation and how his class status has been changed through his business. During our interaction, YSQ is ardent in deconstructing a negative stereotype associated with the Thai Yunnanese in general. From a sociological perspective, he identifies himself with the wider Thai Yunnanese community, which has been attached to a historical social stigma. In this case, the stigma is rooted in drug-trafficking activities that had unfortunately been conducted by the Yunnanese Chinese in the distant past. Due to this deep-seated stigma, YSQ recognises that the process of deconstruction of this negative identity will take a significant amount of time. He opines that if young Thai Yunnanese are willing to contribute proactively to society, then there will be a gradual transformation of others' perspective towards the Thai Yunnanese. To him, this process will take a decade or two. In an analogy he provides, YSQ gives an example of how the PRC is viewed, where there have been frequent reports that products from China are not safe for consumption as they are filled with toxic chemicals. YSQ comments that this is a result of our psychological inclination where we crave for "bad" news. Thus, social media has chosen to exploit such basic instincts by reporting exaggerated, negative social aspects. By inference, YSQ is implying that the local news often drums up the insinuation of Thai Yunnanese as those who are involved with drug-trafficking activities. Nonetheless, YSQ balances his analysis by acknowledging that Yunnanese KMT had indeed trafficked drugs from Myanmar to Thailand in the past. To him, this is an undeniable fact, which Thai Yunnanese can neither refuse, refute, nor reject. Since the source of drug-trafficking in Thailand can largely be traced to the Thai Yunnanese, this has inadvertently given rise to the negative public mindset towards the Thai Yunnanese. As such, YSQ realises that it would require a significant period of time and significant hard work to change this deep-seated notion.

10.4.2 Firm but Painful Decision for Deconstruction

It is crucial to note that the deconstruction process of the "drug trafficker" identity can be a painful one. In the case of YSQ, there are times where he is forced to make the hard decision of articulating his position even at the cost of losing friends. This position is an advocacy of a conscious detachment of the past with an objective revaluation. YSQ painstakingly explained to his friends that one should not judge the Thai Yunnanese simply because of their past drug-related activities. At the same time, he appealed to his friends to empathise and embrace the Thai Yunnanese who have been bearing an unjust social stigma. Nonetheless, he is prepared to lose friends to rectify the stigmatised social identification.

10.4.3 Reconstruction Through Aestheticisation, Professionalism, and Socialisation

Actors are not passive under subjugation despite the long arduous process of the reconstruction of an identity. In this case, YSQ actively reconstructs his identity through his entrepreneurial activities. His company, which initially started as a small apiary with just four beehives has since developed into a larger bee farm and honey manufacturer. Currently, his company is known to be the biggest honey exporter and trader in Thailand. Summarily, his company manufactures a wide range of products including honey, bee pollen, royal jelly, beeswax, honey vinegar, cosmetics, and bee venom balm. From an aesthetic and economic perspective, the company emphasises branding, packaging, and novelty. Apart from the exquisite display of their products, the company is known to be the sole supporter of beekeeping equipment in Thailand. Their aesthetic innovation also includes offering tourists and customers an affective experience in visiting their well-designed museum.

In order to heighten the company's professional standing, YSQ's company acquired certifications from the Food and Drug Administration (FDA) to ensure his products are safe and healthy and are of the best quality. This includes Goods Manufacturing Practice (GMP) and Hazard Analysis Critical Control Point (HACCP) in production process. Moreover, they have also received Thai Halal and Malaysia Halal certificates through the years. In the process of reconstructing a new image and identity, there are times when YSQ is willing to pay an exorbitant price to acquire internationally recognised certificates. For instance, during the time of the interview, YSQ shared that his company was undergoing the process of acquiring ISO2008 so that it could penetrate into the American market. Due to the hefty price, he admitted that there were times when he wondered whether it was worth making such a high financial commitment. Nonetheless, YSQ viewed it as important to remain relevant and credible even though the annual fees are substantially heavy. These certificates are formidable tools in the company's internationally oriented marketing strategy. Hence, their overseas clientele stretches beyond Asian countries, such as Indonesia, Malaysia, China, Singapore, Macau, Taiwan, and Japan, and into the farther global reach of Saudi Arabia, the U.A.E., and Germany.

Besides aestheticisation and professionalism, YSQ has elevated Thai Yunnanese social status in identity reconstruction through his social connections. He consistently builds a good rapport with governmental departments and other public organisations. In reiteration, YSQ nowadays spends only mornings to come into the office while devoting his daily afternoon to play golf. As a socialising sport, this hobby provides him ample opportunities for interaction and bonding with retired military officers, influential governmental officials, and politicians, as well as acclaimed professors. As a result, his company's commodities have often been officially appointed to be sent to other countries as a national representation of Thai products. In summary, aestheticisation, professionalism, and socialisation have effectively propelled his company in the local, regional, and international arena for social and market recognition. From the perspective of identity reconstruction, these performative acts are significant in articulating a new economic identity with renewed currency for Thai Yunnanese community.

10.5 Identifying with Taiwanese in Various Aspects

10.5.1 Distantiation from PRC Chinese Communist Ideology

Born in Thailand, SZW is a second-generation Thai Yunnanese owning a desserts company. Her father left China after witnessing countless atrocities of the PRC communists during the Cultural Revolution, which included the murder of her father's own father. Thus, the negative perception towards PRC communism has been transmitted to SZW and influences her everyday life interactions. When asked about her identity affiliation towards either Taiwan or PRC, her response was clear. She surmised that she would affiliate her identification with Taiwan so that she could distantiate herself from PRC's communist ideology.

10.5.2 Drawing Business Connections

However, there are times when identification with the Taiwanese identity is not necessarily related to a desire for ideological distantiation from the PRC. In the case of SZJ, the sister of SZW, her transnational marriage with a Taiwanese provides her an opportunity to accentuate her Taiwanese identity for business purposes. As a founder and director of an education and tour company, SZJ has often introduced herself as Taiwanese. She has also granted permission for her business associates and colleagues to use her Taiwanese identity for extending business connections. In responding to a question as to whether her Taiwanese identity has helped her in any way, she replied: "Yes, it does help. For example, my business partners will introduce me as a Taiwanese to potential customers. They will say, 'See, our manager here is a Taiwanese!' They do so to draw connections or pull guanxi (拉关系)" (SZJ, Interview, 30 August 2014).

10.5.3 Glamorised Celebration of Taiwan National Day

In terms of a more concrete expression, some Thai Yunnanese celebrate the National Day of Taiwan held on the 10th of October. On this day, a full-course Chinese dinner was served to participants seated in round tables that fill the entire social hall of the Chiang Mai Yunnan Association. Both Thai and Taiwan National Anthems will be sung on this occasion. In terms of decoration, huge Chinese characters depicting this political occasion in Taiwan are prominently displayed on the backdrop of the stage. In addition to three huge Taiwan flags on stage, many small Taiwan flags are strung around the entire dining hall. Furthermore, in a ritualistic manner, speeches were made by representatives from Taiwan, the Overseas Chinese Association (华侨协会) as well as the President (会长) of Chiang Mai Yunnan Association. In addition, cultural performances by students and adults from various villages are also showcased. During the event, apart from individual contributions between 500 and 3000 Baht, the Taiwanese ambassador or representative will also offer sums around USD$500 for the event to be organised. According to SZJ who has witnessed the celebration at both Chiang Mai Yunnan Association and Taiwan itself, there is a distinct difference between the two where the former has a more glamorised version. As a performative act, such celebration serves to galvanise and demonstrate Thai Yunnanese patriotic orientation towards Taiwan.

10.6 Accentuating Thai Identity to Penetrate Unfamiliar Ground

There are times when respondents have chosen to accentuate their Thai identity in order to gain access into privileged business circles. For instance, HZJ, a Thai Yunnanese mobile coffee stall owner and a self-employed marketeer, shared about this aspect in her approach to penetrate the Vietnamese market. Through a Facebook network, she came to know of a group, which is involved in business activities in Vietnam. This group is made up of Thai businessmen who trade in Vietnam and who also know many Vietnamese people. They have started the Facebook group to facilitate Thais who are engaging in business within the Vietnam market. Immediately sensing a business opportunity, HZJ messaged the group with her stylistic Thai language and introduced herself as a Thai national. As a result, she was accepted and assisted by some Thai businessmen who are involved in delivering all kinds of products to Vietnam and Laos. In this case, HZJ has chosen to accentuate her Thai identity so as to gain access to resources within a Thai business group in Vietnam, a country which she is not familiar with.

10.7 Distinguishing Thai Yunnanese Identity from Thai and Thai Chinese

MEI, an owner of a Taiwanese restaurant, has had her fair share of historical marginality as a Thai Yunnanese. In the past, MEI felt that the Thais despised the Yunnanese and rejected those who spoke the Chinese language. Consequently, MEI grew up with the intention to submerge or hide her Yunnanese identity. In no uncertain terms, her Taiwanese husband expressed it in this way: "So we normally do not reveal to outsiders that our wives are Yunnanese because the Thai people hate Yunnanese very much" (ZHM, Interview, 20 September 2014). However, MEI pointed out that she no longer feels that way now. Interestingly, MEI cites the Beijing Olympic Games as a turning point. MEI is proudly distinguishing herself as a Thai Yunnanese who can speak Chinese language in contrast to other Thai Chinese who have lost their linguistic skills to do so.

This linguistic ability to speak in Chinese language has undoubtedly placed the Thai Yunnanese in a strategically advantageous position. With a comparative advantageous edge over other Thai Chinese, it is unsurprising that MEI has chosen to distinguish herself as a Thai Yunnanese. In this vein of thought, MWP, a Thai Yunnanese founder and managing director of a real estate company, shared a similar sentiment in the accentuation of Thai Yunnanese identity in economic perspective. MWP explained that Chiang Mai's economy started to bloom more than a decade ago in 2003. Chiang Mai's tourism industry flourished as many tourists began flooding in. In such a socio-economic landscape, MWP evinced the strengths of being a Thai Yunnanese. To him, Thai Yunnanese can learn and work concurrently. As a result, the Thai Yunnanese embarked and developed on several business projects including factory production, dried longan trade, export, and tourism. Gradually, according to MWP, Thai Yunnanese now have better living standards as compared to the past.

Bilingualism, mobility, endurance, and entrepreneurial skills have been strategically deployed circumstantially by Thai Yunnanese to their advantage over others. In terms of bilingualism, scholars have argued that "the learning of Thai has not led to the demise of the Chinese language" (Tong 2010, p. 43). However, it is noteworthy that the bilingualism of the Thai Chinese majority refers to the Thai language and a Chinese dialect (Tong 2010, p. 42). In other words, the Thai Yunnanese still have an upper hand in this regard. Furthermore, MWP boasted the income of the Thai Yunnanese, particularly for tour guides within the tourism industry who earn a considerable income. Unlike in the past where they used to fetch between 10,000 and 20,000 Baht in monthly salary, Thai Yunnanese involved in the tourism industry now earn at least between 40,000 and 50,000 Baht per month. In summary, these examples illustrate how Thai Yunnanese perform identity by distinguishing themselves over and against other ethnic groups amidst a growing PRC economy in the region.

10.8 Identifying with PRC Yunnanese in Translocalisation

Researchers have discussed translocalism as local to local connections across national boundaries that are created by transnational migrants (Brickell and Datta 2011). Here, translocalisation refers to the performative activities whereby transnational Yunnanese migrants of Thai citizenship affiliate with PRC Yunnanese. In the case of LRJ, he has been raised in Thailand and he has received his Thai citizenship at 28 years of age. He is currently a successful businessman as he is the Thai Yunnanese founder and managing director of a renowned tour company. He is an exemplification of translocalisation for he has chosen to identify himself with PRC Yunnanese.

First, LRJ articulated the historical suffering of Yunnanese as a noble form of sacrifice for the flourishing of PRC at the national level. He stated, "Finally, PRC can experience great rejuvenation. We, as Yunnanese, have also paid a price over one to two generations in exchange for the flourishing of PRC Chinese throughout the nation and I think it's worth it though we are the victims" (LRJ, Interview, 30 September 2014). In no uncertain terms, he bespoke of a seamlessly enjoined identity between him and the PRC Yunnanese through the first-person plural pronoun, "we." Emphatically, the pronoun also galvanised the translocal identity in construction.

In addition, the translocalisation of Yunnanese identity is not a mere verbal discourse. In the case of LRJ, it involves practical actions of identification in performativity. During difficult times caused by natural calamities, LRJ was prompt and generous in charitable donations to Yunnan, PRC. According to a senior official at the Office of Commercial Affairs from Chiang Mai Province, LRJ is the wealthy owner of a reputable tour agency that has sent money to help the needy in China during the flood. According to LRJ, the purpose behind his charitable giving was symbolic in expressing a sense of belonging in the eyes of PRC. In another natural disaster, LRJ contributed financially as well as appealed to others for assistance during the crisis at Yunnan. This has inevitably riveted his translocalised Yunnanese identity as a fellow countryman though located outside of PRC. This performative act possesses symbolic power as he stood in solidarity with the PRC Yunnanese in translocalism as well as from a broader national perspective of PRC in transnationalism.

10.9 Transnationalising Chinese Identity with Multi-Faceted Dimensions

10.9.1 Social Imagination, Ethnic Ascription, Variegated Thai Attenuation, and Exposure Rites

LRJ also engaged identity performativity through the transnationalisation of "Chinese-ness" as part of a seamless, unified, and imagined global overseas Chinese community. In response to my question concerning the apparent shift of patriotism of Thai Yunnanese amidst the economic rise of PRC, LRJ then constructed a transnational affiliation with the overseas Chinese all over the world. In his articulation, LRJ created an invisible transnational bridge, which ideologically connects global Chinese all over the world, including Thai Yunnanese, with PRC. By the same token, MEI transnationalised Thai Yunnanese identity in identification with PRC Chinese. In her case, MEI employed additional ascriptive elements of surname and birthplace as identity markers in transnationalisation. In the same manner, JZB also transnationalised identity through ethnic ascription: "Concerning this Chinese ethnicity, there is no way we can escape. Our Chinese ethnicity is characterised by our looks and physical appearance. There is no escape" (JZB, Interview, 31 August 2014). On the surface, this might seem delimiting, however, at a deeper level, JZB's strategy of ethnic ascription is non-restrictive but flexible. It allows him to draw on cultural resources in construction. In JZB's case, his transnationalised construction of identity with ethnic ascription gave him a culturally rooted tenacity in dealing with life and business challenges.

There are times when transnationalisation of Chinese identity involves attenuating other identities. In deconstructing negative Yunnanese stereotype, YSQ transnationalised his identity between Thai Yunnanese and PRC Chinese while attenuating his Thai identity: "I am a Yunnanese. There is nothing to be ashamed of. I am only a Thai citizen. My ancestors are Yunnanese. I cannot deny my ancestors to say that I have already been changed into a Thai. I am a Thai citizen, but I am actually a Yunnanese. I am Chinese. I am a PRC Chinese. I have never said that I am a Thai" (YSQ, Interview, 2 September 2014). Such attenuation of Thai-ness might reflect the fluidity of identity performativity.

Similarly, ZZJ, another Thai Yunnanese founder and director of a bee-related company, expressed a multi-layered identification where Chinese-ness is transnationalised and Thai-ness attenuated. He said: "We are in Thailand, so we must definitely love and respect the country's land. Yet, the real truth in my heart is that I will forever be a Chinese, the descendants of PRC ethnic Chinese" (ZZJ, Interview, 12 August 2014). Noticeably, ZZJ's patriotic sentiments towards Thailand is secondarily rooted in a geographic residential location in contrast to his identification of transnationalised Chinese-ness. When posed the question concerning whether he had absorbed Thai culture, ZZJ responded: "I don't think I have! No, this is because Thai people are quiet, gentle, and secured. This is unlike our PRC for we grew up through wars. Thailand is a mild place as the Thai people don't know how to suffer" (ZZJ, Interview, 12 August 2014). In addition, resembling JZB's access to cultural resources in ethnic ascription, ZZJ described how he would like to retain Chinese culture. To him, ethnic Chinese are superior in comparison to other ethnicities as they are known to be tenacious, have a long history of enduring suffering, and are diligent. ZZJ's form of transnationalisation of Chinese also has practical aspects. Even though his wife is a non-Chinese, ZZJ frequently brings his wife and children to China and Taiwan for holidays to be exposed to Chinese culture.

However, there are times when transnationalisation of Chinese-ness is concomitant with retaining a Thai identity. Being a third-generation Thai Yunnanese settled in Thailand, MWP did not see a need to attenuate his Thai identity even when he transnationalised his Chinese identity. To him, he is both a Thai and a Thai Chinese simultaneously. MWP's transnationalisation of Chinese identity is inextricably tied to his Thai identity. His "performative rites" of bringing his children to the PRC for exposure tours are preparations for his next generation to extend business into the PRC while maintaining the business base that he has established in Thailand.

10.10 Meeting Different National Representatives with Malleability

From observations in interactions for this research, ZZJ came across as someone who is malleable in politicking. Having served in many associations such as Chinese Business Association, Honey Association, Bee Association, Taiwan Business Association, and Chiang Mai Yunnan Association, ZZJ meets officials from many different countries and organisations. He expressed that in the past he hardly wore his Western suit and as a result, his suit had turned mouldy in his wardrobe. However, nowadays, he has to wear his suit at least once a month. On these occasions, he must welcome and meet representatives from different nationalities. His identity malleability is often stretched as he wears different national suits in meeting these officials. ZZJ reflected his performance of malleable identities in his remark: "There is no separation between politics and economics. It is the field where everyone joins or simply patronise (捧场)" (ZZJ, Interview, 2 September 2014). It is interesting to note how such patronising or literally "flattering in the politic economic field" reflects a flexible performativity of different identity constructions.

10.11 Customised Celebration of Chinese Festivals and PRC National Day

Thai Yunnanese practice the celebration of Chinese festivities including the Lunar New Year (农历新年), Ching Ming Festival (清明节), and Dragon Boat Festival (端 午节). Among these, the most important is the Lunar or Chinese New Year. Similar to many other Chinese, JZB has a family reunion dinner (团圆饭) on the eve of Chinese New Year. After dinner, there are rites conducted for worshipping the God of Wealth (财神爷), Kitchen God (灶神), and Earth God (土地公) to secure prosperity, favour, and protection. In this festival, he also consumes delicacies such as Chinese Dumplings (饺子) and New Year Cake (年糕). To a businessman, eating dumplings shaped similarly to Chinese silver or gold ingots during the Chinese New Year symbolises the wealth he will make in the coming year. Additionally, New Year Cake in Chinese is niangao, which also homophonously means "getting higher year-by-year." For a businessman, the consumption of the New Year Cake represents rising higher in business and becoming more prosperous in the coming year. In addition, JZB gives out red packets (红包), a common custom in Chinese New Year to his parents, children, employees, and others. Culturally, the rite of giving red packets—with money inside—represents sending good wishes, good luck, safety, and peace. Though it is not a public holiday in Thailand, JZB often declares three days off for his employees to render importance to this festival. As for HZJ, she routinely gives herself three to five days of leave as she returns to her hometown in Chiang Rai during the Chinese New Year. Ritualistically, she adorns herself with new clothes, which symbolises new provisions in the coming year. In her case, she performs ancestral worship rites at the eve of Chinese New Year as an act to demonstrate her filial piety to her deceased ancestors while securing blessings from them. As for ZZJ, he has a reunion dinner with delicacies similar to JZB. However, he does not officiate it as a holiday for his company. As a Christian, ZZJ does not have any of the ancestral or Chinese religious worship rites. In a similar way, LRJ, a Muslim, does not conduct any of the non-Muslim religious rites. However, he customarily declares a day off for his employees.

Next, the Ching Ming Festival is a day where the Thai Yunnanese visit the burial sites of their ancestors to "sweep the tombs" in memory of their ancestors. It is a day that they conduct traditional Chinese worship rites. For most Chinese, it normally falls on the fifth of the fifth month of the lunar calendar. However, this is not always the case for Thai Yunnanese transnational entrepreneurs. For instance, LRJ visits his ancestral tomb at Fang District, Chiang Mai Province, at any time of the year that is conducive for him and his family. Carried out without any worship elements, it is a solemn memorial occasion where he and his family commemorate their ancestors.

This also applies for the Dragon Boat Festival, where the Muslim and Christian Thai Yunnanese take rice dumplings but exclude the religious worship elements. Succinctly, these festival celebrations are customised according to their religious affiliation as well as their business preoccupations.

In contrast to the celebration of Taiwan's National Day, the major celebration of PRC National Day on the 1st of October is organised by the Consulate-General of the PRC in Chiang Mai. Mr. Chao XL is the 9th Consul General in Chiang Mai since the Consulate's establishment in northern Thailand in 1991 (Chao 2014). Mr. Chao's consular district covers 12 provinces in northern Thailand, namely, Chiang Mai, Chiang Rai, Lampang, Lamphun, Mae Hong Son, Nan, Phayao, Phitsanulok, Phrae, Sukhothai, Tak, and Uttaradit (Chao 2014). As an organisation, the Consulate is committed to promoting exchange and cooperation between PRC and northern Thailand in the area of politics, economy, trade, culture, and education (Chao 2014). On the day of the national day, the Consulate will organise a cocktail party for their overseas compatriots (侨团), in which the Hainanese, Cantonese, Teochew, Hokkien, and Yunnanese are invited to attend. The main highlight of the event is when the Consul General makes a speech giving some highlights of the recent socio-economic developments in the PRC. In addition, the Chiang Mai Province Governor also presented a speech extolling mutual bilateral cooperation. For the past few years, LRJ has represented the Chinese Chamber of Commerce (中华商会) to give a speech of a similar nature. According to a Thai Yunnanese who attended the session last year, though LRJ did not overtly declare that he is a Thai Yunnanese, his role was a significant statement and conspicuous recognition of his ethnic identity to all. Succinctly, the identity performativity during the celebration of PRC National Day takes on a great symbolic meaning as all different Chinese subethnic groups and political figures are present.

10.12 Political Self-Historicisation of PRC and Taiwan

As a form of everyday politics, Thai Yunnanese transnational entrepreneurs provide a self-styled analysis of past intertwined political relationships between PRC and Taiwan. In a radical fashion, ZZJ attempted to demonstrate the significance of political vigilance through a re-telling of the story of separation between China and Taiwan. As far as ZZJ is concerned, Taiwan belongs to PRC, and there is only one PRC because Taiwan originates from China. As for JZB, he appealed to ethnic characteristics as well as the sibling analogy in describing the relationship between PRC and Taiwan. To him, "Regardless of whether it is from China or Taiwan, we are 'yellow descendants' (炎黄子孙). We are Chinese; Chinese as in Hua (华) in Zhonghua (中华). Chinese can't be split between Taiwan and China. We can't be separated… It's between the two brothers where PRC is the elder brother and Taiwan as the younger brother in the fight" (JZB, Interview, 31 August 2014). Similarly to JZB, MWP described the relationship as brothers, but he added the element of face or mianzi (面子): "Therefore, now, to a Chinese, both Communist Party and Kuomintang are brothers. They can't kill each other. It is only about some political factors and the problem of face: If Taiwan is returned to you, then what face do I have. For example, if I were the president of Taiwan, Ma Ying-Jiu, then when I return it to you, I became a mayor, and I'm no longer the president" (MWP, Interview, 20 September 2014). Further, in the case of MEI, her way to self-historicise PRC and Taiwan is premised on the Chinese language: "Even though they are Taiwanese, they are PRC Chinese people. In my eyes, if they speak Chinese language, then they should belong to PRC Chinese people" (MEI, Interview, 20 September 2014). Finally, as for LRJ, he self-historicised the relationship between PRC and Taiwan through a time-compressed longitudinal perspective by articulating Taiwan's separation as a historical vignette. To him, China's sovereignty is immensely vast in an indissoluble continuum. He described China's immense sovereignty in this way: "If the General Consulate in Chiang Mai asked me this question, I would say I'm Chinese. The future of Taiwan in history is a small episode, and Yunnanese from Yunnan to Burma and then to Thailand is also a small episode in history" (LRJ, Interview, 30 September 2014).

10.13 Transcendental Pragmatism as Mental Reconciliation

The art of negotiating patriotism between multiple sovereignties or entities is no easy feat. Situated in-between these powers, it requires a transcendental approach that is driven by pragmatic rationalisation. First, LRJ conceptualised Taiwan as a province of PRC with the strength in preservation of Chinese culture. Additionally, in his narration, LRJ employed the word "slippers," which refers to a light sandal, typically made of plastic or rubber, with a thong between the big and second toe. Practically, the tight and single thong makes the foot hurt while walking. Metaphorically, it means that a rigid and uni-linear fixation will only cause harm and hurt to oneself. Hence, according to him, one should adopt a transcendental pragmatic paradigm, which serves non-attachment and allows for manoeuvring in accumulating opportunities from multiple places. Going further in this case, LRJ's pragmatic rationale lies in the strength and power of PRC with reference to the outstanding record of national governance in the global arena within a short timeframe. PRC's excellence in recent history becomes a "fortress" for overseas Chinese against any imminent political threats. Finally, the transcendental pragmatism is being brought back to the local context where the interests of the Thai Yunnanese are situated. To LRJ, given their historical marginality, Thai Yunnanese should capitalise on resources from the PRC and even the States to secure a good future.

In another case, MWP's transcendental pragmatism is predicated on the comparative advantage he can leverage on PRC of its military capacity for protection and sheer massive population size for economic prospects. Furthermore, JZB's perspective towards the unification of Taiwan with PRC is another example of transcendental pragmatism. He has hoped to see the unification between Taiwan and PRC from a business concern. Instead of espousing an intellectual or political rationale, JZB emphasised ease in business dealings. Plainly speaking, JZB's premise of seeking unification of Taiwan and PRC stems from the pragmatic benefits found in the transnationalisation of the four currencies. As a form of mental reconciliation in the transnationalisation of Chinese, the various examples mentioned above have demonstrated transcendental pragmatism as the determining factor of analysis.

In summary, this section provides a comprehensive understanding of the negotiation of identity as a practice of skilful performativity by the Thai Yunnanese transnational entrepreneurs in Chiang Mai. In reiteration, identities are crafted with skilful performativity, involving mental, bodily, and object elements in situated multidimensionalities, and are also closely related to economic performance and sustainability. As a negotiating practice, Thai Yunnanese transnational entrepreneurs accentuate and attenuate identities reflexively for pragmatic economic reasons in view of the rise of the PRC. Culturally, glamorisation and customisation are varied strategies for remaking identities too. Hence, this chapter articulates the multiplicity of identities performed by the Thai Yunnanese against different situational contexts.

10.14 Discussion and Conclusion

This discussion and conclusion section engages critical theoretical debates and purports significant contributions to the understanding of identity constructions within the context of Thai Yunnanese transnational entrepreneurs in Chiang Mai. Firstly, this section argues for departure from assimilation and dichotomous explanations of Chinese in Thailand. It also re-introduces polychotomisation and multicomplementarisation in relationality as conceptualisation. Finally, it seeks to show that a new wave of ethnic entrepreneurialism has begun.

10.14.1 Departure from Assimilation and Dichotomous Explanations

Based on the assimilation theory, scholars have postulated that Chinese in Thailand will become indistinguishable in Thai society (Skinner 1963) and will be assimilated due to a lack of formal Chinese education (Amyot 1972; Ossapan 1979). Other scholars such as Wongswadiwat (1973) and Yuesin (1984) have also concurred with these generalised views of assimilation. From the state's perspective, in most parts of modern history, the Chinese are mostly favourably treated and are adopted through pro-assimilationist policy such as integrative national education as early as

1898 (Skinner 1957). However, this research argues against assimilationist model of explanation. Congruent with Tong's (2010, p. 42) argument, this research unravels bilingualism and bicultural education as distinctive ethnic markers. Shu-min (2007, p. 182) demonstrated that Thai Yunnanese have exhibited a determination in retaining Chinese cultural heritage. Furthermore, Shu-min (2007, p. 185) believed that the Thai Yunnanese have successfully converted their Chinese ethnicity into selected cultural markers, which have facilitated them to compete in Thai society and claim distinct social classes associated with professional and middle-class niches (Tong 2010, p. 49). In a similar vein of thought, this research has surfaced other cultural markers such as ethnic ascription, customised celebration of festivities and National Days, effective aestheticisation, professionalism, and socialisation. At the same time, this research further delineates the specificity, complexity, and multiplicity of identity constructions, which are located in interstitial dimensions. It should be noted that non-assimilation does not mean anti-assimilation. This is because Thai Yunnanese transnational entrepreneurs have also accentuated their Thai identity and cultural traits in gaining access to new unfamiliar grounds.

Performance can be construed as a "broad spectrum" and "continuum" of human actions that are framed, enacted, presented, highlighted, and displayed (Schechner 2013, p. 2). At the heart of performativity, human actions are "liveness" where practices, events, and behaviours are studied (Schechner 2013, p. 2). Hence, within this broad spectrum of continuum, "lively" identity constructions in this research are conceptualised as negotiating practice where performativity becomes an analytic tool for scrutiny. Such conceptualisation takes a departure from dichotomistic predomination. Hew's (2011) thesis on Chinese Muslim identities in post-new order explores the difference between public performance and everyday life practice. "Hybridity" has been used to describe and analyse cultural mixing in a broader sense. In particular, "intentional hybridity" and "organic hybridity" have been used to distinguish hybridity as symbolic strategic and everyday practice. Organic hybridity refers to unconscious exchanges and everyday adaption of cultures, meanwhile intentional hybridity is a conscious effort to create a double consciousness of one's identity (Bakhtin 1993; Werbner 1997). However, "hybridity" is problematised as presupposing essentialist authentic form (Asad 1993; Yeoh 2009). Further, in the case of Thai Yunnanese transnational entrepreneurs' identity constructions, every practice is a performance and every performance is a practice. As a case in point, YSQ made a poignant remark with regards to practice as performativity in everyday life: "You are who you are. You can act like Guan Yu in the opera, but you are still who you are (你可以演關公在戲，可是你還是你自己)" (YSQ, Interview, 2 September 2014). Guan Yu is one of the best-known historical figures throughout East Asia. In Chinese opera, his character is often portrayed as the epitome of loyalty and righteousness. Essentially, the context of YSQ's idiomatic expression alludes to a rejection of false pretention, but it is also an advocacy of self-performativity as skilful politicking practice. Hence, employing performativity as an analytic tool removes the artificial dichotomy between "intentionality" and "organic-ality," which can be intricately and inextricably intertwined.

10.14.2 Polychotomisation and Multi-Complementarisation in Relationality

From a circumstantial or situational ethnicity perspective, a man might be tribal in some situations and a wageworker in others (Mitchell 1966). Proponents of situational ethnicity such as Eidheim (1971) introduced dichotomisation and complementarisation to describe fundamental ethnic processes. Dichotomisation refers to the articulation of ethnic relationships through mutual negation, while complementarisation gives an expression of ethnic relationships within a shared language for culturally distinctive groups (Eriksen 2010, p. 280). However, the Thai Yunnanese transnational entrepreneurs in Chiang Mai have multiple sets of interstitial identities in dichotomisation and complementarisation in simultaneity. In other words, within dichotomisation, they are simultaneously distinguishing themselves as Thai, Thai Chinese, and Thai Yunnanese, meanwhile, within complementarisation, they are constructing a shared language with PRC Chinese and PRC Yunnanese in transnationalisation and translocalisation. As such, this chapter demonstrates polychotomisation and multi-complementarisation in complex, multiple, situational circumstances in simultaneity. However, despite its flexibility, situational or circumstantial approach fails to clarify the persistence of ethnicity where there is a certain idea of common identity and loyalty (Gladney 2003, pp. 47–48). In response, Gladney (2003, p. 48) proposed that social relations of power become the focus of attention while adhering to the enduring and mutable nature of symbolic and instrumental aspects of ethnicity. In other words, this chapter argues for a relationality in the processes of identity construction within polychotomisation and multi-complementarisation. For instance, living with her father, SZW grew up with the political ideology of PRC communism as destructive and merciless infused into her. Additionally, her marital relationship with her Korean husband propels her socially to distinguish herself from Thai factory workers. These relationships give rise to polychotomisation in terms of distantiation away from PRC Chinese communism and Thai low-skilled workers' status. On the other hand, the same relationships multi-complementarise her ethnic affiliation with Taiwan as well as with educated Thai middle-upper class respectively and simultaneously. Hence, this chapter articulates a relational approach to polychotomisation and multi-complementarisation in complex, multiple, and situated political circumstances in simultaneity.

10.14.3 Neo-Ethnic Entrepreneurialism

In a classical sense, ethnic entrepreneurship refers to the utilisation of "a set of connections and regular patterns of interaction among people sharing common national background or migration experiences" (Waldinger et al. 1990, p. 3). Ethnic entrepreneurs refer to those who can "succeed in being both, say Indian and African, and switching strategically between these identifications" (Eriksen 2010, p. 279).

However, this chapter problematises its definition in a narrow sense. In the case of Thai Yunnanese transnational entrepreneurs, they go beyond merely sharing common national background or migration experiences to gain success. As the previous sub-section shows, they employ complex polychotomisation and multi-complementarisation in relationality to gain socio-economic advantage. Furthermore, based on this research, respondents possess the entrepreneurial spirit of creating an ethnic identity that reflects a new image. This image is shed on its past stigma while embracing a new professional and successful social status. Apart from YSQ's vehement attempt to deconstruct and reconstruct Thai Yunnanese identity, LRJ's case presents another example to illustrate this creative notion of neo-ethnic entrepreneurialism. LRJ was realistic in describing the lower social status of Thai Yunnanese as compared to other successful Thai Chinese compatriots. However, he did not leave it lying down, he made every effort to create a new image and appealed to his fellow Thai Yunnanese to do likewise. As a successful transnational entrepreneur in the realm of economic business, LRJ has unwittingly joined a new wave of entrepreneurs in the realm of ethnic identity creation. He has also positioned himself as a pacesetter in neo-ethnic entrepreneurialism for other Thai Yunnanese to emulate. Furthermore, from an institutional perspective, the Chiang Mai Yunnan Association provides a mediating platform to facilitate neo-ethnic entrepreneurialism of identity, albeit through a sensitive and gradual negotiation in-between conflicting ideologies.

In summary, this chapter has fulfilled its purpose to explain how identity has been remade as everyday practice and performance through negotiating multiple, interstitial, and relational identities with polychotomisation and multi-complementarisation in neo-ethnic entrepreneurialism. Such an articulation takes a departure from the delimiting nature of assimilation explanation and dichotomous perspective. To reiterate, this chapter contributes to the reification of this volume's value of advocating a transnational paradigm in analysing the critical practices of—in this case—Thai Yunnanese transnational entrepreneurs. Here, transnationality, amidst other prevailing circumstances, dynamically precipitates, innovates, and elucidates the remaking of identities.

References

Amyot J (1972) The Chinese and national integration in Southeast Asia. Institute of Asian Studies Monograph. Chulalongkorn University, Bangkok

Asad T (1993) Genealogies of religion: discipline and reasons of power in Christianity and Islam. John Hopkins University Press, Baltimore

Bakhtin MM (1993) In: Holquist M, Liapunov V (eds) Toward a philosophy of the act (trans: Liapunov V). University of Texas Press, Austin

Biernacki R (2007) Practice. In: Ritzer G (ed) The blackwell encyclopaedia of sociology. Blackwell, Oxford, pp 3607–3609

Boyle P, Halfacree KH, Robinson V (1998) Exploring contemporary migration. Longman, Singapore

Brickell K, Datta A (eds) (2011) Translocal geographies: spaces, places, connections. Ashgate, Burlington
Chang WC (2001) From war refugees to immigrants: the case of the KMT Yunnanese Chinese in Northern Thailand. The International Migration Review 35(4):1086–1105
Chang WC (2004) Guanxi and regulation in networks: The Yunnanese jade trade between Burma and Thailand, 1962-88. Journal of Southeast Asian Studies 35(3):479–501
Chao X (2014) Message from consul general. The consulate general of the People's Republic of China in Chiang Mai, 13 February 2014. http://chiangmai.chineseconsulate.org/eng/news/t1225634.html. Accessed 1 Apr 2015
Eidheim H (1971) Aspects of lappish minority situation. Universitetsforlaget, Oslo
Eriksen TH (2010) Small places, large issues: an introduction to social and cultural anthropology, 3rd edn. Palgrave Macmillan, New York
Gladney D (2003) Ethnic identity in China: the making of a Muslim Minority nationality. Wadsworth/Thomas Learning, California, USA
Guangjin C, Wa Z (2012) Strategic relationship enhanced. China Daily. https://www.chinadaily.com.cn/business/2012-04/18/content_15076281.htm
Hew WW (2011) Negotiating ethnicity and religiosity: Chinese Muslim identities in post-new order Indonesia. Dissertation, Department of Political and Social Change, School of International, Political and Strategic Studies, College of Asia and the Pacific of the Australian National University
Hill AM (1998) Merchants and migrants: ethnicity and trade among Yunnanese Chinese in Southeast Asia. Yale SEA studies Monograph 47, Yale University Southeast Asia, New Haven
Mitchell JC (1966) Theoretical orientations in African urban studies. In: Banton M (ed) The social anthropology of complex societies. Tavistock, London, pp 37–68
Ossapan P (1979) The Chinese in Thailand. In: Dhiravegin L (ed) Reader on minorities in Thailand. Phraephittaya, Bangkok
Reckwitz A (2002) Towards a theory of social practices: A development in culturalist theorizing. European Journal of Social Theory 5(2):243–263
Ritzer G (2008) Sociological theory, 7th edn. The McGraw-Hill, New York
Schechner R (2013) Performance studies: an introduction, 3rd edn. Routledge, London and New York
Shu-min H (2007) A Chinese diasporic community in Northern Thailand: contested political loyalty and shifting ethnicity. In: Tan CB (ed) Chinese transnational networks. Routledge, London and New York, pp 172–190
Skinner GW (1957) Chinese society in Thailand. Oxford University Press, London
Skinner GW (1963) The Thailand Chinese: assimilation in a changing society. Lecture presented at the Thai Council of Asia Society
Tong CK (ed) (2010) Identity and ethnic relations in Southeast Asia: racializing Chineseness. Springer, Singapore
Tong CK (2014) Chinese business: rethinking Guanxi and trust in Chinese business networks. Springer, Singapore
Waldinger R, Aldrich H, Ward R (1990) Ethnic entrepreneurs: immigrant business in industrial societies. Sage Publications, Newbury Park, London and New Delhi
Wang L (2006) Hui Yunnanese migratory history in relation to the Han Yunnanese and ethnic resurgence in Northern Thailand. Southeast Asian Studies 44(3):337–358
Werbner P (1997) Introduction: The dialectics of cultural hybridity. In: Werbner P, Madood T (eds) Debating cultural hybridity: multi-cultural identities and the politics of anti-racism. Zed Books, London, pp 1–28
Wongswadiwat J (1973) The psychological assimilation of Chinese university students in Thailand. Dissertation, Urbana-Champaign, University of Illinois
Yan X (2001) The rise of China in Chinese eyes. Journal of Contemporary China 10(26):33–39
Yeoh SG (2009) For/against Hybridity: religious entrepreneurships in a Roman Catholic pilgrimage shrine in Malaysia. Asian Journal of Social Science 37(1):7–28

Yuesin S (1984) Communication behavior and Thai cultural assimilation: a study of Chinese in the Chiang Mai municipal area. Dissertation, Department of Public Relations, Chulalongkorn University

Kian Cheng Lee holds a Ph.D. (Social Sciences), M.A. (Southeast Asian Studies), M.Th. (Asian Christianity), M.Div., M.A. (Biblical Studies), and B.Sc. (Physics). He is an Assistant Professor in the Faculty of Political Science and Public Administration while being affiliated with the School of Public Policy and Faculty of Social Sciences at Chiang Mai University. He is also a member of the board of the Asian Pastoral Institute Ltd. in Singapore. Kian Cheng's research interests concern Chinese business practices, transnational entrepreneurship, politics, cultural and citizen diplomacy, and Thai religious networks. He can be contacted at kiancheng.lee@cmu.ac.th.

Chapter 11
Chinese Reactions to Indonesia's Nationality Law: An Analysis of Chinese Language Newspapers

Matsumura Toshio

Abstract This study discusses the issue of Chinese nationality in Indonesia dating back to the creation of the Indonesian state in the 1950s. The Chinese previously had a transnational identity and formed an extensive network covering diverse regions, including China and neighbouring colonies, frequently travelling back and forth between them. However, Indonesia's newly established nation-state regarded the transnational framework as a 'problem' in the process of identifying the members of the Indonesian state. This study aims to understand how the Chinese attempted to deal with this new situation during this period. During the decolonisation process, Indonesia enacted a new nationality law in 1958 that aimed to clarify the legal status of the Chinese people in the country. Since the beginning of the twentieth century, the Chinese had had dual legal status as the nationals of China and subjects of the Dutch East Indies. However, the implementation of the new law required the Chinese people to choose between the Chinese and Indonesian nationalities. Using the Chinese-language newspapers published in Indonesia as the main source of information, this study analyses the reactions of the Chinese community in Indonesia to the implementation of the nationality law.

Keywords Chinese · Indonesia · Nationality · Chinese-language newspapers · Nationalism in the 1950s · Nationalisation of the economy

11.1 Introduction

This study aims to understand how the Chinese people responded to the newly introduced system of nationality in Indonesia in the 1950s. The origins of the Chinese in the present-day Indonesia can be traced back to a long line of immigrants from Southern China, dating from before Dutch colonisation. Those who localised among

---

M. Toshio (✉)
Graduate School of Humanities, Osaka University, Osaka, Japan
e-mail: tmatsu.hmt@osaka-u.ac.jp

H. M. Y. Ho and Y. Chan (eds.), *Transnational Southeast Asia*, Asia in Transition 29,
https://doi.org/10.1007/978-981-96-3608-2_11

them were called *Peranakan* and developed their businesses (for example, tax collection contracts and opium trade) in a way that was adapted to the political, economic, and social systems of the Dutch colonial backdrop. The second half of the nineteenth century also saw a huge wave of immigration. Unlike the *Peranakan*, they were labourers who came to the Dutch East Indies at that time in search of a livelihood due to factors such as rapid population growth in Southern China, opening of the Chinese ports after the Opium War, and disturbances in the country caused by the Taiping revolt. They were distinguished from the *Peranakan* and called *Sinkeh,* meaning newcomers.

Even in the twentieth century, the presence of the Chinese (both local-born Chinese or *Peranakan* and newcomers or *Sinkeh*) continued to be a problem for the Dutch rule, and in the late nineteenth and early twentieth centuries, the Qing dynasty tried to incorporate the overseas Chinese. In the meanwhile, in the early twentieth century, the Chinese in the Dutch East Indies responded to this, and they founded the *Tiong Hoa Hwee Koan* 中華会館to promote their identity as Chinese. To a certain extent, this movement influenced the Qing dynasty's policy. Consequently, it established a nationality law based on the *jus sanguinis* or bloodline principle (1908), making the Chinese overseas part of the Qing dynasty. Meanwhile, the Dutch also established the principle of granting status as Dutch subjects or *Nederlandsch onderdaanschap* to the Chinese, especially those born in the Dutch East Indies, and established a Dutch-style education for them. Thus, there was a dual attribution of the Chinese to their place of residence and to China (Sadayoshi 2006, pp. 9–12).

Meanwhile, the system of governance in the Dutch East Indies from the late nineteenth century onwards was based on the division of the population. The Europeans were placed at the top, with the local population, the object of domination, below them, and the Chinese and other foreign residents of Asian origin (including Arabs and Indians) given unique status as intermediaries between the two and were placed below the European division.

According to Sadayoshi, the status of the Chinese at that time was not limited to China and the place of their residence, but some acquired rights as Japanese subjects by acquiring Taiwanese status because Taiwan was a Japanese colony at the time, and the status of its inhabitants was that of Imperial Japanese subjects. Others acquired status as subjects of the British colony. He emphasises that the Chinese at that time had 'multiple nationalities' rather than 'dual nationality'. Sadayoshi and Kudo also emphasise that the reality of the Chinese at the time was not so much that they mainly belonged to the Dutch territory, but that they had established transnational business networks and did not necessarily live exclusively in the Dutch territories (Kudo 2023; Sadayoshi 2006).

Kudo (2023) reveals that within this system of resident classification, some Chinese aiming for European citizenship sought to raise their status within the Dutch East Indies to the same status as Europeans by acquiring Japanese citizenship. How, then, was this possible?

This was premised on the enactment of the Japanese Act (1899) as a precedent. Prior to the enactment of this law, the Japanese living in the Dutch East Indies had the same status as the Chinese. However, as a result of the discussions between Japan and the Netherlands as part of the revision of unequal treaties with the Western countries, the Japanese in the Dutch colonies, including Indonesia, were given the same status as the Europeans. This upgradation of the Japanese also affected the Chinese, and following the example of Japan, modernisation based on the Western model emerged as an important issue among the Chinese in the Dutch East Indies (Sadayoshi 2006, pp. 13–14).

Under these circumstances, there have been cases of individuals aiming for and successfully obtaining European citizenship at the individual level. According to Kudo, this was the path to obtaining 'Japanese' status through gaining Taiwanese nationality (or originally being a Taiwanese national). As a concrete example of this, she discovered Kwik Djoen Eng 郭春秧. He was born in Fujian in 1859 and went to Semarang in Central Java in 1876, where he worked in the trading business from 1894 onwards. He entered the trade business between Taiwan and Java, and established relations with the Japanese companies and government, while also cooperating with the Dutch companies. In addition to his Chinese and Dutch subject nationalities, he also used his Taiwanese and Japanese nationalities, as well as his European equivalent legal status. He also acquired British nationality (British Empire subject status), which enabled him to move his business to Hong Kong in the 1920s. However, this multiple nationality situation became subject to restrictions in the 1920s.

11.2 Localisation of the Chinese

These points may force a reconsideration of the conventional perception of Chinese migration to a foreign country, and of assimilation and localisation into the society as irreversible changes. It can also be understood as an illustration of the recent 're-migration' of the Chinese as a more universal model.

According to Yamashita (2023, pp. 35–38), re-migration refers to the phenomenon in which the Chinese who have moved abroad from China move again from their place of residence to a third region: the migration of the Chinese refugees from Vietnam after the 1975 reunification of Vietnam to the USA, or to France, the former colonial suzerain of Vietnam. Considering each concrete case of re-migration, it is clear that the idea of Chinese migration as a one-directional, one-off, and irreversible move is inconsistent with the facts.

Not considering their migration as a 'one-time only move' but as a circulation between various regions is important. This point is important because it forces a reconsideration of the fundamental view of studies on the overseas Chinese, which has long been concerned with the localisation of the Chinese people. Chinese migration has been understood as a unidirectional migration and settlement from China to other parts of the world, or localisation. In terms of Southeast Asia, this concept is often expressed as '*xia nan yang* 下南洋 (going down to the Southern Seas)' or change from '*luo ye gui gen* 落葉帰根 (leaves fall in the country of residence, but roots return to China) into *luo di sheng gen* 落地生根 (settling in the country of residence and taking root there)'.

In addition, Liu and Liao (2006) argue that previous studies on the overseas Chinese have been too heavily weighted towards belonging to the nation-state. Whether emphasising the aspect of being overseas Chinese 華僑 or the post-war aspect of being 'not overseas Chinese華僑 but Chinese華人', there has been too much focus on belonging and loyalty to the nation-state, and too much emphasis on how the Chinese people behave as members of the nation-state. Indeed, it is not difficult to imagine that in the twentieth century, the role of the nation-state soared, and the issue of belonging to it was recognised by people as a significant problem. However, other aspects such as belonging to multiple regions other than the nation-state, such as belonging to the locality of Quemoy Island 金門島 (Kawashima 2011), or to the West Kalimantan region rather than Indonesia (Matsumura 2017), a sense of closeness to multiple regions and actual movement and migration, are not as highlighted above. It is also possible that feelings of familiarity with multiple regions, actual movement, and migration (or circulation) were discarded in the concentration on perceptions of belonging to the nation-state.

11.3 Nation-State and Nationality

The abovementioned situation began to change significantly, synchronising with the global shift in the situation regarding nationality. Previous studies (e.g. Torpey (1999)) focuses on the birth of passport system) trace the origins of this concept of nationality as membership of a state, particularly in the West, and it was brought to each colony through colonial rule.

Regarding post-independence Indonesia, defining the shape of the state and membership was regarded as important. The fact that ethnic division was constructed during the Dutch period is worthy of special mention. This placed the Chinese in a special status that is discerned from indigenous inhabitants. This social structure was also observed after Indonesia's independence, creating a social perception of the Chinese as different from that of the local population. Due to this socioeconomic background and nationality issue, the status of the Chinese was the first issue to be raised after independence.

Nevertheless, at the beginning of its independence, Indonesia applied the *jus soli* or place of birth principle, thus presenting the concept of Indonesian citizenship to encompass those living in its territory. This allowed the Chinese to acquire Indonesian citizenship if they did not voluntarily renounce it.[1] Some people renounced their Indonesian citizenship, which later led to controversy regarding their nationality.

However, a provision that would later become problematic also emerged. It was a distinction between genuine Indonesians or *orang-orang bangsa Indonesia asli* and others who were permitted by law to become Indonesian citizens or *orang-orang bangsa lain yang disyahkan dengan undang-undang sebagai warganegara*. This distinction led to a debate in the 1950s on the category in which the Chinese were included.

Another important factor was the social context of the 1950s, when the status of the Chinese was defined by law, that is, the economic superiority of the Chinese over the locals. In the context of the 'nationalisation of the economy', policies favouring 'national protagonist' local people's enterprises were adopted; in the meanwhile, various restrictions were placed on those run by the Chinese nationals alongside the European-owned enterprises. The issue of national membership—who to accept as citizens—and the 'nationalisation of the economy', which was an important political agenda at the time, were inseparable.

An important development in the nationality issue occurred in 1955. In April 1955, during the Asian-African Conference, China and Indonesia confirmed their policy of abolishing the previous dual nationality of the Chinese nationals residing in Indonesia. This resulted in the enactment of the Nationality Law of 1958 and its implementation in 1960, which set clear guidelines on the issue of nationality, specifically forcing the Chinese nationals to choose between the Chinese or Indonesian nationalities. The climax of the nationalisation of the economy, Presidential Decree No. 10 or *Peraturan Presiden No.10*, occurred at the same time as this law came into effect. It prohibited foreign nationals (most of whom were Chinese) from engaging in commercial activities in rural areas, thereby depriving many Chinese of their livelihoods. The decree came into force in January 1960. In the aforementioned extremely severe situation for the Chinese people, a debate on nationality within the Chinese community was initiated.

11.4 Main Sources

The main source material used in this study was the Chinese-language newspapers published in various parts of Indonesia. The readers of Chinese-language newspapers were Chinese people who understood the language. It was the Chinese-language newspapers of the time that led such people to think and presented them with an image of what the Chinese community should be like. Therefore, reading Chinese-language newspapers was the best way to understand what was being discussed in the Chinese community at the time in relation to nationality and what their interests were.

The newspaper media, in which the Chinese became central, is said to have played a central role in Indonesia's media history in general (Yamamoto 2008). They began publishing various newspapers at the beginning of the twentieth century.

*Da Gong Sheng Bao*大公商報, used in this chapter, was first published in 1908 and was the earliest newspaper to be published. It was banned during the Japanese military rule but was revived in the 1950s and gained wide readership as a pro-PRC newspaper (Li 2003, p. 353; 357). *Su Men Da La Min Bao* 蘇門答臘民報 was first published in 1914 and widely read by the Chinese, especially in Medan, North Sumatra (Li 2003, pp. 353–354; 357). *Xin Bao* 新報was the most influential newspaper in the Chinese community. It was launched in 1921 as a sister edition to the Malay-language edition and was first published in 1910. It advocated patriotism for the Chinese nation, respect for the Chinese culture, and opposition to colonialism, and had a circulation of about 5000 in the 1930s. After Indonesia gained its independence, it supported the People's Republic of China. *Sheng Huo Bao* 生活報was first published in 1945 Jakarta by patriotic Chinese leaders and carried articles relating to the international politics, and supporting the Indonesian nation's war for independence. It supported the People's Republic of China after the war. Its circulation rose rapidly from 30,000 in 1958 to 50,000 in 1960 (Li 2003, pp. 358–361).

In addition to these, *Yin Hua Jing Ji* 印華経済 and *Mei Ri Dian Xun* 毎日電訊, both published in Jakarta, are quoted in this study.

11.5 Analysis of the Articles Published in the Chinese Newspapers

From this section onwards, I examine the Chinese people's views on nationality during the abovementioned troubled period using the contemporary Chinese-language newspapers, as described above. First, I introduce an article that appeared in the editorial in the daily *Xin Bao* (30 May 1959) entitled *tan san hao*談三好 meaning 'talking on three good choices'.

> PRC's Policy is 'three good choices三好'. It is good if the overseas Chinese adopt the nationality of their country of residence, it is equally good if the overseas Chinese withhold their Chinese nationality. It is also good if the overseas Chinese return to China and participate in the construction of their homeland. China supports and understands the Chinese nationals acquiring Indonesian citizenship. It has also been indicated that they are welcome to remain Chinese citizens. When the Chinese make their choice of nationality, they first need to fully understand the spirit of this Chinese 'three good choices' policy. Many would claim, 'It is true that all three are good', but it is impossible to select all three. You[2] have to choose one of these three. You must decide on yourself. Of course, choosing your own nationality is not the same as deciding whether or not to drink a cup of coffee. Therefore, thinking over this issue is necessary. When considering the nationality issue, one must account for the actual situation of the individual and the family before making the final decision. In this process, it is important not to bring up *daguo sixiang* 大国思想 or Great Patriotic Idea when choosing nationality because it can lead people to wrong choice. One should be wary of the arrogant idea of regarding oneself as a national of a great state and others as citizens of a small state.

Particular mention is made of *daguo sixiang*, perhaps because such a tendency was observed among the Chinese. We also note that the reason for choosing the Indonesian nationality was to stay in the country for a longer period. Peranakan was born in Indonesia and grew up there; thus, they did not want to be treated as foreigners, which could be a reason influencing their choice of nationality. Apart from this, this article emphasises rights and obligations as members of a nation, as seen in the following quote:

> Where there are rights, there are obligations. When one becomes a member of a nation, one naturally enjoys the rights that the state provides, but one inevitably has obligations as well. If you retain your Chinese nationality, you must comply with the policies of the Chinese government. If you make such a choice, you must not complain about the politics

> of the country in which you reside. You must respect the customs and manners of the people of the country in which you live and try to coexist with them. If they choose to become Indonesian citizens, they must swear allegiance to the nation, enjoy their rights as citizens, and fulfil their obligations. These obligations include the obligation to have their children receive Indonesian education and to perform military service.

Here, it is particularly emphasised that there is only one nationality per person and that while a half-national way of being was tolerated, this is not acceptable today. This indicates the transnational nature of nationality and belonging in earlier times.

11.5.1 In Relation to Military Service Issues

The issue of military service has often been discussed in the Chinese-language newspapers. At that time, the Indonesian military was increasing its build-up to put down a local rebellion in Sumatra.[3] In this context, a law on emergency military service, Law No. 66 of 1958 on Military Service (*Undang-Undang No. 66 Tahun* 1958 *tentang Wajib Militer*) was issued. This law stipulated that the Indonesian nationals who refused to register for military service at a certain time or who did not participate in military service for any reason shall be punished with imprisonment or fines (Article 63). For the Chinese who had not settled on the nationality issue, the question was raised as to whether they had refused to serve in the military.

On this theme, *Su Men Da La Min Bao* (19 January) reported that foreign-born 'overseas Chinese' who are not Indonesian citizens are not required to register for military service. This indicates that there was a sense of doubt and uneasiness among the Chinese community, at least in Medan, regarding whether they had to serve in the military. Meanwhile, the same newspaper (27 January) stated that 'undocumented persons' must also register for this emergency military service because they are considered Indonesian nationals, except for those who willingly renounced their Indonesian citizenship in the early 1950s.

Mei Ri Dian Xun (29 May) and *Xin Bao* (27 May) reported that the initial training of soldiers in West Java had begun. It also introduced the Army's statement: 'In order to solve the domestic security problem, the army has decided to increase its forces in a short period of time. This is aimed at solidifying territorial integrity, restoring the normal state of the government, and rebuilding the economy'.

Sheng Huo Bao (27 May) published an article, which explained the drafting process step-by-step. The reason for such detailed articles was probably due to the high level of interest among the Chinese in how the conscription system would be introduced.

Da Gong Sheng Bao (11 March) said that Baperki[4] supported the emergency military service law. According to this article, a member of Baperki in Surabaya said, 'The participation of Chinese in military service will dispel the misunderstandings held by a handful of people about Chinese, namely that Chinese who are Indonesian citizens enjoy only rights but do not fulfil their duties as citizens'. *Da Gong Sheng Bao* (4 March) reported that healthy Indonesian nationals were required to serve two years of military service and that failure to register for such services would result in three months in prison or a fine. Penalties for failing to fulfil this 'obligation as an Indonesian citizen' were also reported in *Sheng Huo Bao* (3 March) and *Xin Bao* (3 March). Emergency military service was an opportunity for the Chinese to realise that, to obtain Indonesian citizenship, they must naturally fulfil their obligations and enjoy their rights.

11.5.2 The Case of Riau Islands (Difficulties in Becoming an Indonesian National)

In Java, restrictions on the Chinese are more strictly applied, and in West Java, there have been forced evictions of Chinese people due to the implementation of the Presidential Decree No. 10. However, people's awareness of nationality selection is lower outside Java, and restrictions may not have been applied as strictly as in Java. In my own research in West Kalimantan, I found no evidence that the Presidential Decree No. 10 was applied as strictly as in Java (Matsumura 2017). This section examines the case of Riau, which is close to Singapore and has been described in newspapers.

Sheng Huo Bao (10 June) reported as follows:

> 'Few people informed the local court in Riau of their intention to renounce their Indonesian nationality between 27 December 1949 and 27 December 1951. The reason for this may be that there was a low level of awareness among the Chinese about their nationality, and they had doubts about Indonesia.
>
> In particular, people living on the small island of Riau are poor and do not have a concept of nationality. In response, Baperki established a branch in Tanjung Pinang, Riau. The head of the branch went to great lengths to explain nationality issues in the small island villages. Since then, the residents have become more aware of their nationalities. Baperki helped them prepare documents on their place of birth, marriage certificates, and other documents needed to obtain Indonesian citizenship. Currently, the Chinese in Riau are required to have a birth certificate and a certificate of renunciations of Chinese nationality from the Chinese side to obtain Indonesian nationality. Those who were born in Riau and did not voluntarily renounce their Indonesian nationality are also considered "overseas Chinese." If a woman born in China or elsewhere marries an Indonesian national, then even if she has a marriage certificate, she can only have the same Indonesian nationality as her husband after obtaining a certificate of renunciation of Chinese nationality from the Chinese consulate'.

What is clear from this case is that the person must fulfil strict conditions to obtain Indonesian nationality. First, the Chinese must issue a document proving that the person has renounced his or her Chinese nationality, and then the person must go to the Indonesian District Court with that document. The conditions for naturalisation were also very strict; compared with the early 1950s, all those who did not renounce their Indonesian nationality (those who were passively recognised as Indonesian citizens by *jus soli*) were also considered foreigners.

In addition, the relative lack of awareness among the Chinese people regarding nationality is also described. The entire section here highlights the nation-state's attitude of not tolerating transnational ways of being. Tensions between transnational and national logic can be seen. The logic of nationalism is that Indonesian nationality can only be acquired through the act of renouncing one's nationality, which is a symbol of ties to other states as proof of political allegiance and a sense of belonging to Indonesia. This reflects a give-and-take situation.

11.5.3 Dispute Over Nationality Provision in Semarang

Xin Bao (17 January) reported that the District Court in Semarang had begun accepting applications for reinstatement. *Da Gong Sheng Bao* (20 March) stated that there were many people in Semarang who wanted to obtain Indonesian nationality; statistics showed that between February and March, there were 276 applicants, the largest number of whom were Chinese, with approximately 250. Under these circumstances, an incident stirred the Semarang Chinese, who were highly aware of the need to obtain Indonesian nationality. This situation was triggered by a notice issued by the Semarang District Court. *Sheng Huo Bao* (2 June) reported the following:

> The notice of the District Court of Semarang was as follows: 'With regard to children born after the transfer of power from the Netherlands to Indonesia on 27 December, 1949, they are Indonesian nationals. According to the provisions of the 1958 Nationality Law, children born after the transfer of power on 29 December, 1949 are automatically Indonesian nationals. The nationality of the parents is dependent on children born before 27 December, 1949. As for children born after the transfer of power, if they want to choose their nationality, they have the right to choose their nationality after they reach the age of 21, in accordance with the provisions of Articles 2 and 13 of the Nationality Law No. 62 of 1958'.

This statement appeared to have caused controversy, and many articles on this topic have been published since then. *Da Gong Sheng Bao* (4 June) reported, 'On 25 May, the Semarang District Court issued a statement that children born after the transfer of power automatically became Indonesian nationals and were considered Indonesian nationals unless their parents or guardians applied for their de-registration'. However, the situation changed after June. On 8 June, the same newspaper reported the following:

> The statement issued by the District Court of Semarang, namely the notice that children born after the transfer of power would become Indonesian nationals, stirred anxiety among overseas Chinese in Central Java; on 4 June, the District Court abruptly corrected this, stating that the previous announcement had been a mistake. It stated that children of overseas Chinese born after the transfer of power from the Netherlands to Indonesia would also inherit their fathers' nationalities.

Initially, the court applied the place of birth principle or *jus soli* to the children born after 27 December when political power was transferred to Indonesia and held that children were automatically recognised as Indonesian nationals even if their fathers were Chinese nationals. However, in practice, the patrilineal lineage principle or *jus sanguinis* was consistently applied, and all children born even after the transfer of power to Indonesia were also considered Chinese nationals if their fathers were Chinese nationals, and the children were not automatically granted Indonesian nationality. This article clearly shows that the Chinese in Surabaya had high hopes for the Indonesian interpretation of *jus soli*, which was betrayed.

Sheng Huo Bao (10 June) subsequently reported:

> Regarding the announcement by the Semarang District Court in the case, many overseas Chinese in Semarang gathered at the District Court to demand their children's 'Certificate of Deregistration of Chinese Nationality' and to have their new nationality examined again. After 10 days, the district court told them to go to the Immigration Department to obtain a 'certificate of de-registration'. They also approached the tax bureau to have the children's expatriate tax removed. However, the Immigration Department disagreed and refused to issue the certificate.
>
> The head of the Foreign Overseas Chinese Registration Department of the Semarang Immigration Department told the head of the Chinese Assembly that the view of the Semarang District Court was a misunderstanding. The announcement of this view of the Semarang District Court has caused anxiety among overseas Chinese. The reporter spoke to the head of the Nationality Department of the Semarang District Court. He said that their interpretation was completely wrong and that the correct interpretation was that 'children born after the transfer of power on 27 December 1949 automatically inherit their father's nationality, just like children born before that date'. He demanded that this correction be made in the newspaper.

In *Yin Hua Jing Ji* (13 June), under the headline, 'Chinese living in Central Java, be careful'! reported on the mistake of the court in Semarang 'This notification of the court in Semarang has caused confusion among the Chinese people and they have made this an unprecedented and deplorable thing'.

This perplexing misinformation from the District Court confused the Chinese, who were seeking Indonesian citizenship and left them more than slightly disillusioned with the administration. These newspaper articles also showed that many people in Semarang wanted an Indonesian nationality and were eager to gather information about the procedure.

What did this high interest of Surabaya Chinese in acquiring Indonesian citizenship represent? It was the awareness that they could never survive in the transnational way they had been. The Chinese fully understood that they would have to abandon their Chinese nationality to acquire Indonesian nationality, and they acted on this understanding of the importance of fully belonging to Indonesia, where they were born and raised, and having their rights guaranteed by the state. These articles show that the state was not unilaterally forcing them to acquire nationality, but that the residents were also responding.

11.6 Conclusion

The cases I have examined so far re-examined how the relationship between the transnationalism of the Chinese people and the attitude of the nascent nation-state was established. In post-war Southeast Asia, whether in previously colonised countries or in countries that had maintained their independence, such as Thailand, there was a common debate on how the Chinese, who were considered to have ties to China, should be integrated into the nation-state. The Chinese had previously enjoyed the status of subjects of the empire (colonies) and were included within it; however, after the colonial power was eliminated, their belonging became a political issue.

In Indonesia, in particular, the Chinese had strong links with the Dutch, and the local people believed that the Chinese exploited the local people as pawns of the Dutch. Therefore, it was difficult for the Indonesian state as it was to accept the Chinese. The Indonesian policies we have seen so far (forcing the Chinese to choose their nationality, depriving the Chinese of their livelihood in the context of economic nationalisation) reflect the vulnerable position in which the Chinese have been placed since the Dutch period. The State did not tolerate their transnationalism and tried to encompass their entire existence (or partially exclude them as outsiders).

From the details of the newspaper articles examined, we can understand how the Chinese acquired their nationality. What can be seen from the high level of interest in the issue of nationality, as seen in the military service issue and the case of Semarang, is that they have abandoned their previous transnational way of being, shifted towards belonging to a single state (Indonesia), and have their lives protected by it.

Furthermore, I point out that the issue of Chinese nationality is significant to the Indonesian society as a whole. The issue of Chinese nationality stems from the existence of two categories of citizens in the Indonesian Constitution of 1945: original or native Indonesians (*Indonesia Asli*) and those who acquired Indonesian nationality through legal procedures. The process of acquiring Indonesian citizenship, as stipulated by the Nationality Law of 1958, was a legal procedure that distinguished the Chinese from native Indonesians. This law demonstrated that the Chinese were not native Indonesians.

During the Suharto regime (1968–1998), the assimilation policy towards the Chinese was adopted to thoroughly break the ties between the Chinese and China, and the localisation of the Chinese progressed. However, despite the increased localisation, they continued to be discriminated against.

This principle was reviewed after the Suharto regime fell. The new nationality law enacted in 2006 confirmed the principle that people born and raised in Indonesia, including the Chinese, are native Indonesians (*Indonesia Asli*) (Verdiansyah 2007). The fact that the Chinese born and raised in Indonesia are regarded as native Indonesians, which guarantees them equal status with the non-Chinese Indonesians, can be interpreted as the terminal point of the localisation of the Chinese. As we have seen, for the Chinese, whose place of life is in Indonesia, belonging to Indonesia was the most realistic option, but it took a long time for the Chinese to be accepted as full citizens of Indonesia because the state was aware of the danger of their transnational potential.

Notes

1. Article 26, paragraph 1 of the Constitution of the Republic of Indonesia 1945.

2. In the original text of this quote, the word 'you' is used with the nuance of directly addressing the reader.
3. The politicians who broke away from the left-leaning Sukarno declared the formation of their own government, the PRRI (Revolutionary Government of the Republic of Indonesia), and confronted the central government, which was briefly suppressed by the Indonesian National Army.
4. Baperki (Badan Permusjawaratan Kewarganegaraan Republik Indonesia, the Indonesian Nationality Consultive Association) was established in 1954 by Siauw Giok Tjhan. He was born in Surabaya in 1914 and immersed himself in political activities under Dutch rule, sympathising with the ideals of the founding of Indonesia. After Indonesia's independence, he established Baperki, which appealed mainly among the Chinese people to share their destiny with Indonesia and assisted them in acquiring Indonesian nationality.

References

Da Gong Sheng Bao (19 January, 4 March, 11 March, 20 March, 4 June, 8 June 1959)

Kawashima S (2011) Kyoukyou to shiteno kinmon: rekishi teki haikei (Quemoy as a place of origin for overseas Chinese: historical background). Chiiki kenkyu 11(1):43–61

Kudo Y (2023) Oranda ryou higashi indo no taiwanseki min to nihon no keizai shinshutsu (Taiwanese nationals in the Dutch East Indies and Japanese economic expansion). Southeast Asia History and Culture 52:5–27

Li, ZH (2003) Yin hua xian qu ren wu guang hui sui yue: yin ni hua ren bao gan he du li xian xian shi hua (The glorious years of Indonesian Chinese pioneers: a history of newspapers and magazines published by Indonesian Chinese and leaders of independence of Indonesia). Lian tong huawen shu ye you xian gongsi, Jakarta

Liu H, Liao C (2006) Network, identity, and ethnic Chinese studies: towards a re-examination of regional orders in 20th century east Asia. Tounan asia kenkyuu 43(4):346–373

Matsumura T (2017) Indonesia kokka to nishi kalimantan kajin (Indonesian state and West Kalimantan Chinese). Keio University Press, Tokyo

Mei Ri Dian Xun (29 May 1959)

Sadayoshi Y (2006) Ranryoo ki indonesia kajin no tajuu "kokuseki" to houteki chii no jissou (More than dual "nationalities" and legal status of Chinese in colonial Indonesia). Kindai 96:1–34

Sheng Huo Bao (3 March, 27 May, 2 June, 10 June, 24 June 1959)

Su Men Da La Min Bao (19 January, 27 January 1959)

Torpey J (1999) The invention of the passport: surveillance, citizenship and the state. Cambridge University Press, Cambridge

Undang-undang dasar negara Republik Indonesia 1945 (Constitution of the Republic of Indonesia 1945)

Undang-undang no. 62 tahun 1958 tentang kewarganegaraan Republik Indonesia (Law no. 62 of 1958 on citizenship of the Republic of Indonesia)

Undang-undang no. 66 tahun 1958 tentang wajib militer (Law no. 66 of 1958 on military conscription)

Verdiansyah C (ed) (2007) Jalan panjang menjadi WNI (The long road to becoming an Indonesian citizen). Penerbit buku Kompas, Jakarta

Xin Bao (17 January, 18 January, 3 March, 19 April, 27 May, 30 May 1959)

Yamamoto N (2008) Higashi indo ni okeru jaanarisuto kuukan to kajin (Journalistic space and the Chinese in the Dutch East Indies). Political Science and Area Studies at Keio University, pp 223–258

Yamashita K (2023) Kakyou, kajin wo shirutame no 52 shou (The 52 chapters for understanding Chinese overseas). Askashi shoten, Tokyo

Yin Hua Jing Ji (13 June 1959)

Matsumura Toshio is a lecturer at Graduate School of Humanities, Osaka University, Japan. He has been teaching courses in East and Southeast Asian history, political science, and Indonesian language. For his doctorate he focused on the Chinese community in Indonesia and wrote his dissertation on the history of the Chinese in West Kalimantan based on his own field research.

Chapter 12
Transnational Muslim Identity Crisis: A Comparative Analysis of *the Accidental Malay* and *Seasonal Adjustments*

Shafinur Nahar and Hannah Ming Yit Ho

Abstract This chapter provides a comparative analysis of the representation of Muslim identity in the novels *The Accidental Malay* by Karina Robles Bahrin and *The Seasonal Adjustments* by Adib Khan, focusing on the crisis of Muslim men in the context of transnationalism. The study employs a qualitative approach, using literary analysis and thematic analysis to examine the key themes and motifs related to Muslim identity in both novels. The findings suggest that while both novels explore the complexities of Muslim identity, they do so from different perspectives, with *The Accidental Malay* focusing on the internal struggles of a Muslim man whose Muslim identity has been shaped by his short-term stay in the UK, in Malaysia, within a national border and *Seasonal Adjustments* examining the external challenges faced by Muslim men in diaspora and home, Bangladesh, as well. The chapter concludes by discussing the implications of these representations for understanding Muslim identity in contemporary literature and its relevance to transnational contexts.

Keywords Muslim men · Identity crisis · Transnationalism · Diaspora · Home

12.1 Introduction

Transnationalism encompasses the "social, cultural, economic and political relations which are between, above or beyond the nation-state, interconnecting, transcending" (p. 864) and often surpassing what has traditionally been the central focus for the last two centuries (Grillo 2004). This concept is particularly relevant in the context of Muslim identity, which is increasingly debated globally, especially regarding the identity and practices of Muslims in Western countries. Grillo (2004) highlights that the heightened visibility of Islam has connected discussions worldwide on issues

S. Nahar (✉) · H. M. Y. Ho
Faculty of Arts and Social Sciences, Universiti Brunei Darussalam, Gadong, Brunei Darussalam
e-mail: shafinurnahar@gmail.com

H. M. Y. Ho
e-mail: hannah.ho@ubd.edu.bn

H. M. Y. Ho and Y. Chan (eds.), *Transnational Southeast Asia*, Asia in Transition 29,
https://doi.org/10.1007/978-981-96-3608-2_12

with Muslim practices and identities coming under scrutiny from both Muslim and non-Muslim societies.

Roose (2021) further explores the transnational dimension of Muslim identity by describing how Islam provides a sense of belonging to Muslims in Western countries. Such belonging is important for Muslims when they navigate and accept both local and Western cultures as well as practise their Islamic faith regardless of social challenges. Moreover, Roose's study points out an increasing number of self-assured Muslim men who are blending regional and Western ways with Islamic practice. Given the transnational identities and homes of Bangladeshi Muslims (Ho 2023), which will be a focus of analysis in this chapter, the resilience of Muslim characters in managing their religious beliefs in the transnational and diaspora contexts will be examined through a comparative analysis of Anglophone novels by Bangladeshi and Malaysian authors.

Muslim male identity is defined via religious and ethnocultural factors but is also shaped and influenced heavily by transnationalism, migration, and multiculturalism. A literary study by Ho (2021) examining a migrant's conversion to Islam contends that Muslim men assume the role of "gatekeepers managing the (non-) entry of others into Malayness." (Ho 2021, p. 149). In conflating Malayness with Muslim identity that fulfils the demands of Malay-Muslim synchronicity in the Malay world, Ho addresses an important conversation about the process of migration and transnational determinants of Muslim identity. Furthermore, Islamaphobia that has singled out religious patriarchy to create a perception of Muslims as intractable has also shaped responses in literature (Ho 2022). Contemporary literature, thus, offers representations that are focused on the challenges that Muslim characters encounter between diaspora and homeland, as well as their integration of Islamic values in the modern and transnational world. A complex negotiation of Muslim identity is highlighted and warrants further examination in the transnational context.

In addressing a crisis of Muslim men in the transnational context, this chapter analyses *The Accidental Malay* by Karina Robles Bahrin (2022) and *Seasonal Adjustments* by Adib Khan (1994), where the two fictional works offer insights into the lives of Muslims in the diaspora and national home. The former is a Malaysian novel, while the latter is a novel by a Bangladeshi diaspora writer. A comparative analysis of the novels reveals challenges that arise from living in distinct multicultural settings and the conflict between maintaining tradition and modernity. While tradition is often equated with cultural values associated with Muslim practices, modernity is conventionally understood as Western liberal beliefs that do not necessarily align with Muslim tenets. This chapter aims to explore the way that Muslim men in these novels navigate this binary model of identity when negotiating between the demands of tradition and modernising their Muslim identity. In the national and transnational contexts, Muslim male characters challenge and subvert the juxtaposing structures of religious tradition and modernity to elicit their multicultural realities.

Karina Robles Bahrin's debut novel, *The Accidental Malay*, features the protagonist Jasmine, a Malaysian businesswoman who did not know about her Malay-Muslim identity before, and later faces the revelation of her Muslim identity that places her in a dilemma to accept her newly discovered religious identity. A subplot of the novel revolves around her complex relationship with Iskandar, her Malay-Muslim boyfriend, focusing on his inner conflict while trying to reconcile personal desires with societal expectations. Iskandar is the novel's second main character and is described through Jasmine's perspective. The main struggle that Iskandar encounters is having to balance traditional values and modern beliefs, which is evident in his romantic relationship with ethnic Chinese Jasmine Leong that defies social expectations within the Malay-Muslim dominant nation. In relation to Iskandar, the novel portrays a crisis of Malay Muslim men resulting from their religious and cultural orientations that are delimited in a Malaysian nation.

Adib Khan's *Seasonal Adjustments* centres around Iqbal, a Bangladeshi immigrant living in Australia who returns to Bangladesh after eighteen years. His transnational life has shaped Iqbal's identity. The novel explores his struggles of acceptance and integration in different cultures. Through his complex transition from a conservative and traditional Bangladeshi culture to a modern and liberal Australian culture, Iqbal experiences internal conflicts between individual choices and cultural expectations as a man. The novel also accentuates the struggle of balancing cultural and religious identities, while embracing new cultures and ideas in the host country.

Both male characters, Iqbal from *Seasonal Adjustments* and Iskandar from *The Accidental Malay* have experienced life in the diaspora. Both are heterosexual Muslim males. Iqbal migrated to Australia for a permanent living, while Iskandar migrated for a short term to Malaysia for education in the United Kingdom. Their transnational experiences influence the development and evolution of their male identities. Despite the temporal gap in publication and cultural differences between the characters, both novels address issues of life in the diaspora, intercultural marriages, and religious identity, for which both novels are relevant for discussing contemporary discourses about the crises of Muslim men in the transnational context. By analysing the two male characters, this present study affords significant insights into personal struggles, societal and cultural expectations, and religious boundaries, providing profound reflections on identity construction in a global world.

These novels invite readers to rethink the ways that traditional notions of faith and culture may lead to a crisis in Muslim men's existence in the modern world. The main objective of the research is to examine the representations of the Muslim male identity crisis in *The Accidental Malay* and *Seasonal Adjustments* and highlight the common but also different traits of Muslim identity that are embodied through their male protagonists. In comparing and contrasting these characters, the chapter aims to promote an understanding of a literary discourse of Muslim masculinity in contemporary Anglophone Southeast and South Asian works and, in taking the transnational as an operative framework, reveal identity tensions and conflicts that Muslim men experience.

12.2 Muslim Identity in Malaysia and Bangladesh

In Malaysia, Muslim identity is vigorously associated with Malay ethnicity since Islam is the official religion of the state. Malaysian norms of Muslim identity draw from Islamic teachings, Malay culture, and colonial influences. These prescriptions give a framework for defining Malaysian Malay Muslim identity. In the Malay world, Malay Muslims engage in the Malay language and Islamic religion as components of a local cultural value system that aligns with a notion of national identity (Ho 2019, 2021). In Malaysia, to be Malay is also to be Muslim per se (Federal Constitution 1957, 160.2), connecting both identity features. Thus, as it advances the Islamic agenda as the Malay agenda, enforcement of a measure of "being Malay" and "being Muslim" becomes the paradigm of "good subjects" among Malay Muslims (Peletz 2021). Noritah Omar (2006) found that temporary Islamic contracts and the voluntary submission of Malay women in light of religious and cultural norms establish Malay Muslim men's power. The perceived norms for the Malays are that men are expected to be strong and financially capable of providing for the family. However, all these factors of power relations, religion, and cultural values influence the dynamism and fluidity of Malay-Muslim masculinity in Malaysia.

On the other hand, in Bangladesh, Bengali Muslim men's identity has been formed as an outcome of centuries of cultural intersection of Islam into Bengali society. Bengali Muslim men's masculine identity is conceptualised within different historical periods and institutional settings, marked by gender power differentials (Hasan et al. 2017). Islam strongly influences the perception of masculinity in the context of Bangladesh, and being a good Muslim is an integral part of being a man (Hasan et al. 2017). However, they also form diverse and non-hetero norms of masculinity based on religion, politics, history, and cultural globalisation (Hasan et al. 2017).

The conceptualisation of Malaysian-Muslim male identity in literature, especially concerning transnationalism, has been tricky and compounded. The evolution of Malaysian Anglophone literature has seen various shifts and struggles, and authors have had problems articulating religion, culture, and identity about the different global forces and transnational affiliations. The political separation of Singapore from Malaysia in 1965 affected many writers who wrote in English and created the disassociation of many writers from Malaysian literature; this shows how this movement transcended beyond the country to form the Malaysian Anglophone literary traditions (Quayum 2020). Muslim-Malay writers, such as Rehman Rashid, Dina Zaman, Amir Muhammad, and Karim Raslan, have chosen English as a tool for writing for the global audience while grappling with the Malay-Muslim identity to assert modernity (Ng 2009). The narratives are characterised by ambivalent and gender role conflicts and employ irony and humour to subvert the transnational Malaysian paradigms of conformity. Their works explore the problematic relationship between modern perspectives on Islam and lingering conservative forces, presenting narratives that simultaneously advocate for and critique certain aspects of Islamic practices in Malay life in the context of transnationalism. Nevertheless, the portrayal of the development of Muslim men's identities in Malaysian Anglophone writings is mainly addressed with sensitivity as authors navigate the delicate balance between religion, culture, and sociopolitical issues in the face of transnational forces (Mukherjee 2019).

Bangladeshi Anglophone fiction writing explores male identity with themes, such as liberation war and its aftermath, nationalism, religion, gender issues, and identity politics (Quayum and Hasan 2022) in both domestic and diasporic spaces. Male characters are usually the main protagonists in Bangladeshi Anglophone fiction, and their actions significantly impact how the plot unfolds (Quayum and Hasan 2022). Bengali Muslim men in Bangladeshi Anglophone fiction enact and maintain their masculine roles and identities, exercising control over their wives, daughters, and household decisions, highlighting the intersection of local and global norms (Töngür 2013). Bangladeshi Anglophone fiction writers often depict Bengali Muslim men in the diaspora, highlighting their complexities and challenges in navigating their identity as Muslim and relationships across borders, which have been expressed by Hasanat (2017) when stating the way that fictional works "intensify the diasporic subject's staggering quest for identity" (p. 58). Characters grapple with the complexities of their identity, while seeking to establish their definitions of home and identity (Hasanat 2017), which reflect tensions between national and transnational Muslim identities.

The concept of transnationalism has garnered increasing scholarly attention in understanding Muslim identity within the contexts of migration and globalisation. Transnationalism involves exchanging cultural, economic, and social elements across national boundaries (Vertovec 2001), significantly influencing the perception and expression of Muslim identity. Research emphasises that while transnational influences may reshape Muslim identities, national contexts continue to play a pivotal role in shaping how Islam is practised and experienced (Grillo 2004).

Studies, such as those by Mercanti (2012) and Faruque (2022), on *Seasonal Adjustments* focused on the representation of identity crises in the diaspora, highlighting the complexities faced by Iqbal in negotiating a sense of belonging at home and in the diaspora. Conversely, *The Accidental Malay* offers a novel perspective on the Muslim male identity crisis that remains underexplored in existing scholarly works.

12.3 Transnationalising Muslim Identity

Vertovec (2001) defines transnational identity as the construction of self across social contexts that transcend national borders. To unveil the crisis of Muslim male identity in the context of transnationalism, this chapter compares the male characters from *The Accidental Malay* by Karina Robles Bahrin (2022) and *Seasonal Adjustments* by Adib Khan (1994). Hence, this analysis implements a qualitative research methodology that combines literary and thematic analytical techniques. Regarding literary analysis, a close reading has been done through the character development of Iqbal and Iskandar in their transnational context. Additionally, thematic analysis involves the identification and cross-comparison of themes, including identity crisis and adaptation to the new culture, as well as intercultural relationships in both novels.

This study draws on previous studies on transnationalism and Muslim masculinities to connect the thematic analysis. Thus, this approach may offer a nuanced understanding of Muslim men within their literary representations and advance literary discourses on their identity crisis in the transnational context.

12.4 The Accidental Malay and Seasonal Adjustments

The Accidental Malay explores the complex identity of Muslim men through Iskandar, a non-practising Malay Muslim involved romantically with the female Chinese protagonist, Jasmine Leong. Iskandar's character development and identity crisis highlight the tensions that arise when individuals do not conform to traditional Muslim identity expectations in Malaysia. His experiences and actions throughout the narrative exemplify his transnational Muslim identity.

The novel introduces Jasmine in the beginning chapters, subtly highlighting the underlying tension in her relationship with Iskandar. Jasmine is described as the "ideal mistress—not a wife," reflecting her disappointment over Iskandar's absence because "His wife has taken ill with a cold" (Bahrin 2022, p. 1). This early portrayal reveals the complexity of Iskandar, a Muslim man with a Muslim wife and a Chinese lover. Jasmine's waiting and his ignorance are depicted through her actions: "She checks her phone for the thousandth time. No sign of Iskandar. Not even a text" (Bahrin 2022, p. 21).

Iskandar's experience in London exemplifies how Muslim identities are negotiated and reshaped through transnational experiences, highlighting the intersection of local and global influences on Muslim identity (Mandaville 2001). As a Malay Muslim, he encountered unique challenges and opportunities in a Western metropolitan context. In London, where he first met Jasmine, he was placed within specific racialized and cultural frameworks that shaped both his self-perception and how others viewed him. Western countries often process different men differently, categorizing them into hierarchical racialized boxes based on factors like race, ethnicity, skin colour, education, and religiosity (Hearn 2017). For Iskandar, this meant navigating his identity about "normative" white masculinity and experiencing a position of both privilege and disadvantage (Hearn 2017). Although he returned home and settled down, his stay in London profoundly influenced his identity, reflecting the interplay of local and global cultural forces on his sense of self.

Transnational identities involve living across multiple cultural and national boundaries, which can "shape and influence people's behaviour and daily practices" (Metcalf 1996, p. 19). They also involve balancing personal desires, cultural expectations, and religious obligations. Iskandar's character embodies these complexities. His initial encounter with Jasmine in London marks a pivotal moment where he begins to navigate between his Malaysian-Muslim heritage and Western cultural influences. Adopting the name "Alex" in his Western social circles illustrates his adaptation and negotiation of multiple identities: "I'm Iskandar, but they call me

Alex here. Short for Alexander. I guess the Arabic version is too difficult for white tongues" (Bahrin 2022, p. 22).

Iskandar's drinking habits present a significant complexity in his transnational identity. As a Muslim, consuming alcohol is prohibited, yet Iskandar engages in drinking, reflecting his internal conflict and the influence of his Western lifestyle. This behaviour signifies the tension between his religious identity and his attempt to fit into a different cultural context. This is illustrated when "a half-empty bottle of whisky sat on the coffee table" (Bahrin 2022, p. 25), symbolising his struggle to reconcile his faith with the lifestyle he has adopted abroad.

Despite his integration into Western society, he retains his cultural and religious identity as a Muslim. Iskandar's engagement with religious practices, such as during Ramadan, and his attempts to adhere to Islamic teachings amidst lapses into behaviours like drinking further underscore his negotiation of religious and cultural boundaries. When he states, "Trying to cut down. Ramadan is almost here. Gotta face the music soon" (Bahrin 2022, p. 32), it shows his acknowledgement of Ramadan as a time for spiritual cleansing contrasts with his indulgences, revealing a tension between his faith and his lifestyle influenced by global contexts.

The pressures of familial and cultural expectations become apparent when Iskandar returns to Malaysia. In an interview with Wee (2022), the author Karina Robles Bahrin said, "Being Malay and Muslim impacts a lot of a person's life; there are many expectations set." Her statement is very true in the case of Iskandar, who once said to Jasmine, "To be Malay today means to be Muslim, little else" (Bahrin 2022, p. 146), which means contemporary Malay identity is largely defined by Islamic beliefs. This religious centrality influences social practices, including marriage, where Muslim-Malay individuals are often expected to marry within the faith, reinforcing the connection between Malay identity and Islam. The obligation to marry within his faith poses such a significant challenge, as seen in Jasmine's reluctance to convert to Islam and Iskandar's adherence to this requirement: "Conversion to Islam was not an option for her. Grandmother would never allow it. And he could only marry a Muslim; the law said so" (Bahrin 2022, p. 26). His relationship with Jasmine, who is a non-Muslim, challenges these norms and prompts a crisis in his Muslim identity. His family's expectations further emphasise this challenge after his father's bypass surgery and his mother's wish for him to marry a family friend (Bahrin 2022, p. 27). Iskandar's guilt over his simultaneous involvement with Jasmine while married to a hijabi Malay woman reflects his internal struggle between religious adherence and personal fulfilment. His confession that he "should never have married" (Bahrin 2022, p. 31) his wife and his intense emotional connection with Jasmine underscore his internal turmoil and a quest for personal fulfilment amid societal and religious constraints.

Iskandar's emotional turmoil is evident again in his confession to Jasmine: "I keep trying, but you're still there. Even if I haven't seen you. Even if we haven't spoken. You're here, in my head" (Bahrin 2022, p. 28). It reveals the lasting impact of their relationship and the unresolved nature of his feelings, highlighting the complexities in his male identity. Iskandar's experience of secrecy and compartmentalisation in his relationship with Jasmine speaks to broader themes of identity negotiation in transnational contexts: "Over the last four years, they have kept things going by gliding over the little details. Telling each other just enough, without letting on too much" (Bahrin 2022, p. 33). Their decision to conceal their affair reflects a need to protect their relationship from societal judgement and potential consequences, illustrating the challenges faced by individuals in navigating multiple identities across cultural and religious borders.

Iskandar's decision to divorce his wife and pursue a relationship with Jasmine reflects his willingness to challenge societal norms for love and fulfilment. Nevertheless, his insistence on Jasmine's conversion underscores his internal struggle with his Muslim identity and acceptance. The struggle Iskandar faces as a Muslim is a result of the profound influence that religious identity has on the autonomy and choices made by Muslim men in different societal contexts. Jasmine's statement further evidences this, "It's not his fault—he was born a Muslim" (Bahrin 2022, 65). Iskandar acknowledges, "If I weren't Muslim by default, I probably wouldn't be anything" (Bahrin 2022, p. 173), reflecting the hegemony of Muslim identity over Malay identity. But, this intersection of ethnic and religious identity with his gender identity gives him privilege. Though, Iskandar further critiques the community as "hypocrites," emphasizing, "It's not the religion, it's us" (Bahrin 2022, p. 173) that attributes the internal conflict to human failings rather than religious teachings, highlighting a larger societal crisis of Muslim men where religious expectations often take precedence over individual desires, including marriage.

The novel's conclusion reveals Iskandar's vulnerability and dependence on Jasmine as a source of stability amidst societal pressures. His disappointment when Jasmine decides against marriage underscores his reliance on her as a refuge from his tumultuous life. This dependence is further complicated by Jasmine's uncertainties and decisions, which ultimately challenge Iskandar's perceptions of control and stability. His intention to marry Jasmine, his decision about "divorce gets finalised next month" (Bahrin 2022, p. 172), and his move to Mont Kiara denotes a profound personal transformation and a decisive rejection of societal Muslim norms for which Jasmine describes him as "a right rebel" (p. 174). Finally, his pursuit of personal interests, as stated in his plans for "pick[ing] up a couple more gigs" (Bahrin 2022, p. 172), signifies a break from traditional expectations and a shift towards embracing personal autonomy, which is explained his own words:

> "But am I happy? In parts. Not that I know how to be different, without being Muslim. The Eid, the fasting, knowing I am doing something rather naughty every time I break the rules. But do I feel guilty? Not really. It's just...it's the only way I know how to be." (Bahrin 2022, pp. 173-174)

In essence, Iskandar's character embodies the broader tensions within Malaysian society regarding cultural authenticity, religious identity, and individual autonomy. Iskandar's narrative is a poignant exploration of the challenges individuals face navigating the intersections of tradition and modernity in a globalised world.

Unlike Iskandar, Adib Khan's novel *Seasonal Adjustment*'s protagonist, Iqbal, permanently moves to Australia. Khan explores the complex life of Iqbal, a Muslim male migrant, highlighting his experiences with religious traditions, nationalism, and cultural barriers within a transnational context (Mercanti 2012). Iqbal's journey underscores the importance of embracing diversity and mutual recognition in shaping his unique Muslim identity, reflecting the intricate nature of transnational identity formation (Mercanti 2012). Iqbal's temporary return to Bangladesh from Australia after eighteen years is driven by a need to rediscover his identity. His identity, which he perceives in Australia, is always living "in a perpetual state of conflict, torn between what was and what should have been. There is a consciousness of permanent loss" (Khan 1994, p. 143). This conflict happens regarding his cultural and religious boundaries. His Muslim identity as a male has evolved due to his experience as a migrant. Even though he is not religious or practicing Muslim anymore, he cannot ignore his Muslim identity. Now, he wants to rediscover his identity to reconcile the sense of loss and displacement by embracing the inclusive and compassionate aspects of Islam aligned with the globally shared values of humanity. His letter to Clair encapsulates this quest where he explains that he has returned to "find the past" associated with his cultural and religious identity and to see himself "in its reflection" (Khan 1994, p. 2) to know who he is.

This journey of self-discovery is a common challenge for transnational Muslim individuals who seek to reconcile their fragmented identities. He settles down in Australia driven by the desire to escape the turmoil of his homeland, which results in deep feelings of guilt and alienation as described as his "moral failing" (Khan 1994, p. 8) due to his family's perception of his move as selfish in leaving after nine months of the Bangladesh Liberation war in 1971, which resulted in a newly independent Bangladesh that Iqbal describes as "a nation born in agony" (Khan 1994, p. 8). Religion played a divisive role in the Bangladesh Liberation War as Muslims and Hindus engaged in physical conflict. After this war, the Muslim democratic nation-state of Bangladesh needed people like him to build up the new nation, so Iqbal's immigration to Australia was taken as a self-interest relocation. Upon his return to Bangladesh, Iqbal's sense of dislocation becomes prominent as he realises the Muslim intensification of his family, which deepens his identity crisis. He notices their judgmental and alienating attitude towards him for not being a practising Muslim.

Iqbal's Muslim identity with a modern view creates significant tension within his traditional family and the broader Muslim community. His modern views are often at odds with the religious norms expected by his family in Bangladesh and his in-laws in Australia. He remains "a *non* […] Outside the human race" (Khan 1994, p. 149), showing his feeling as an outsider in the West and in his native culture as well (Faruque 2022).

Levitt (2003) argues that migration contexts foster the emergence of diasporic, global, and transnational religious expressions through recreating "global religions by making them local and then starting the process anew" (p. 849). This process involves global religious institutions shaping migrant experiences while migrants adapt and reinterpret these traditions to suit local contexts, thereby evolving religious practices within transnational communities. He faces discrimination from Australians and his Muslim community due to his Bangladeshi background or Bengali ethnicity (Tan 2003). His first visit to the mosque in Preston reveals divisions within the Muslim community based on ethnicity and nationality (Tan 2003). This diversity underscores how Islam "has travelled in diaspora" (Mandaville 2001, p. 115) and transcends geographical and cultural boundaries, highlighting its transnational nature among its adherents.

Iqbal experiences racism and xenophobia in Australia, which further underscore the challenges of maintaining a cohesive Muslim identity in a globalised world. He is acutely aware of the "racism and the misconceptions against Muslims" (Khan 1994, p. 139) in Australia. Initially optimistic about Australia, his hopes are shattered by the reality of racial prejudice: "When I first arrived in Australia, I thought it was an ideal sanctuary[...] I suppose if you get to know anything well enough, you discover its flaws" (Khan 1994, p. 123).

His disappointment with Australia starts with the racist behaviour he experienced on his first visit to a mosque in Melbourne. He found in the footpath a red-haired man physically pushing a "burqa-clad woman against a fence and was taunting her with racist abuse" (p. 196), reflecting a sense of entitlement and aggression towards.

> "dark-skinned foreigners were probably meant to subject themselves passively to verbal insults and physical abuse in return for the privilege of living in his country. They were not supposed to retaliate and yell, 'Fuck off, you diseased convict dick!' at the tall, white residents of the land. He will never know how scared I was of his size and how I loathed myself for stooping to his level of obscenity" (Khan 1994, p. 196)

Iqbal's confrontation, although defensive, underscores the fear and indignation provoked by overt racism and the struggle of feeling compelled to respond in kind. He is often frustrated with his in-laws' conservative Christian views, which leads to a sense of "sadness of isolation" (Khan 1994, p. 150). In a family gathering, Keith, Iqbal's father-in-law, insists Iqbal and Michelle baptise their daughter Nadine because, according to him, "She is a Catholic. It's an essential part of our religion" (p. 82). Keith further explains his viewpoint that "[t]here is a commonality of unshakable belief among Catholics which gives us strength and a clarity of vision about ourselves. It is all about sharing and participation in the richest communal tradition in the world" (p. 84), thus viewing Christianity as essential for Nadine to know "who she is and where she belongs" (p. 85) as well as for her integration into mainstream "Australian life" (p. 85). Iqbal resists it, saying she was not "born with a religious tag around her wrist" (p. 82), which fuels a heated argument with Keith for whom Christian rituals are "essential land-marks" of their "faith" (p. 83). Keith is not ready to accept Iqbal's non-Christian view. Iqbal's response of "How did you determine she is a Christian and not a Muslim?" (p. 84) sparks a tense debate about religious choices. While Keith advocates for Nadine's baptism into Christianity, Iqbal argues for Nadine's right to inherit his Muslim identity. This clash marginalises the immigrant Muslim perspective as Keith prioritises Christian norms and cultural assimilation, highlighting broader tensions around cultural identity and religious differences in a multicultural society like Australia. These experiences underscore the layered discrimination that transnational Muslims encounter in cultural contact spheres.

Iqbal's marriage to a white Catholic woman, Michelle, exemplifies the cultural and religious challenges that he faces. His family's disapproval of his marriage underscores the clash between his personal choices and their cultural expectations. To his family and relatives, his wife is always "a consumer of pork, would have been treated as a catastrophe destined to ruin my life" (Khan 1994, p. 150). Iqbal recounts how his mother unexpectedly started talking about his "life in a foreign land" (p. 51) where he was "trapped into marriage" (p. 51) for what he remains "blameless" (p. 51), but his wife is considered the "culprit" (p. 51). This also demonstrates how Iqbal is perceived as a victim of marriage to a foreign non-Muslim woman. To his family, this marriage is an act of manipulation that undermines his ability to make autonomous decisions. They believed he should have chosen someone from the Muslim community and Bengali society to preserve his religious and cultural lineage. This view is rooted in a patriarchal mindset in which Muslim men perceive themselves as meek targets of non-Muslim women. It enforces the social belief that Muslim men should adhere to the imposed cultural and religious rules concerning marriage in order to maintain the unity, identity, and purity of their community's traditions.

This reflects the pressures in Bengali society that transnational Muslim men face to conform to traditional norms. Such demands create "the contestation between the reinforcement of cultural identity and its new construction and its new construction represented, respectively, through Bengali/Bangladeshi identity and the Muslim identity" (Riaz 2009, p. 96). Regarding this, Iqbal's resistance highlights the struggle for personal agency and a more inclusive identity, navigating family expectations and societal norms. For example, his reluctant participation in his daughter's "aqeeqa" programme in Bangladesh, contrasted with his discomfort with halal slaughter, symbolises his struggle with his cultural heritage because of his exposure to diverse cultures. As he reflects, "There are occasions when I regret my exposure to the diversity of cultural radiation which has bleached my individuality" (Khan 1994, p. 9). The term "cultural radiation" metaphorically suggests that different cultural experiences have impacted him, which erased parts of his cultural self in a transnational setting which eventually causes an internal conflict.

This internal conflict sheds light on the persistent challenges that transnational Muslims face in balancing personal convictions with communal norms. However, his disdain for religion is evident in his negative view of figures like Moulana Khawaja Rahmatullah Azad, a "self-appointed communicator with Allah" (p. 44) who exploits people's faith for personal gain. Iqbal's childhood encounters with religious men, including Muslim mullahs and Catholic brothers, instilled a deep resentment towards religion, shaping his desire to raise his daughter, Nadine, in a multicultural environment free from religious imposition. This is particularly evident in Iqbal's advocacy for his daughter Nadine's autonomy in determining her religious identity and life choices: "Nadine will be among a slowly growing minority which will learn how to combine traditions […] By the time she is an adult, the narrowness of life within a single tradition may become a handicap" (Khan 1994, p. 85).

Iqbal's belief in combining traditions and embracing diversity suggests a progressive approach to identity formation that transcends conventional boundaries. Also, his family in Bangladesh expects him to raise his daughter Nadine "as a Sunni Muslim" (Khan 1994, p. 35); there emerges a significant societal pressure on transnational Muslim men to uphold and transmit religious and cultural traditions to the next generation. By challenging traditional notions of Muslim male dominance and control over family members, he emphasises individual agency and the fluidity of Muslim male identity in a transnational context. His resistance to dominant cultures and traditions reflects the complexities of cultural adaptation and identity negotiation of diasporic individuals.

As a transnational Muslim man, his identity crisis is evident when Iqbal proclaims himself as "a free floater" (Khan 1994, p. 249). However, his father rejects this idea when reprimanding him: "You are still a Muslim. You were born as one. You do not have a choice. You may have moments of private doubt about religion. We all do. But you cannot shake off your heritage" (Khan 1994, p. 249). His father's words underscore the way that Iqbal has not conformed to traditional Muslim masculine roles as he jettisons a rigid Muslim identity to embrace "a diversity of experiences" (p. 249). His positionality within an out-group is clear when admitting, "Yes, there are moments when I get lonely" (p. 249). Navigating familial and national expectations of Muslim identity, Iqbal embraces transnational perspectives as a Muslim choosing individuality rather than communality.

Iqbal's journey in *Seasonal Adjustments* provides a nuanced perspective of transnational Muslim identity crisis. He grapples with traditional power structures that are governed by patriarchal norms and influenced by his family background as well as societal expectations. His rejection of traditional religious authority signals his struggle to meet normative expectations as a Muslim man (Mercanti 2012). Through Iqbal's experiences of cultural dislocation, religious tension, and familial demands, Khan highlights the challenges of embodying a Muslim identity in a globalised world. The novel underscores a plea to inclusivity to transcend cultural and religious boundaries, thus advocating for a more interconnected and inclusive Muslim male identity.

12.5 Discussion

The global mobilisation of Muslims based on shared identity and values is not a new phenomenon (Mandaville 2009). However, Muslim migrants in Western societies continue to face intersectional discrimination, as seen through Iqbal's experiences with racism and xenophobia, as well as misconceptions about Islam in Australia. Iqbal's identity as a transnational Muslim is shaped by his migration to Australia, where he confronts modernism and cultural dislocation.

Transnational identities emerge from complex negotiations between homeland traditions and host country dynamics, significantly influencing personal identity (Esteban-Guitart and Vila 2015). Iqbal's experience in Australia represents a pivotal shift as he negotiates between Bangladeshi heritage and Australian societal norms. This process involves adapting to new cultural norms while maintaining connections to his religious and cultural roots, reflecting the fluidity inherent in transnational identities (Vertovec 2001). Similarly, Iskandar's short-term migration underlines his commitment to his Malaysian-Muslim identity while abroad. Despite his brief stay in a non-Muslim dominant nation, Iskandar's adherence to traditional values and integration within his community underscores the resilience of transnational identities in preserving cultural and religious continuities.

Iqbal's marriage to Michelle, a non-Muslim white woman, challenges traditional norms within his Bangladeshi-Muslim community, reflecting broader trends where personal choices clash with communal expectations in multicultural societies. This highlights the transformative potential of individual agency in reshaping identities amidst religious and cultural diversity. Conversely, Iskandar's preference for a spouse within his cultural and religious community emphasises his conformity with social expectations of preserving cultural heritage and maintaining religious continuity. His extramarital affair with Jasmine leads to feelings of guilt. While he will not fully commit to non-Muslim and non-Malay Jasmine, his Islamic values are not served by entering into an illicit sexual relationship with her that involves committing adultery while he is legally married to another woman. Iskandar's sexual promiscuity can be explained by Banks' (1990) observation about Malay Muslim men who "do not find Islam as sexually restrictive as Westerners apparently find Christianity" (p. 541). Islam also permits polygamy as it condones marriage between a man and up to four women, so it is possible for Iskandar to take Jasmine as his second wife if he so chooses. Iskandar's action in divorcing his Muslim wife instead signals his hypocrisy and lack of courage to face the truth about his own dilemma as a modern and transnational Muslim man living within the familial and national confines of Malaysia. Both protagonists' experiences illustrate the way that their Muslim identity crises stem from their relationships with non-Muslim women that conflict with societal norms within the Muslim national community.

In both cases, Muslim identities intersect with transnationalism through a negotiation of cultural and religious practices and values at home and in the diaspora. This dynamic interaction underscores the way Muslim men like Iqbal and Iskandar navigate and redefine their Muslim male identities within global contexts while maintaining their ties to religious and cultural heritage shaped by familial and national prescriptions.

Thus, the Muslim masculine identity crisis suffered by Iqbal and Iskandar is a result of the tension created by the national hegemonic discourse of Muslim identity and their transnational experiences. National hegemonic discourse prescribes strict patterns that combine religious beliefs and cultural norms, including the restriction of marriage among different faiths for which the spouse must convert to Islam and the adherence to certain religious practices, such as abstinence from *khalwat* or an unlawful sexual union outside of marriage. This is contrasted with transnational experiences that encourage an exercise of freedom to choose, embrace, and accept cultural diversity without jettisoning Muslim identity.

Iqbal and Iskandar embody the dilemma of Muslim men through the personal choices that they make and the challenges they experience. Iqbal's marriage to and Iskandar's extramarital relationship with non-Muslim women is a deviation from social norms, thus producing a conflict between what they choose and what is expected of them. Their dilemma exemplifies demands of accountability to maintain religious standards while dealing with interpersonal relationships in multicultural spheres of contact within and beyond the nation. Ultimately, their identity crisis epitomises a struggle for religious individualism against conformism, even while their religiously deviant acts may be understood as religious hypocrisy that leads to their sense of loneliness.

12.6 Conclusion

Karina Robles Bahrin's *The Accidental Malay* and Adib Khan's *Seasonal Adjustments* offer profound insights into the Muslim male identity crisis within trans/national contexts. While exploring its complexities, both novels do so by means of accounting for different national contexts. Each of their narratives highlight the social and personal predicaments encountered by Muslim men in navigating traditional values and modern influences within a multicultural society.

In challenging preconceived notions of Malay Muslim identity, *The Accidental Malay* ultimately portrays the male character's self-discovery and cultural integration in modern-day Malaysia. Its exploration of Iskandar's dilemma reflects struggles of cultural assimilation and adaptation that Muslim men face in the national setting while also influenced through their transnational experiences. Similarly, *Seasonal Adjustments* explores Iqbal's experiences as a Bangladeshi immigrant in Australia, depicting his quest for identity amid the cultural clash between his homeland traditions and adopted Western values. The choices that both protagonists make demonstrate the outcome of their transformations. A resolution of their identity crises is offered in their rejection of societal norms, which results in their exercise of personal autonomy that carries the caveat of their social alienation within the familial and national Muslim community.

Both novels illustrate the evolving dynamics of Muslim male identity that are influenced by transnational experiences of their characters. The literary works underscore the way that globalisation and migration shape identities, presenting nuanced narratives that resonate with the themes of cultural authenticity, religious identity, and individual autonomy in a globalised world. The authors do not deny that the Muslim identity of their protagonists is shaped by upbringing and cultural backgrounds. Nonetheless, Muslim masculinity undergoes a continual transformation through their interactions within a multicultural and transnational society, reflecting the dynamic evolution of Muslim identity in global contexts. In this way, the novels enrich an understanding of a negotiation of Muslim male identity crisis amid complexities of contemporary transnationalism.

References

Bahrin KR (2022) The Accidental Malay. Epigram Books.

Banks DJ (1990) Resurgent Islam and Malay rural culture: Malay novelists and the invention of culture. American Ethnologist 17(3):531–548. http://www.jstor.org/stable/644860

Esteban-Guitart M, Vila I (2015) The voices of newcomers: a qualitative analysis of the construction of transnational identity. Psychosocial Intervention 24:17–25. https://doi.org/10.1016/j.psi.2015.01.002

Faruque F (2022) The question of "writing back" in Adib Khan's "Seasonal Adjustments". Praxis: J Dept English Rajshahi Univ 13:123–136

Federal Constitution. (1957) Article 160. http://www.agc.gov.my/images/Personalisation/Buss/pdf/FederalConsti(BItext).pdf Accessed 10 July 2024

Grillo R (2004) Islam and transnationalism. Journal of Ethnic and Migration Studies 30(5):861–878. https://doi.org/10.1080/1369183042000245589

Hasan MK, Aggleton P, Persson A (2017) The makings of a man: social generational masculinities in Bangladesh. Journal of Gender Studies 27(3):347–361

Hasanat F (2017) Religion, diaspora and the politics of a homing desire in the writings of Zia Haider Rahman, Tahmima Anam and Monica Ali. Asiatic: IIUM Journal of English Language and Literature 11(1):57–68. https://doi.org/10.31436/asiatic.v11i1.962

Hearn J (2017) Men around the world: global and transnational masculinities. In: Armengol JM, Bosch M, Vilarrubias À et al (eds) Masculinities and literary studies. Routledge, pp 39–55. https://doi.org/10.4324/9781315230733

Ho HMY (2019) Women doing Malayness in Brunei Darussalam. Southeast Asian Review of English 56(2):147–165

Ho HMY (2021) Localisation of Malay Muslim identity in Brunei Darussalam: a modern nation's cultural and economic goals. In: Ho HMY, Deterding D (eds) Engaging modern Brunei: research on language, literature and culture. Springer, Singapore, pp 127–143

Ho HMY (2022) Beyond intractability: Muslim women negotiating identities in Brunei Darussalam. In: Jussawalla F, Omran D (eds) Muslim women's writing from across South and Southeast Asia. Routledge, London and New York

Ho HMY (2023) Resilient women: transnational homes and identities in Tahmima Anam's a golden age. South Asian Diaspora 15(2):255–270

Khan A (1994) Seasonal adjustments. Allen & Unwin, Sydney

Levitt P (2003) "You know, Abraham was really the first immigrant": religion and transnational migration. The International Migration Review 37(3):847–873. https://doi.org/10.1111/j.1747-7379.2003.tb00160.x

Mandaville P (2001) Transnational Muslim politics: reimagining the umma. Routledge, London and New York

Mandaville P (2009) Muslim transnational identity and state responses in Europe and the UK after 9/11: Political Community, Ideology and Authority. Journal of Ethnic and Migration Studies 35(3):491–506. https://doi.org/10.1080/13691830802704681

Mercanti S (2012) "No better or worse than anyone, but an equal": negotiating mutuality in Adib Khan's "Seasonal Adjustments". Journal of the Association for the Study of Australian Literature (JASAL) 12(2):1–13

Metcalf BD (1996) Introduction: sacred words, sanctioned practice, new communities. In: Metcalf BD (ed) Making Muslim space in North America and Europe. University of California Press, California, pp 1–27

Mukherjee D (2019) Malaysian English Writing Today. World Literature Today. https://www.worldliteraturetoday.org/blog/cultural-cross-sections/malaysian-english-writing-today-dipika-mukherjee

Ng AHS (2009) Islam and modernity in the works of two contemporary Malay Anglophone writers: Che Husna Azhari's "Mariah" and Karim Raslan's "Neighbors." The Journal of Commonwealth Literature 44(3):127–141

Omar N (2006) Colonialism and Malay masculinity: Malay satire as observed in the novel "Kawin-Kawin". Sari 24:107–120
Peletz MG (2021) Hegemonic Muslim masculinities and their others: perspectives from South and Southeast Asia. Comparative Studies in Society and History 63(3):534–565
Quayum MA (2020) Malaysian literature in English: A critical companion. Cambridge Scholars Publishing
Quayum MA, Hasan MM (2022) Bangladeshi literature in English: A thrice born tradition. Journal of Postcolonial Writing 58(6):733–743
Riaz A (2009) Interactions of "transnational" and "local" Islam in Bangladesh. In: Mandaville P, Noor FA, Horstmann A et al (eds) Transnational Islam in South and Southeast Asia: movements, networks, and conflict dynamics. National Bureau of Asian Research, pp 79–100
Roose JM (2021) Masculinity and Muslims: contemporary debates. In: Lukens-Bull R, Woodward M (eds) Handbook of Contemporary Islam and Muslim Lives Springer Nature Switzerland AG, pp 889–906 https://doi.org/10.1007/978-3-030-32626-5_46
Tan YEH (2003) Discourses of multiculturalism and contemporary Asian-Australian literature. Dissertation, University of Adelaide, Adelaide
Töngür AN (2013) Chanu's thwarted dreams and dislocation in Brick Lane by Monica Ali. Journal of Faculty of Letters 30 (3):249–262
Vertovec S (2001) Transnationalism and identity. Journal of Ethnic and Migration Studies 27(4):573–582 https://doi.org/10.1080/13691830120090386
Wee J (2022) On menopause and being Malay: author Karina Robles Bahrin opens up on award-winning debut novel "The Accidental Malay". Malay Mail. 2 September. https://www.malaymail.com/news/life/2022/09/02/on-menopause-and-being-malay-author-karina-robles-bahrin-opens-up-on-award-winning-debut-novel-the-accidental-malay/26088 Accessed 10 July 2024

Shafinur Nahar currently pursuing her Ph.D. at Universiti Brunei Darussalam, is dedicated to advancing the discourse on Muslim masculinities in South Asian and Southeast Asian Anglophone literature through her thesis. Her scholarly contributions include research articles published as chapters in esteemed books by academic publishers, such as Rutgers University Press and Routledge (Taylor & Francis). Her primary research interests encompass South Asian Diaspora writing, gender and religion, media and language, and the study of gender-based violence. Alongside her academic pursuits, Shafinur serves as the editor of the literary magazine *Prachya Review*.

Hannah Ming Yit Ho is an Assistant Professor at the Faculty of Arts and Social Sciences, Universiti Brunei Darussalam. She is an Associate at the Asia Research Institute at the National University of Singapore. Her research lies at the intersection of race, gender, and nation with an interdisciplinary focus on literature, culture, and education. She has been awarded grants to study cultural and literary representations of the Chinese in Brunei Darussalam. Her articles have been published in *Asiatic, International Journal of Asia Pacific Studies, Kritika Kultura, South Asian Diaspora, Suvannabhumi,* and *The Wenshan Review*. She coedited *Engaging Modern Brunei: Research on Language, Literature and Culture* (Springer, 2021).

Chapter 13
"South of Hong Kong, Almost as Big as Singapore": Transnational Identity and International Visibility in Contemporary Anglophone Bruneian Novels

Kathrina Mohd Daud

Abstract Despite a high level of literacy, widespread proficiency in English and a rich oral tradition, it was only in 2009 that the first Bruneian novel in English was published. Although the Malay language novel has a longer history dating back to 1951, the Bruneian literary ecology as a whole to date can be described as still nascent, with low levels of awareness, production, publication and dissemination of local literature across the country. It is not surprising then that the majority of media and literature consumed by Bruneians still originates from outside the country, with popular sources being East and Southeast Asia, the USA and the UK. Nevertheless, the landscape of Anglophone Bruneian literature is changing rapidly, with over 15 novels and novellas in English published since 2009, in addition to numerous collections of poetry, literary zines and original scripts for theatre and new media published and produced. Using Driscoll, Fletcher and Wilkins's notion of genre worlds to map out the "social, industrial and textual practices" (Wilkins et al. in Genre worlds: popular fiction and twenty-first-century book culture, University of Massachusetts Press, 2022) of contemporary Anglophone Bruneian novels as a genre, this chapter will show how the genre has constructed itself for international visibility through transnational rather than local identity construction, ranging from the material production of the texts through to their thematic and narrative contents. This chapter will map how the genre, influenced by an awareness of representations of Brunei from outside the country, has employed various narrative and paratextual tactics in order to carve out and (re)claim a space in the global imaginary.

K. Mohd Daud (✉)
Universiti Brunei Darussalam, Bandar Seri Begawan, Brunei Darussalam
e-mail: kathrina.daud@ubd.edu.bn

H. M. Y. Ho and Y. Chan (eds.), *Transnational Southeast Asia*, Asia in Transition 29,
https://doi.org/10.1007/978-981-96-3608-2_13

13.1 Introduction

"We wanted to **put Brunei on the map**,"—KYSB on producing an EDM music album, *Sounds of the Brotherhood*, a compilation of Bruneian musicians.

"I really want to put **Brunei on the map** in the area of music [.]"—singer Maria Grace Koh.

"One of my main goals is to **put Brunei on the map** of boxing."—Muhammad 'Irfan bin Haji Abdul Razak, a Bruneian boxer.

"The main thing for us is still to **put Brunei on the esports map** [...]"—Yee, manager of the Goodfellas Gaming team.

"Wu Chun's contribution to the country has been massive all these years and has literally **put Brunei on the map** [...]"—Rano Iskandar on the actor Wu Chun.

"Superfish Growers wants to be the largest producer of Japanese premium muskmelons in Southeast Asia, hoping to **put Brunei on the map** with its newly introduced product brand, Melts."—Wardi Wasil

"LiveWIRE Brunei Programme Manager Hj Abd Yusrin Dato Hj Mohd Salleh said The shortlisting helps **put Brunei on the map** for innovation and entrepreneurship."—Aaron Wong

As observed previously (Kathrina 2019), the phrase "putting Brunei on the map" is one that appears with regularity in local media and social media, particularly in reference to achieving an international standing as a barometer of excellence or achievement. In the examples above, it is frequently used aspirationally, in the context of the aspiring individual or organisation articulating their motivations and *raison d'etre.*

The premise that underlies the use of this phrase is that of Brunei Darussalam's relative or perceived invisibility in the international realm. This perception ranges from the fields of sports, to music, to agriculture and innovation. International visibility can thus be said to be an aspirational measure for many Bruneians seeking validation of their competency and expertise, both on an individual and a representative level.

The Bruneian arts critic and commenter Faiq Airudin identified another aspect of this aspiration in the creative industries, as "firstism" (2016), in which creative organisations, products and/or individuals claim to be the first at achievement or existence in order to, amongst others, garner local support and attention, frequently by identifying firstness in an international arena (Kathrina 2019).

This aspirational desire for a global presence is central to the work of this chapter, which proposes that this preoccupation has shaped the transnational identity construction of the Anglophone Bruneian novel as a genre. By considering the contemporary Anglophone Bruneian novel as a genre, and understanding this genre as part of a literary ecology, it becomes possible to identify the transnational traits, influences and outlooks which thus shape the contents of the genre, in particular the way that Brunei itself has been imagined and articulated. Given that, as elsewhere, the novel form in Brunei has historically been intertwined with the nation-building project, its imaginings of the nation provide significant insight into the primary discursive tensions which are at stake around the concept of Bruneian identity, particularly in the context of Bruneian-ness in relation to the world.

13.2 The Contemporary Anglophone Bruneian Novel

The history and development of Anglophone Bruneian literature have been amply covered elsewhere (Ampuan et al. 2023; Chin 2016; Deterding and Ho 2021; Kathrina et al. 2016). Separate from the development of Malay language literature in Brunei Darussalam, yet echoing its developmental trajectory and traits, Anglophone Bruneian literature only really begins to manifest as a discernible body of work in the 2000s, with the most significant medium being that of the novel.

In this chapter, I define the Anglophone Bruneian novel accordingly: any novel or long-form work of fiction written in English which meets at least two of the following criteria:

(1) Has been produced or published in Brunei;
(2) Is set in or is about Brunei, or an analog of Brunei;
(3) Has been written by someone who explicitly either identifies themselves as Bruneian or as Bruneian-based in the paratextual material.

The aim of this definition is not meant to be prescriptive, or restrictive, but to offer useful working parameters for the genre, within which to reflect on novels which are explicitly and/or directly invested or engaged in the arena of Bruneian literary development; that is, novels which identify or position themselves as Bruneian novels in some way. Using the criteria above, I exclude from the genre of the contemporary Anglophone Bruneian novel, works such as Anthony Burgess' *Devil of a State* (1961). Burgess, who was shortlisted for the Nobel Prize in Literature in 1973, and possibly best known for his dystopian novel *A Clockwork Orange* (1962), confirms in his autobiography, *Little Wilson and Big God* (1986), that *Devil of a State* was about Brunei, where Burgess had worked as a teacher in the 1950s until health issues forced his return to the United Kingdom. Thus I categorise the novel as a novel about Brunei, rather than being a Bruneian novel.

As such, this definition accepts what previous studies and scholars have generally identified the first Anglophone Bruneian novel, Selamat Munap's *The Wild Men of the East* in 2009, with approximately twenty novels and novellas which can be considered Anglophone Bruneian novels having been published since then—an average of approximately one text a year. This average is slightly misleading, however—five novels were published in 2022, an unprecedented high demarcating a clear growth in the industry.

About half of these have been self-published through platforms such as Raider Publishing or CreateSpace and are available both digitally and in print; the other half have been traditionally published through five domestic publishers and two Singaporean publishers; the genres range from literary, satirical, thriller, speculative, fantasy and coming-of-age, amongst others. This diversity is indicative of the relative youth of the industry—many of these novels are the first of their kind in the sphere of Anglophone Bruneian literature and thus can be plausibly said to be drawing not (solely) from a domestic literary tradition but from a global one.

Subsequently, studies of Anglophone Bruneian literature, including my own work, have generally focused on individual texts, situating them in the global context through the use of critical lenses such as ecocriticism (Rinni 2021; 2022), postcolonial discourses about identity, nation, gender, and home (Chin 2016, 2018; Ho 2021; Ho and Rinni 2021; Kathrina 2017), including Chinese diasporic identity (Ho 2021), and in relation to popular genres such as romance (Kathrina 2021), global Muslim fiction (Kathrina 2019), horror (Kathrina 2016), and speculative fiction. These close-reading approaches, and the transnational focus on the texts as part of global book culture and discourses are a natural response to two factors: first, the seeming lack of connectivity between the novels themselves as a body of work, and secondly, the self-evident participation of the novels in "the transnational processes of contemporary book culture," both discursively and materially. Nevertheless, the limitation of the close-reading approach is that it provides an incomplete understanding of the literary landscape as a whole, which then obscures the way in which the processes of production and circulation, amongst others, influences and shapes the texts as narratives.

This chapter offers a way to understand these texts collectively, proposing that as a body of work, they constitute a genre within a singular literary ecology, which governs the way they construct themselves for an audience.

13.3 Literary Ecologies and Genre Worlds

Alexander Beecroft's typological model of ecologies to understand the "modes of literary reception and circulation throughout human history" in *An Ecology of World Literature* (2015) is predicated on the understanding that.

> "It is impossible to understand any given literature qua literature solely through an analysis of the texts read through it. Rather, any given literature must, I believe, be understood as being in an ecological relationship to other phenomena—political, economic, sociocultural, religious—as well as to the other languages and literatures with which it is in contact" (p. 59)

This framework emphasises the "distinct and mutually interactive nature" (p. 59) of the various ecological phenomena, in which any changes to one phenomenon has the potential to alter another, and vice versa. To understand Anglophone Bruneian texts as part of a distinctive literary ecology is to interrogate how the specific phenomena within that ecology have in fact created and shaped the genre of the contemporary Anglophone Bruneian novel.

Methodologically, I draw on the work of Lisa Fletcher, Beth Driscoll and Kim Wilkins, who have proposed the genre worlds approach as a corrective to the overemphasis on close-reading genre texts. Fletcher et al. suggest that "while close reading genre texts is an important part of understanding their cultural meaning, a broader view is enabled by attending to the collaborative networks in which they arise" (2018, p. 4), that is, the elements which make up the genre world. According to Fletcher et al., the genre world describes "the collective activity that goes into the creation and circulation of genre texts, and is particularly focused on the communities, collaborations and industrial pressures that drive and are driven by the processes of these socio-artistic formations."

As a methodology, it is "an analysis of the interrelated parts of the genre world that can be achieved by identifying the cast of characters, examining the cooperative links between them, and investigating what influence they exert on the final work of art, that is the genre fiction text." (Fletcher et al., *Genre Worlds and Popular Fiction,* 2018). These interrelated parts can be broadly categorised as the social, textual and industrial elements which act on the text.

Taking this approach to the genre world of the contemporary Anglophone Bruneian novel offers a way of understanding the genesis, development and status of the genre. As Wilkins et al. note, "genres are formed not only textually (that is, by recognisable and categorisable textual features) but also socially and through industrial processes" (2022, p. 7). The next section will chart how these elements, which are transnational in scope, have worked upon the Anglophone Bruneian novel as a genre to produce texts which are themselves transnational in orientation and construction.

13.4 The Transnational Identity of Contemporary Anglophone Bruneian Novels as a Genre

In a previous section, I outlined the parameters by which I identify contemporary Anglophone Bruneian novels as a genre. Of these parameters, only one has to do with the aesthetics or thematics of the text (set in/about Brunei or an analog of Brunei). The other two parameters have to do with the paratextual element of authorial self-identification ("Has been written by someone who explicitly either identifies themselves as Bruneian or as Bruneian-based in the paratextual material") and the means of production of the text as a cultural, material artefact ("Has been produced or published in Brunei"). These criteria can also be broadly understood as being drawn from the social, textual and industrial facets of the literary ecology in which this genre is enmeshed. This section will briefly sketch the transnational features and influences of the social and industrial aspects of the genre; and how this has influenced the textual aspects, in the following section.

The social aspect of the genre world refers broadly to the ways that people "interact with books, including writing them, talking about them, studying them, reviewing them, attending conventions, interacting with fandoms and so on" (Wilkins et al. 2022, p. 26). In the context of the contemporary Anglophone Bruneian novel, the social aspect refers to not just the writers and the readers, but also to the writers' awareness of their audience, and how this shapes the ways that the writers position their works in the paratextual material, including author biographies, interviews, forewords and afterwords and so on. The contention of this chapter is that writers in the genre demonstrate that the expected and aspired-for audience or readership is transnational and international rather than local through both their choice of publishing platforms as well as their paratextual statements; thus that they are not just Bruneian writers, but Bruneian writers in the world. These paratexts offer us an insight into the perceptions of the genre world as it exists.

An example is this afterword by KH Lim, in the 2014 novel *Written in Black*. Lim writes "to the reader interested in learning more about Brunei, there is plenty of information available on the web for your perusal. Or better yet, you could go over and see for yourself." Lim's novel was the first internationally traditionally published Anglophone novel in the genre and was heavily influenced by the American writer Mark Twain's *The Adventures of Huckleberry Finn*. This afterword demonstrates an expectation of a non-local audience.

The next novel to be traditionally published, Aisha Malik's halal romance *Jewel* in 2019, had previously been published as an Amazon ebook in 2017. It was not available in print until it was published by the small Bruneian publisher Heartwrite Co. In personal communications with the writer, Malik acknowledged the difficulties that Bruneian-based readers would have accessing the book, given the structure of the Amazon digital publishing platform, which requires a credit card (Kathrina 2019, p. 202). Additionally, at the time, not all local banks allowed purchases on Amazon. Malik, who identifies as a Muslim writer but not a Bruneian writer, situates herself and her readership clearly as transnational—"young Muslims anywhere; not particularly Brunei or anything like that" (p. 202). As above, despite the different genres (coming-of-age and Muslim romance), these "early" Bruneian writers shared an expectation of readership as non-local.

The novels from 2020 to present, including Kathrina Mohd Daud's *The Fisherman King* (Epigram 2020), Tina Afiqah's *The Bubble Princess and the Stone Heart* (Qasrun Nafis 2021) and Jeemah Jefri's *The Tale of the Unsung Tailor* (Qasrun Nafis 2022) marked a new development in the genre world, as the authors articulated a desire for their works to be read by a domestic as well as an international audience. There was a shift in emphasis from an international readership to a domestic and international readership. In an interview with the *Borneo Bulletin*, Tina Afiqah noted that the promoting "acknowledgement of traditional stories" (Adib 2022) was a central impetus to the publications produced by ParaBelle Studios, including *The Bubble Princess and the Stone Heart*. Tina also noted that "similar to Disney, we strive to find new and inventive ways to ensure that our stories continue to be embraced and appreciated by locals and international friends alike" (Adib 2022).

This shift in desired readership marks an orientation inward, that is reflected in the narratives produced within this first generation of Anglophone Bruneian novels (the textual element) and made clear in the changing industrial aspect of the genre.

The industrial aspect of genre worlds refers broadly to those "arenas of activity that are shaping the twenty-first-century production and circulation of genre fiction: editing, distribution, self-publishing, and algorithms" (Wilkins et al. 2022, p. 26).

A clear shift from seeking an international audience to a regional and domestic one can be observed based on the publishing platforms chosen by the writers. The majority of Anglophone Bruneian novels from 2009 to 2020 were self-published, and available digitally to an international readership. From 2020 onwards, there has been a marked preference for traditional publishing, with domestic publishers, in particular, Qasrun Nafis Publishing House (QNPH) and M Content Creations leading the way. This is despite the fact that domestic publishers have a very limited distribution network for print internationally, and of these, only M Content Creations makes its novels available in digital form as ebooks (i.e. feasibly accessible by an international audience). By and large, the other domestic publishers make novels available only in print, and primarily for the domestic market through local bookstores,[1] meaning that these novels are only accessible by a local readership.

Examining the existence of these domestic publishers also offers an insight into the development of the industry, and further illuminates the difference between what I shall call the two "phases" of the Anglophone Bruneian novel (the first phase being primarily international in orientation (2009–2019), and the second phase shifting focus to both domestic and international (2020-present), signaling a shift in its engagement with the transnational. Of the four domestic publishers which have published novels (QNPH, M Content Creations, Heartwrite Co and Dewan Bahasa dan Pustaka), only one is older than a decade (the Dewan Bahasa dan Pustaka).[2] The production of the Anglophone Bruneian novel from 2020 is dominated by young domestic publishers.

This turn towards a domestic audience is particularly interesting for a medium that Brennan calls "naturally cosmopolitan" (1990, p. 56). As Brennan points out, historically, "[...] under conditions of illiteracy and shortages, and given simply the leisure-time necessary for reading one, the novel has been an elitist and minority form in developing countries when compared to poem, song, television and film. Almost inevitably it has been the form through which a thin, foreign-educated stratum (however sensitive or committed to domestic political interests) has communicated to metropolitan reading publics, often in translation" (1990, p. 56).

This orientation towards the domestic, however, does not preclude the transnational nature of the industry. As per Tina Afiqah's comments, there is still a strong sense of awareness of the potential for a regional and international audience. The shift to a more domestic focus is a shift in emphasis rather than an exclusion; certainly, as the next section will show, the textual elements of the genre are, in influence and inception, unmistakably transnational.

13.5 Imagining Brunei for a Global Audience

"Outsider" representations of Brunei, both in news media and fiction, are dominated by an Orientalist fascination with the perceived decadence and hedonism of the monarchy and with Brunei's Islamic identity. For example, as Kathrina (2019) notes,

> "In 2013, when it was officially declared that Brunei would be introducing and implementing the Syariah Penal Code, there was an international outcry, with international media and activists calling the move a return to 'the dark ages'. (Buncombe 2014)" (p. 193)

Indeed, Western media's coverage of the announcement was marked by sensationalist rhetoric that drew on Islamophobic anxieties about fundamentalism, radicalism and extremism. In "Representations of Islam in Anglophone Bruneian fiction" (2019), I mapped out how Bruneian Islam as represented in Anglophone Bruneian fiction, demonstrated an awareness of these external representations of Brunei and offered correctives to these negative stereotypes through their portrayal of Bruneian Islam as part of a global (transnational) and "moderate" Islamic identity (p. 202, 204).

Outside of Brunei's Islamic identity, popular media has focused on the decadence, hedonism and excess of the monarchy, with a representative article in the American magazine *Vanity Fair* titled "The Prince Who Blew Through Billions" in 2011 (Mark Seal). The Sultan of Brunei has become a byword for extreme wealth and excess in pop culture. Some examples include:

> **Miguel Syjuco *Ilustrado*** (2010) "The Manila Polo Club, where he and Jerald learned to play tennis, always swinging far over the back fence to see the ball boys climb like monkeys after the balls. The large home of the U.S. Ambassador, with its concerns and high walls. The large home of the Sultan of Brunei's brother's Filipina mistress…" (p. 118) "Or many become domestic or contract workers in Manila. Maybe Brunei or Saudi." (p. 97)
>
> **Kevin Kwan *Crazy Rich Asians*** (2013) "I used to see them all the time in Brunei. The Sultan of Brunei only uses Gurkhas as his private protection force." (p. 122) "privacy obsessed Brunei princesses" (p. 272) "the Sultan of Brunei's convoy" (p. 277)

Earlier studies (Kathrina 2019, 2021) have shown how early Anglophone Bruneian novels have clearly demonstrated awareness of these external representations and provided correctives by explicitly engaging and refuting stereotypes about Brunei's monarchy, emphasising its paternalistic, familial aspect. Other texts such as Aammton Alias' Bunian Conspiracy series have pushed back against the perception of Brunei as a primitive, invisible "jungle island" (Amir 2013) by emphasising the value of the rainforest biodiversity in the age of the Anthropocene, reframing what has historically been a point of denigration into a point of pride.

Novels from 2020 onwards, however, in their pivot away from a global audience, have also seen a narrative change. A new focus on re-tellings and reclamations of folktales—*The Fisherman King* (2020), *The Bubble Princess and the Stone Heart* (2021) and *Pulau Jong Batu* (2022) have emerged, moving away from engagement with sensationalist external media narratives about Brunei, towards a fresh "refraction of an imagined nation" (Nwakanma 2008, p. 8) through their re-articulations of local myths, legends and folklore. Given that these articulations have, as established in the previous section, been in the main, brokered through a domestic publishing ecosystem, it can be said that the intended audience is primarily local. The second phase of the development of the Anglophone Bruneian novel has turned away from rebutting international perceptions of Brunei, focusing instead on establishing narratives which do not have the primary focus of "writing back" to the gaze of non-resident readers. This new phase from 2020 signals a decolonial turn in the contemporary Anglophone Bruneian novel, which is engaged with the transnational in a different way from the earlier phase.

How this re-orientation has influenced local narrative-making can be seen in the preference not just for narrative reclamations but also in the preference for historical and speculative Bruneian settings, with only two Anglophone Bruneian novels to date that have chosen the realistic mode to represent contemporary Brunei: KH Lim's *Written in Black,* 2014 and Jeemah Jefri's *Not Your Alai,* 2023, from this second "phase."

In the non-realistic mode, novels have offered narratives either from historical (Kathrina Mohd Daud's *The Witch Doctor's Daughter* (2022) and Jeemah Jefri's *The Tale of the Unsung Tailor* (2022)) or futuristic Bruneian timelines (Amir Falique's *The Forlorn Adventure,* 2013); or of alternative, speculative versions of Brunei, such as Hamizan Yasin's *Baffling Journey of Bino and Jana* (2022), set in a fictional fifth district in Brunei. These portrayals of Brunei which mix fact, fiction and the fantastic in a way that would not necessarily be discernible to a reader unfamiliar with Brunei, suggests a new unconcern with corrective accuracy, towards an imaginative expansion, that while inevitably still affected by global discourses (for example, around climate change and geo-political tensions, as well as the regional turn towards speculative fiction) is not primarily concerned with rebutting negative or incomplete perceptions of Brunei. This also suggests less focus on the international visibility that was central to the thematics of the novels from the first phase.

Even while in the realistic mode, for example, the engagement with the transnational in the second phase is demonstrably less a corrective response than it is a carving out and signaling of how global concerns have affected local sensibilities. In *Not Your Alai* (2023), for example, Jeemah Jefri posits local considerations of femininity as part of global feminist discourse, while keeping the focus firmly on the local.

This re-orientation has also manifested in a turn away from popular transnational genres such as romance and sci-fi, towards what can be understood as literary fiction, suggesting that the Anglophone Bruneian novel is re-balancing within what literary scholar Franco Moretti triangulated as a compromise between "foreign *plot*; local *characters*; and then, local *narrative voice*" (2000, p. 65) that he identified as part of a culture's movement towards the modern novel (Kathrina, forthcoming). While literary fiction as a genre also has clear transnational influence and traits, these are less prescriptive than popular fiction and allow local narratives to develop a novel form that is more representative, perhaps, of the conventions, norms and aesthetics of local stories as they have developed from their oral literary past.

13.6 Going Forward

This chapter began by considering the prevalence of the desire in the domestic creative industries to "put Brunei on the map" as an aspiration and measure of success, with the phrase denoting a desire for international recognition and acclaim. This chapter will conclude by suggesting that it has been the work of what I will designate as these two phases or generations (2009–2019, 2020 onwards) of Anglophone Bruneian novels to consider seriously the versions of Brunei which enter the global space, and that they do this "self-portraiture" (Izzuddin 2023, p. XI) from a transnational framework socially, textually and industrially.

Literary scholar Eddie Tay noted of the Singaporean context that "close readings of literary texts are necessary, but alone do not provide a sufficient understanding of the contemporary literary landscape" (2017, p. 121). In a country like Brunei Darussalam, where the production of literature has historically been bound with the project of nation-building, constrained by cultural policy and influenced by genres and literature from outside the country, using a genre worlds methodology to understand the literary landscape allows for repeated tropes, rhetoric and narrative and aesthetic choices to be understood as part of a genre pattern and convention.

The nature of genres is evolutionary, and even within a genre as young as the Anglophone Bruneian novel, there have been shifts and developments since 2009, with texts responding not just to each other, but to non-textual elements such as audience reception, political pronouncements and industry developments.

An examination of the social, industrial and textual aspects of the genre world of the Anglophone Bruneian novel shows that it must be understood as a transnational rather than a local product. While the inequalities of power in contemporary transnational book culture have limited the mobility and visibility of Bruneian novels beyond the local, those same inequalities have influenced and shaped not just the orientation of Bruneian novels towards a first global and then domestic audience, but also their aesthetic and narrative form. These flows may be understood in Beecroft's terms as multi-directional as well as reciprocal.

In this chapter, I have proposed understanding the Anglophone Bruneian novel as a genre world in itself, a way of understanding the novels as a body of work with shared characteristics, that illuminate the transnational way that Brunei as a nation is imagined and articulated by Bruneian artists, in response to the specific social, textual and industrial elements of this genre world at this particular point in time. This chapter has elucidated the transnational identity of the genre world (materially and ideologically), and how there are limitations to its mobility which reflects the state of global contemporary book culture; a transnational identity without international visibility, and increasingly a domestic orientation.

Going forward, it is crucial to remember that in the context of national identity, and identity formation, the contemporary Anglophone Bruneian novel occupies a different position in the construction of the social imaginary than other genres, e.g. the Malay language novel, theatre, film, short stories and poetry. The genre worlds methodology allows us to interrogate these other genres and mediums more effectively, offering a fuller understanding of how, why and to whom a nation is narrated in an era of globalisation.

Notes

1. With minor exceptions.
2. Which published its first books in 1963 (Nellie DPHS 2012, p. 5).

References

Adib N (2022) Traditional tales for a new generation. https://borneobulletin.com.bn/traditional-tales-for-a-new-generation/ Accessed 1 July 2024

Aisha M (2017/2019) Jewel. Ebook/Heartwrite Co, Bandar Seri Begawan

Amir F (2013) The forlorn adventure. Trafford Publishing

Ampuan Haji Brahim AHT, Kathrina MD (2023) An overview of Bruneian literature: history, historiography and critical approaches. In: Ooi KG, King VT (eds) Routledge handbook of contemporary Brunei. New York: Routledge, New York, pp 189–217

Beecroft A (2015) An ecology of world literature: from antiquity to the present day. Verso, London and New York

Brennan T (1989) Salman Rushdie and the third world: myths of the nation. Palgrave Macmillan, London

Brennan T (1990) The national longing for form. In: Bhabha HK (ed) Nation and narration. Routledge, London and New York, pp 44–70

Buncombe A (2014) International outcry as Brunei introduces Sharia Law and takes country back to dark ages. https://www.independent.co.uk/news/world/asia/international-outcry-as-brunei-introduces-sharia-law-and-takes-country-back-to-the-dark-ages-9308088.html

Burgess A (1961) Devil of a state. William Heinemann Ltd, London/Melbourne/Toronto

Burgess A (1962) A clockwork orange. William Heinemann Ltd, United Kingdom

Burgess A (1986) Little Wilson and big god. Weidenfield and Nicholson, United Kingdom

Chin GVS (2016) Bruneian women's writing as an emergent minor literature in English. In: McLellan J, Chin GVS (eds) Special issue: English in Brunei Darussalam. World Englishes 35(4):587–601

Chin GVS (2018) Counter-narratives of the nation: writing the modern Brunei Malay woman. In: Chin GVS, Kathrina MD (eds) The Southeast Asian woman writes back: gender, identity and nation in the literatures of Brunei Darussalam, Malaysia, Singapore, Indonesia and the Philippines. Springer, Singapore, pp 129–148

Deterding D, Ho HMY (2021) An overview of the language, literature and culture of Brunei Darussalam. In: Ho HMY, Deterding D (eds) Engaging modern Brunei: research on language, literature and culture. Springer, Singapore, pp 1–17

Faiq A (2016) Creative industry needs support. https://faiqairudin.com/writing/creative-industry-needs-support/ Accessed 19 June 2024

Fletcher L, Driscoll B, Wilkins K (2018) Genre worlds and popular fiction: the case of twenty-first-century Australian romance. J Popular Culture 51(4)

Helgesson S (2018) General introduction: the cosmopolitan and the vernacular in interaction. In: Stefan Helgesson S, Alling AM, Lindqvist Y, Wulff H (eds) World literatures: exploring the cosmopolitan-vernacular exchange. Stockholm University Press, Stockholm, pp 1–14

Hamizan Y (2022) Baffling journey of Bino and Jana. M Content Creations
Ho HMY (2021) Localisation of Malay Muslim Identity in Brunei Darussalam: a modern nation's cultural and economic goals. In: Ho HMY, Deterding D (eds) Engaging modern Brunei: research on language, literature and culture. Springer, Singapore, pp 127–144
Ho HMY, Rinni HA (2021) Constructs of the modern home: negotiating identity in anglophone Bruneian literature. In: Ho HMY, Deterding D (eds) Engaging modern Brunei: research on language, literature and culture. Springer, Singapore, pp 163–176
Izzuddin R (2023) Editor's note. In: Izzuddin R (ed) Ink-Cantations: written self-portraits from Borneo. Buku IBDE, Kuala Lumpur, pp I–XIII
Jeemah J (2022) The tale of the unsung tailor. Qasrun Nafis Publishing House
Jeemah J (2023) Not your alai. Qasrun Nafis Publishing House
Kathrina MD (2016) The ghosts of Islam and the unseen world in Bruneian horror stories. World Englishes 35(4):602–611
Kathrina MD (2017) Articulating female citizenship in Norsiah Gapar's Pengabdian". In: Chin GVS, Kathrina MD (eds) The Southeast Asian Woman writes back: gender, identity and nation in the literatures of Brunei Darussalam, Malaysia, Singapore, Indonesia and the Philippines. Springer, Singapore, pp 41–54
Kathrina MD (2019) Packaging MIB: representations of Islam in Anglophone Bruneian fiction. In: Fischer J, Jammes J (eds) Muslim piety as economy: markets, meaning and morality in Southeast Asia. Routledge, New York, pp 193–210
Kathrina MD (2020) The fisherman king. Epigram, Singapore
Kathrina MD (2021) Translating Islam: conversion and love in Bruneian fiction. In: Chin GVS (ed) Translational politics in Southeast Asian literatures: contesting race, gender and sexuality. London and New York: Routledge, London and New York, pp 132–150
Kathrina MD, Chin GVS, Maslin J (2016) Contemporary English and Malay literature in Brunei: a comparison. The use and status of language in Brunei Darussalam: a kingdom of unexpected linguistic diversity. Springer, Singapore, pp 241–251
Kwan K (2013) Crazy rich asians. Doubleday
Lim KH (2014) Written in black. Monsoon Books
Moretti F (2000) Conjectures on world literature. New Left Review
Nellie DPHS (2012) Empowering a reading culture: a Brunei Darussalam's perspective. In: Paper presented at the TK conference on reading, Bangkok, 10–11 May 2012
Nwakanma O (2008) Metonymic eruptions: Igbo novelists, the narrative of the nation, and new developments in the contemporary Nigerian novel. Research in African Literatures 39(2):1–14
Rinni HA (2021) Examining the imagined environments in contemporary Bruneian fiction: developing Southeast Asian ecocriticism. In: Ho HMY, Deterding D (eds) Engaging modern Brunei: research on language, literature and culture. Springer, Singapore, pp 145–162
Rinni HA (2022) Into the haunted forest: reading anti-extractivism in Aammton Alias's the Last Bastion of Ingei. SARE 59(1):70–87
Seal M (2011) The prince who blew through billions. Vanity Fair
Selamat M (2009) The wild men of the east. Raider Publishing International
Syjuco M (2010) Ilustrado. Tantor Media, Incorporated
Tay E (2017) The Singapore Writers Festival: global identity, cultural policy and the post-independence literary landscape. Southeast Asian Review of English 50(1):121–132
Tina A (2021) The bubble princess and the stone heart. Qasrun Nafis Publishing House
Wilkins K, Driscoll B, Fletcher L (2022) Genre worlds: popular fiction and twenty-first-century book culture. University of Massachusetts Press
Yen TB (2022) Pulau Jong Batu. [Jong Batu island]. Qasrun Nafis Publishing House

Kathrina Mohd Daud is an Assistant Professor in the Faculty of Arts and Social Sciences, Universiti Brunei Darussalam. Her research focuses on the intersections of popular fiction, Bruneian fiction, and representations of religion in literature. She co-edited *The Southeast Asian*

Woman Writes Back: Gender, Identity and Nation in the Literatures of Brunei Darussalam, Malaysia, Indonesia and the Philippines (Springer, 2017) with Grace V.S. Chin, and her work has appeared in several edited volumes published by Routledge and Springer, as well as in the *Journal of Commonwealth Literature* and *World Englishes*.

Chapter 14
Between the Imaginary and Reality of Dunia Melayu: MABBIM as a Case Study of Transnational Cooperation in Language Planning

Badriyah Yusof, Elmy Maswandi, and Wafiqah Asnola

Abstract The Majlis Bahasa Brunei, Indonesia, Malaysia (MABBIM) was established in 1985 with a shared interest to develop, fortify and promote Bahasa Melayu/Indonesia as the official language, language of the region, and one of the dominant languages in the world. In hindsight, the establishment of MABBIM did not just mark the shared geopolitical location of the countries within MABBIM, it also highlights the shared historical and cultural values of these nations as embedded in the term *bahasa serumpun*. MABBIM's establishment also demonstrates an aspiration to relive an apex moment of Malay Civilisation history in the ninth and seventeenth centuries, in which Bahasa Melayu/Indonesia was a lingua franca. At the nexus of this aspiration and the reality of the translanguaging community, this chapter critically reflects on the role and challenges of MABBIM as a regional language council in contemporary Southeast Asia by analysing to what extent this aspiration is realised, negotiated, and challenged in each participating country's language policy and planning amid sociopolitical and sociocultural practices; and at the regional level (particularly considering the effort to make Bahasa Melayu an official ASEAN language). This chapter will shed light on the strengths and limitations of transnational cooperation in language planning in the current setting and how this information can be utilised to improve both state and regional policies.

Keywords Transnational cooperation · MABBIM · Language policy and planning · Imagined community · Dunia Melayu

B. Yusof (✉) · E. Maswandi · W. Asnola
Faculty of Arts and Social Sciences, Universiti Brunei Darussalam, Gadong, Brunei Darussalam
e-mail: badriyah.yusof@ubd.edu.bn

H. M. Y. Ho and Y. Chan (eds.), *Transnational Southeast Asia*, Asia in Transition 29,
https://doi.org/10.1007/978-981-96-3608-2_14

14.1 Introduction

For decades, transnationalism has been traditionally discussed in terms of flows and exchanges of ideas, people, and objects across nation-states. Owing to globalisation brought about by migration, transnationalism transcends political, cultural and economic activities and permeates the thoughts and behaviours of individuals, society and state at large. Due to its multidimensional, dynamic and complex nature, Portes et al. (1999, p. 220) propose the need to have a certain unit of analysis in delimiting the domain of discussion on transnationalism. This unit, comprising 'communities, economic enterprises, political parties', amongst others, methodologically determines the system of networks and peripheries circumventing the phenomenon. Understanding this limitation is not only imperative in determining the scope of discussion but also a point of inquiry. Thus, studies on transnationalism are often approached from the grassroots level, or 'from above' (Smith and Guarnizo 1998), where focus is given to the initiatives operated by institutional actors including state leaders, policymakers, or gatekeepers, to name a few. Accordingly, this 'from above' approach resonates with the aim of this chapter, i.e. to reflect on the function of transnational cooperation as an instrument to promote the political and cultural ideals of Nusantara or the Malay World through Bahasa Melayu. Epistemologically, the establishment of this statutory body in the first place is also seen as a way to relive the language's idyllic supremacy as a lingua franca in the region—as a language of religious, cultural, economic and political importance.

14.1.1 Transnational Cooperation Across Domains

Since the dawn of civilisation, transnational cooperation has adorned the scene of politics and international relations. Transnational cooperation can come in the form of bilateral or multilateral exchanges of resources between participating states, often mutually agreed upon to serve the interest of the local, regional, and, to an extent, international community. At the local forefront, this cooperation translates further into governing mechanisms in the form of policies, regulations, or activities in general, operationalised in calibration within the political and sociocultural context of the country. By the same token, this cooperation also brings stability, economic progress, and a positive political image to the region as states can better address shared challenges such as terrorism, trade barriers, and environmental issues in a collaborative manner rather than independently.

Examples of transnational cooperation can be found in economic partnerships, such as those within the European Union (henceforth, EU) and its policies of facilitating the free movement of goods, services and labour, leading to increased economic stability and growth for member states. Regional cooperation in security matters is exemplified by initiatives under the North Atlantic Treaty Organization (NATO), which involves enhancing collective defence capabilities and deterring potential threats. Cultural exchange programs and educational collaborations also enrich societies by promoting mutual understanding and tolerance, which can lead to harmonious international relations. Ultimately, the pooling of resources and expertise through transnational cooperation enables states to achieve common goals more efficiently, contributing to regional stability and prosperity.

In the realm of language planning, the European Union serves as a prime example of how transnational cooperation can benefit member states with shared interests. Rooted in the principle of linguistic diversity, the EU's language policy recognises 24 official languages, and supports numerous regional and minority languages. This policy aims to ensure that EU citizens can access information and services in their own languages, fostering inclusivity and participation in the democratic process (European Commission 2024). The establishment of the Erasmus + program in the education sector further facilitates student and staff mobility across member states, enhancing communication and promoting intercultural understanding (European Commission 2023).

Similarly, UNESCO's language programs also provide a global perspective on transnational language planning, focusing on the preservation and promotion of linguistic diversity worldwide. Recognising languages as key to cultural identity and sustainable development, the organisation's initiatives such as the 'Atlas of the World's Languages in Danger' raises awareness about endangered languages and mobilises efforts for their revitalisation (Moseley 2010). Similarly, UNESCO's Education for Sustainable Development or ESD program also promotes multilingual education as a means to enhance learning outcomes and cultural awareness, particularly in multilingual societies (UNESCO 2017). By fostering international cooperation and providing frameworks for effective language policies, UNESCO maximises its role in preserving linguistic diversity and promoting intercultural dialogue on a global scale.

In Southeast Asia, the Association of Southeast Asian Nations (ASEAN) exemplifies transnational language planning by promoting English as a lingua franca to facilitate regional integration while respecting the linguistic diversity of its member states. Therefore, English is used as the working language for ASEAN's official documents and meetings, enabling effective communication and cooperation among countries with diverse linguistic backgrounds (Kirkpatrick 2010). At the same time, ASEAN recognises the importance of maintaining national languages and cultures, which are integral to the identity and heritage of its member states. Initiatives such as the ASEAN Language and Culture Program aim to preserve and promote the region's linguistic and cultural diversity by supporting language education and cultural exchange activities (Lim et al. 2019). This dual approach to language planning helps ASEAN balance the need for a common language for regional cooperation with the preservation of its rich linguistic heritage. A much smaller but closer example is the Majlis Bahasa Brunei-Indonesia-Malaysia or MABBIM, a trilateral language council comprising of three Malay-speaking states with Singapore as an observer. Examining the development and impact of this smaller-scale, regional initiative provides insights into a more localised efforts to maintain linguistic unity and at the same time balance diversity. It also provides a microcosmic view of how transnational cooperation is realised and challenged across time in achieving its objective.

14.2 The Linguistic Landscape of Bahasa Melayu/Bahasa Indonesia in Southeast Asia

In the present context, Bahasa Melayu and/or Bahasa Indonesia hold a significant position in the active member states of MABBIM. Bahasa Indonesia, for instance, officially became the state language of Indonesia on 18 August 1945, and has since served as a unifying force and the official language of the Republic of Indonesia (Nursalim et al. 2020). The journey of the language's development has witnessed various spelling modifications, including changes introduced by van Ophuijsen, the Republic Spelling or *Soewandi Spelling* and the *Ejaan Bahasa Indonesia yang Disempurnakan* or widely known as EYD. These adjustments have aimed to refine and standardise the language, ensuring its effectiveness as an inter-ethnic language.

In extension, the production of *Kamus Indonesia-Malaysia-Inggris* (Basri et al. 2010) also serves as a practical tool designed to help speakers of both Bahasa Melayu and Bahasa Indonesia in navigating and understanding the nuanced vocabulary distinctions inherent to each language. Additionally, Bahasa Indonesia plays a crucial role in educational institutions as it is the language of instruction for teaching and learning throughout the education system, from elementary school to university level (Kozok 2012). On a larger scale, this language also serves as a symbol of unity and national identity, reflecting the nation's commitment to inclusivity among the diverse population of Indonesia, preservation of heritage, and cultural sustainability.

Whereas in Malaysia, Bahasa Melayu (also referred to as Bahasa Malaysia) has also been developed as a modern and comprehensive medium through various initiatives especially within education. While academic institutions house linguists and scholars specialising in Malay language studies, educational professionals including curriculum developers shape language proficiency in schools and higher education. Similarly, language associations and organisations provide platforms for collaboration and knowledge exchange among language professionals, fostering a thriving linguistic ecosystem in Malaysia. One of the advocates championing this effort is the *Dewan Bahasa dan Pustaka* (DBP) or the National Language and Literature Bureau which plays a crucial role in language planning and research for the country. Some of the commendable efforts the bureau has made include the documentation of Bahasa Melayu through the publishing of the monolingual dictionary known as *Kamus Dewan* in the 1970s and the introduction of *Sistem Ejaan Rumi Baharu Bahasa Malaysia* in 1972. This standardised system was launched simultaneously in Malaysia and Indonesia and later adopted globally, including in Singapore and Brunei (Seong 2009). To date, DBP is actively involved in bilingual dictionaries, which cover both English and French, as part of its role as a prominent language development institution. More than 552 dictionaries from different publishers have contributed to the evolution of the language, especially since 2002. The bureau's focus on creating new terms, especially in scientific fields, also demonstrates the integration of Malay as a language for scientific purposes. DBP's extensive efforts, ranging from improving spelling to establishing standard pronunciations and developing terminology, highlight the codification and advancement of Malay as a modern language that can effectively express contemporary communication (Mohd Naim 1991; Seong 2009).

On the other hand, the linguistic landscape of Brunei prior to Western colonisation was mainly shaped by Bahasa Melayu due to the trade and the expansion of Islam (McLellan et al. 2016). With the introduction of the British residency system in the early 1900s, English was introduced into administrative and educational domains, while Bahasa Melayu remains the primary language for daily communication and cultural activities (Jones et al. 1993). The Resumption of Independence in 1984 marked the beginning of formal language planning efforts at preserving Bahasa Melayu while promoting English as a global language, resulting in the implementation of bilingual education policies, where both Bahasa Melayu and English are used as mediums of instruction in schools (Ministry of Education 2015). The government has also introduced the 'Dual Language Programme' which aims to enhance students' proficiency in English, particularly in science, technology, engineering, and mathematics (STEM) subjects (Ministry of Education 2018). Similarly, key educational reforms, such as the National Education System for the twenty-first Century or SPN21, emphasised the importance of bilingualism within the curriculum framework (Sharbawi and Jaidin 2020), to strike a balance between proficiency in Malay and English and incorporating elements of Melayu Islam Beraja as its national philosophy. The government has also launched the 'National Language Policy and Planning' initiative, which aims to promote the use of Malay and enhance language education in schools (McLellan et al. 2016).

14.2.1 The Historical Dimension of Bahasa Melayu as a Lingua Franca

Historically, Bahasa Melayu has played a crucial role as a lingua franca in the historical Southeast Asia, serving as a common medium of communication among individuals who speak different languages. In his book The Indianized States of Southeast Asia, Coedès (1964) details the rise of Srivijaya and its role in promoting cultural and linguistic exchange, including the use of Old Malay. As trade and cultural exchanges flourished, Bahasa Melayu quickly spread throughout the region, becoming the predominant language for trade, commerce, and diplomacy (Andaya 2006). In the fifteenth century, the Malaccan sultanate further solidified this language's position as a lingua franca through its extensive trade networks and cultural exchanges with other Southeast Asian kingdoms and China (Miksic 2013). The strategic location of the sultanate along the Strait of Malacca facilitated trade and cultural interactions between India, China, and Southeast Asia, leading to the widespread use of Malay as a common language for trade, communication and diplomacy not only among Southeast Asian kingdoms but also beyond (Reid 2015). On the same axis, Bahasa Melayu has also played a crucial part in the dissemination of Islam in Southeast Asia, with Islamic scholars and traders utilising Bahasa Melayu as a means of instruction and communication (Andaya 2006). Moreover, the status of Bahasa Melayu as a lingua franca has fostered the growth of a collective Southeast Asian identity and cultural legacy. The widespread use of Malay as a lingua franca has facilitated mobilities, exchange and cooperation among these countries, promoting integration across national borders. In the current setting, it is this same ideology that underlies the establishment of MABBIM as a transnational cooperation in language planning.

Bahasa Melayu's simplicity and versatility, paired with its borrowing of vocabulary from other languages, make it an excellent choice for communication across linguistic boundaries (Adelaar 2004). According to Sneddon (2003), the grammatical structure of Bahasa Melayu is relatively straightforward, with a simple phonology and a lack of tonal distinctions, making it easy to learn and use. At present, despite differences in nomenclature, Indonesia predominantly employing Bahasa Indonesia (BI), and Malaysia, Brunei Darussalam, Singapore and parts of Southern Thailand utilising Bahasa Melayu (BM). Critics and scholars often posit that these two languages essentially represent a single linguistic entity with divergent stylistic elements (Asmah 2008). Linguistically speaking, these languages, referred to as cognates, share a linguistic heritage that allows speakers of one to understand the other generally. This linguistic similarity forms the basis for mutual intelligibility between Bahasa Melayu and Bahasa Indonesia. According to Zaim (2014), the relationship between BM and BI is intricate, as more than 80% of their vocabulary is shared. However, the remaining 20% exhibits differences that have emerged over time due to the evolving usage of language within respective native-speaker communities. While speakers of both languages may generally comprehend each other, certain limitations arise due to disparities in lexicons, word structures, and grammatical constructions. These distinctions emerge from unique historical, political and linguistic contexts that have influenced the treatment of language in each region. The resultant impact extends to variations in grammar, vocabulary and pronunciation, creating a linguistic tapestry shaped by the divergent influences of Dutch in Indonesia and English in Malaysia and Brunei Darussalam (Zaim 2014). The absorption of Dutch in Bahasa Indonesia, alongside the influence of the English language in Malaysia, Brunei Darussalam and Singapore, can be attributed to the historical contexts of colonial occupation in these respective countries. Moreover, these languages have evolved and adapted by incorporating terminologies from various sources. This linguistic borrowing reflects the historical and cultural interactions that these regions have experienced over time. The inclusion of terms from Arabic suggests influences from Islamic traditions, while words borrowed from Dutch, Portuguese and Spanish may be remnants of colonial periods. Additionally, incorporating elements from indigenous languages underscores the rich cultural tapestry that has shaped the linguistic landscape of these languages. Befitting the development, understanding these distinctions in the contemporary setting is crucial for effective communication, knowledge production and collaboration. It is also with this understanding that Malaysia, Indonesia and Brunei established MABBIM in 1985.

14.3 MABBIM: Of Genesis, Aspirations and Diverging Interests

The establishment of Majlis Bahasa Brunei Indonesia dan Malaysia or MABBIM in 1985 is a testament to member states' willingness to work closely in standardising the Malay language in the region. This initiative began with the standardisation of spelling between Indonesia and Malaysia and later expanded to encompass various other linguistic aspects, including terminology and grammar. Originally known as Majlis Bahasa Indonesia-Malaysia (MBIM), MABBIM was founded on 23 May 1972. The groundwork was laid when the Minister of Education of Malaysia, Hussein Onn, consulted with the Minister of Education and Culture of the Republic of Indonesia, Mahsuri S.H., in Jakarta. This meeting resulted in a Joint Statement that emphasised collaborative efforts in developing their respective national languages, including spelling, terminology, and grammar, thereby providing a solid foundation for linguistic cooperation between two countries. After completing the harmonisation of spelling, MBIM focused on standardising terms in various fields of knowledge, particularly in the basic sciences such as biology, chemistry, and physics, as well as in the social sciences (Asmah 2008). Careful study and research by linguists from both countries resulted in the publication of *Pedoman Umum Pembentukan Istilah Bahasa Melayu* in 1975 and *Panduan Penyusunan Kamus Istilah* in 1982. On 8 November 1985, Brunei Darussalam joined the 24th MBIM Conference in Jakarta, leading to the renaming of the council to Majlis Bahasa Brunei Darussalam-Indonesia-Malaysia or MABBIM.

This collaborative framework provides a platform for regular dialogues, exchanges, and joint decision-making processes aimed at fostering linguistic unity. With the aim of fostering a stronger sense of unity and camaraderie among its member countries by enhancing their national or official language as tools for broader communication, MABBIM also focuses on language construction and development, prioritising language coordination through scientific research, creative writing, and the establishment of comprehensive guidelines and standards. To achieve this, MABBIM conducts regular meetings to ensure effective coordination and alignment of linguistic efforts among the member countries. Meetings are held twice a year, with the location alternating between member countries (Sugono et al. 2000). The annual General Meeting or *Sidang Majlis* discusses fundamental issues such as language guidelines and MABBIM's Five-Year Plan, while the yearly Expert Meeting or *Sidang Pakar* debates the harmonisation of terms in specific fields or subfields. Additionally, initiatives such as *Siri Kongres Bahasa Melayu, Pidato Kebangsaan, Syarahan MABBIM,* and *Sambutan Bulan Bahasa* involving scholars and language enthusiasts alike contribute even more to the collaborative language efforts within MABBIM. These collaborative ventures not only highlight active participation among participating countries but also demonstrate a shared commitment to achieving linguistic harmony within MABBIM. In complementing this effort, institutions such as the *Badan Pengembangan dan Pembinaan Bahasa Indonesia*, *Dewan Bahasa dan Pustaka Malaysia*, and *Dewan Bahasa dan Pustaka Brunei Darussalam* have been actively engaged in preparing educational materials that delve into grammatical disparities associated with word formation, including derivation and inflection affix administration in both Bahasa Melayu and Bahasa Indonesia. Although grammatical differences between the two languages may be minimal, a meticulous examination and clarification on these aspects are essential. Inaccuracies in applying words and phrases, even if grammatically similar, could pose significant obstacles to seamless communication, particularly between member states and language learners aspiring to master these languages.

Despite this promising development, this transnational entity has faced numerous challenges in recent decades. These challenges include financial constraints, differences in political views, and diverging ideologies. In today's global landscape, where social, political and economic stability is often in flux, maintaining a cohesive spirit and managing differences within MABBIM has become increasingly difficult. These adversities, evident in general meetings and the media, have significantly impacted academic discussions and public discourse. Over the years, these challenges have evolved into noticeable divisions and animosity. An example of this was when the proposal put forward by Malaysian former Prime Minister Ismail Sabri Yaakob in 2022 to designate Bahasa Melayu as the second language of ASEAN was met with rejection by Indonesia's Minister of Education, Culture, Research, and Technology, Nadiem Makarim. Nadiem instead advocated for further study and discussion at the regional level, arguing that Bahasa Indonesia or BI is a more suitable choice for ASEAN's official language due to its historical, legal and linguistic advantages. Nadiem highlights that BI has become the most widely spoken language in Southeast Asia, extending its reach to 47 countries globally. The widespread availability of Bahasa Indonesia for Foreign Speakers (BIPA) courses was offered by various institutions, including the Language Development and Development Agency of Indonesia's Ministry of Education and Ministry of Culture, Research, and Technology. It has also been taught to foreigners by 428 institutions, and also as a subject in a number of campuses in Europe, the United States, and Australia, as well as at universities in Asia (Wardhana et al. 2020).

As a counterproposal by Malaysia's Muslim Youth Movement of Malaysia (ABIM), which welcomed the initiative of making Malay language"as ASEAN's second language, suggested the term 'Bahasa Melayu—Nusantara' to recognise the diverse Malay-speaking communities across the archipelago, aiming to counterbalance and navigate potential conflicts. In their argument, this initiative was befitting considering Bahasa Melayu's significance during the era in which the Malay world served as the centre of a Southeast Asian cosmopolitan civilisation. They also highlighted the evolution of language in Indonesian discourse, where the term 'Bahasa

Melajoe' was initially used before adopting the name 'Indonesian' after the historic *Sumpah Pemuda* or Youth Pledge in 1928. This discussion points to Decision Number 8 of the Second Indonesian Language Congress in 1954, which acknowledges the Malay language as the origin of Indonesian adapted to contemporary Indonesian society (Pusat Pembinaan dan Pengembangan Bahasa 1995), validating that the term itself was widely embraced across regions with Malay-speaking communities worldwide and has the potential to unify the Malay world.

The linguistic debate surrounding the second language of ASEAN and the proposed recognition of Malay as a second language highlights the complexity of language dynamics in the region. The rejection by Indonesia, rooted in historical and linguistic considerations, sets the stage for further discussions among ASEAN member states about the linguistic direction that best serves the interests of the diverse and interconnected Southeast Asian community. Furthermore, the recent announcement on the designation of BI as an official language of the General Conference of UNESCO (2023), would not only potentially impede the vision of reliving Bahasa Melayu as the region's lingua franca, but also casts doubt on past and future initiatives aimed at realising this vision. This contentious scenario raises critical questions about its implications for the Malay World region and its active state participants.

14.4 Transnationalism and Reliving the Imagined Community

Benedict Anderson's 'Imagined Communities' concept posits that nations are socially constructed communities imagined by those who perceive themselves as part of that group (Anderson 2020). In the case of MABBIM, the historical and linguistic unity underpins the contemporary notion of a collective identity, essential not only for forming an imagined community but also as a basis to rearticulate transnational cooperation in language planning. Despite MABBIM's ideals to preserve and celebrate a shared linguistic heritage and foster linguistic unity and cooperation among participating countries, any initiatives must also be critically examined in light of member countries' political, social and economic differences. For instance, sustaining activities ranging from organising regional meetings, workshops and competitions to undertaking collaborative research, joint publications, and terminology-building efforts can be resource-strenuous over time. In the 2023 council meeting, state members were urged to respond to challenges brought about by the rapid expansion of digital technology and world dynamics, realising the multifaceted potential of such technology and dynamics in a creative manner. For example, renewed collaborative efforts can leverage technology to develop language resources such as online dictionaries, language learning platforms, and translation tools, democratising access to language education, which in turn promotes linguistic equality (Pimienta et al. 2009) and enhancing knowledge production in Bahasa Melayu.

At the same time, the language council must also navigate the complexities of national identities and regional geopolitics. For instance, Malaysia's proposal and Indonesia's rejection underscores the complexities of regional language politics and the challenge of achieving linguistic unity amid national interests and historical legacies. Therefore, the promotion of a common linguistic identity by MABBIM is not merely a cultural endeavour but also a political one, reflecting the aspirations and tensions inherent in regional integration efforts. The historical role of Malay as a lingua franca highlights the political dimensions of these efforts, as the language's historical prestige and utility are leveraged to promote regional cooperation. However, the political and economic disparities among member states necessitate careful navigation to ensure that language planning efforts are inclusive and equitable. At the same time, common linguistic identity must also be carefully threaded with caution to avoid the potential homogenization of linguistic and cultural diversity, marginalising the local linguistic varieties and expressions (Blommaert 2010). One way of addressing this challenge is by adopting a more inclusive approach that acknowledges and incorporates local linguistic varieties within its standardisation endeavours, as reflected in the production of *Kamus Nusantara* (dictionary) and the *Pusat Rujukan Persuratan Melayu* (an online reference portal). These references encompass not only Bahasa Melayu/ Bahasa Indonesia but also the dialects and varieties of linguistic expressions of local and indigenous communities. Beyond developing resources, MABBIM could also extend this effort to develop policies that support similar initiatives (Hornberger and Johnson 2007).

On the same axis, tensions or conflicts among state members may affect their future cooperation, especially in language planning. In extreme cases, conflicts over language policy can strain diplomatic relations and weaken regional alliances, as exemplified by the disagreements over language use in the European Union, which have led to tensions between member states, affecting broader political cooperation and integration efforts (Krzyżanowski and Wodak 2011). It can also disrupt trade and investment, as language barriers and policy disagreements create uncertainty and reduce the efficiency of cross-border business operations (Grin 2003), not to mention the ethnic and linguistic tensions within and between countries, leading to social fragmentation and undermining efforts to promote multiculturalism and social cohesion (Skutnabb-Kangas 2013).

In handling this predicament, MABBIM is faced with a tough decision ahead should it want to maintain its status quo as the preserver and gatekeeper of the Malay World's linguistics assets, or should it be dissolved to pave the way for efficiency, prestige and relevance. In light of this, adopting the former would require a paradigm shift not just in ideational and aspirational sense but also in terms of operationalisation of mechanisms and collaboration. Furthermore, sustaining its operations would reinforce Southeast Asia's commitment to preserving cultural diversity through a common linguistic framework, fostering deeper regional integration and mutual understanding. Moreover, MABBIM's continued presence could serve as a catalyst for innovative approaches in education, research and economic collaboration, positioning the Malay language as a cornerstone of regional identity and cooperation. On the other hand, the dissolution of MABBIM in Southeast Asia would not just be the end of an era in regional linguistic planning, but also a significant loss of a pivotal institution that has long fostered regional unity and cultural cohesion. The potential disruption of decades of collaborative efforts in knowledge production, hindrance to social mobilities and cultural exchanges, and diminishment of economic opportunities across member states, all underscore the urgency of this issue. This potential dissolution raises profound questions about the future trajectory of Southeast Asian cooperation in navigating linguistic diversity and fostering a shared regional identity.

14.5 Conclusion

This MABBIM case study provides valuable insights into transnational cooperation, particularly in terms of linguistic planning and cultural integration. It demonstrates how transnational entities can navigate historical, political and linguistic complexities to foster unity among diverse nations. By standardising Bahasa Melayu across Indonesia, Malaysia, and Brunei, MABBIM highlights the potential for a shared linguistic framework to contribute to regional cohesion. Transnationalism, as seen in the context of MABBIM, involves nations collaborating beyond geopolitical boundaries to address common challenges and leverage shared cultural assets. The initiative aims to harmonise language norms and serve as a platform for enhancing educational exchanges, scientific research and cultural dialogue. This cooperative framework not only supports a vision of Southeast Asia where linguistic diversity coexists with a common understanding, but also inspires hope for a future characterised by mutual respect and solidarity among member states.

MABBIM's approach to language planning reflects a nuanced understanding of transnationalism's role in shaping regional identities. By acknowledging the historical legacy of Bahasa Melayu as a lingua franca and adapting it to contemporary needs, MABBIM contributes to strengthening a collective Southeast Asian identity that transcends national boundaries, emphasising shared cultural values and histories that unite rather than divide. At the same time, it also raises concerns about the potential displacement of local meanings and marginalisation of indigenous languages. However, the challenges facing MABBIM serve as a reminder of the complexities of transnational cooperation. Financial constraints, diverging political interests, and varying priorities among member states underscore the need for urgent and sustained diplomatic efforts and strategic investments in institutional capacity. These challenges reflect the broader issues within transnationalism, where balancing national sovereignty with regional integration remains a delicate and pressing endeavour.

For what it is worth, the experiences of MABBIM suggest several considerations for enhancing transnational cooperation. Fostering inclusive decision-making processes that respect each member state's sovereignty and priorities is essential to ensure that initiatives like language planning resonate with local contexts while advancing regional goals. Additionally, it also contributes to a vision of regional integration anchored in shared cultural heritage and mutual understanding by standardising Bahasa Melayu and promoting linguistic unity. As Southeast Asia navigates the complexities of globalisation, MABBIM exemplifies how transnational cooperation can harness linguistic diversity to build bridges and foster a cohesive Dunia Melayu, setting a precedent for collaborative efforts in addressing broader global challenges.

References

Adelaar KA (2004) The Austronesian languages of Asia and Madagascar: a historical perspective. In: Adelaar A, Himmelmann N (eds) The Austronesian languages of Asia and Madagascar. Routledge

Andaya BW (2006) The flaming womb: repositioning women in early modern Southeast Asia. University of Hawaii Press

Anderson B (2020) Imagined communities: reflections on the origin and spread of nationalism. In: Seidman S, Alexander JC (eds) The new social theory reader. Routledge, pp 282–288

Asmah HO (2008) Ensuring linguistic harmony within MABBIM. Dewan Bahasa dan Pustaka Malaysia (Language and Literature Bureau of Malaysia), Kuala Lumpur

Basri R, Boesira SD, Amiruddin Y et al (2010) Kamus bahasa Indonesia–Malaysia–Inggris. KBRI (Dictionary of the language of Indonesia–Malaysia–English KRBI). Kuala Lumpur

Blommaert J (2010) The sociolinguistics of globalization. Cambridge University Press

Coedès G (1964) The Indianized states of Southeast Asia. East-West Center Press

European Commission, Directorate-General for Education, Youth, Sport and Culture (2023) Erasmus+: EU programme for education, training, youth and sport. European Commission

European Commission, Directorate-General for Education, Youth, Sport and Culture (2024) About multilingualism policy: the importance of multilingualism. European Commission

Grin F (2003) Language policy evaluation and the European Charter for Regional or Minority Languages. Palgrave Macmillan UK, London. https://doi.org/10.1057/9780230502666

Hornberger NH, Johnson DC (2007) Slicing the onion ethnographically: layers and spaces in multilingual language education policy and practice. Tesol Quarterly 41(3):509–532

Jones G et al (1993) Multilingualism and bilingual education in Brunei Darussalam. Journal of Multilingual and Multicultural Development 14(1–2):39–58. https://doi.org/10.1080/01434632.1993.9994519

Kirkpatrick A (2010) English as a lingua franca in ASEAN: a multilingual model. Hong Kong University Press

Kozok U (2012) How many people speak Indonesian?. Indo-Pacific Languages and Literatures Website

Krzyżanowski M, Wodak R (2011) Political strategies and language policies: the European Union Lisbon Strategy and its implications for the EU's language and multilingualism policy. Language Policy 10(2):115–136. https://doi.org/10.1007/s10993-011-9196-5

Lim MA, Anabo IF, Phan ANQ et al (2019) The state of higher education in Southeast Asia. Jakarta

McLellan J et al (2016) The language situation in Brunei Darussalam. In: Othman NAH, McLellan J, Deterding D (eds) The use and status of language in Brunei Darussalam. Springer Singapore, pp 9–16. https://doi.org/10.1007/978-981-10-0853-5_2

Miksic JN (2013) Singapore and the Silk Road of the sea, 1300–1800. NUS Press. https://doi.org/10.2307/j.ctv1nthqk

Ministry of Education (2015) Education system in Brunei Darussalam

Ministry of Education (2018) Brunei Darussalam education statistics. Bandar Seri Begawan

Mohd Naim D (1991) Perkembangan ejaan rumi Bahasa Melayu: bahagian II (Development of roman spelling for Bahasa Melayu: part 2). Sekata (A Word) 9(1):28–48

Moseley C (ed) (2010) Atlas of the world's languages in danger. UNESCO

Nursalim MP et al (2020) Bahasa Indonesia. Universitas Pamulang (Pamulang University), Banten

Pimienta D et al (2009) Twelve years of measuring linguistic diversity in the internet: balance and perspectives. UNESCO Publications for the World Summit on the Information Society
Portes A et al (1999) The study of transnationalism: pitfalls and promise of an emergent research field. Ethnic and Racial Studies 22(2):217–237. https://doi.org/10.1080/014198799329468
Pusat Pembinaan dan Pengembangan Bahasa, Departemen Pendidikan dan Kebudayaan (1995) Resolutions of Indonesian Language Congress II (1954). In: Resolutions of Indonesian Language Congress I-VI. Pusat Pembinaan dan Pengembangan Bahasa (Centre for Language Development and Expansion), Departemen Pendidikan dan Kebudayaan (Department of Education and Culture), Indonesia
Reid A (2015) A history of Southeast Asia: critical crossroads. Wiley-Blackwell
Seong TK (2009) Bahasa Melayu dan komunikasi dalam dunia globalisasi: persoalan keupayaan Bahasa Melayu, keutuhan jati diri bangsa Melayu dan ketakatan budaya Melayu (Bahasa Melayu and communication in the world of globalisation: questioning the capability to use Bahasa Melayu, the integrity of the Malay identity and the strength of the Malay culture). Jurnal Bahasa (Language Journal) 9(1): 100–117
Sharbawi S, Jaidin JH (2020) Brunei's SPN21 English language-in-education policy: a macro-to-micro evaluation. Current Issues in Language Planning 21(2):175–201. https://doi.org/10.1080/14664208.2019.1657267
Skutnabb-Kangas T (2013) Linguistic genocide in education–or worldwide diversity and human rights? Routledge
Smith MP, Guarnizo LE (eds) (1998) Transnationalism from below, vol 6. Transaction Publishers, New Brunswick
Sneddon JN (2003) The Indonesian language: its history and role in modern society. University of New South Wales Press
Sugono D et al (2000) Pembinaan Bahasa Indonesia/Melayu dalam dunia pendidikan (Development of Bahasa Indonesia/Malay in the world of education). Departemen Pendidikan Nasional Jakarta (Jakarta National Department of Education)
UNESCO (2017) Division for inclusion, peace and sustainable development, education sector. In: Education for sustainable development goals: learning objectives. UNESCO
UNESCO (2023) Recognition of bahasa indonesia as an official language of the general conference of UNESCO. UNESCO
Wardhana IP et al (2020) Local historical learning in: multicultural education in the era of merdeka learning. Social Human Educ Stud (SHEs) Conf Ser 3(2):143–148. https://doi.org/10.20961/shes.v3i2.46233
Zaim M (2014) Unparalled lexicon of Malay and Indonesian Languages: problems in developing learning materials. In: Bridging the unbridge: changing paradigms in Malay-Indonesian Studies, Hankuk University of Foreign Languages, 14–16 May 2014

Badriyah Yusof is Assistant Professor of Malay Language and Linguistics at the Faculty of Arts and Social Science, Universiti Brunei Darussalam, specialising in Critical Discourse Studies and Translation Studies. Her research interests include discourses of politics and governance, Malay manuscripts, and the translation industry. She is also one of the appointed members of the National Committee for Bahasa Melayu in Brunei Darussalam (JKTBMBD), and an associate editor for *Southeast Asia: A Multidisciplinary Journal*.

Elmy Maswandi is a postgraduate student in Malay Language and Linguistics at the Faculty of Arts and Social Sciences, Universiti Brunei Darussalam. His current research project focuses on the critical discourses of marginalised groups in Brunei, particularly former drug users. Elmy has been published in the *International Journal of Linguistics and Indigenous Journal* (IJLIC), *Jurnal Bahasa* (Malaysia*), Jurnal Pengajian Melayu: Jurnal Antarabangsa Dunia Melayu, PENDETA*, and more.

Wafiqah Asnola is a Malay Language and Linguistics postgraduate at the Faculty of Arts and Social Sciences, Universiti Brunei Darussalam. Her current research explores the phenomenon of social inclusion in 1950s Brunei's Seven Committee Report. Wafiqah has published in *Southeast Asia: Multidisciplinary Journal, Jurnal Bahasa* (Malaysia), *Jurnal Pengajian Melayu, Melayu: Jurnal Antarabangsa Dunia Melayu,* and *PENDETA,* among others.

Part III
Cultures

Chapter 15
Traversing Transnationalism: Enrique's Malayness in Historical Fiction from Southeast Asia

Hannah Ming Yit Ho

Abstract This chapter employs the transnational as the operative framework to analyse various literary representations of a Malay historical figure who was christened Enrique by the Portuguese circumnavigator Ferdinand Magellan in the sixteenth century, which was a period of European colonial expansion in Southeast Asia. Contemporary novels engage an imaginative reworking of Enrique's personal life and historical times in this early colonial period. Briefly discussing Harun Aminurrashid's *Panglima Awang* (1958) and Carla M. Pacis' *Enrique El Negro* (2002), this chapter will focus mainly on examining the fictionalisation of Enrique in Danny Jalil's *Enrique the Black* (2021). The Malay protagonist's transnational migration and mobility can be seen to facilitate his learning and unlearning of a colonial and Orientalist discourse of his Malay identity. Economic and cultural globalisation have led to transnational contestations of race and religion in Southeast Asia, including the Malay and Muslim identities of Enrique. By analysing his transnational travels, a critique of colonial and post-postcolonial transnational processes of globalisation is observed in the literary renderings of Enrique's experiences of traversing national boundaries and cultural landscapes that produce in him a sense of shifting homes.

Keywords Malay identity · Southeast Asian historical fiction · First Malay circumnavigator · Transnational race and religion · Migrant homes · Anglophone novels

15.1 Introduction

Historical fiction as a genre has been oft employed to explore the transnational impact of historical processes of colonisation and globalisation that have shaped discourses of identity. While the term of transnationalism is attributed to Ian Tyrrell to promote a critical approach to historical writings (*American Historical Review* 1991), it has

H. M. Y. Ho (✉)
Faculty of Arts and Social Sciences, Universiti Brunei Darussalam, Gadong, Brunei Darussalam
e-mail: hannah.ho@ubd.edu.bn

H. M. Y. Ho and Y. Chan (eds.), *Transnational Southeast Asia*, Asia in Transition 29,
https://doi.org/10.1007/978-981-96-3608-2_15

since gained traction not only in the field of history but also in literary humanities in which fiction and history merge (Frassinelli et al. 2011). Transnational literature comprising fiction produced by authors characterised as "post-postcolonial" (*Chronicle* 2000), as Pakistani writer Mohsin Hamid articulates those who are engaged in economic and political flows of culture through contemporary modes of globalisation, are strategically poised to engage, unpack, and interrogate the politics of identity contestations within and across national borders that have been conceived as rigorous and impervious. The displacement, migration, and mobile flows that continue in the post-postcolonial period can be traced back to geographical flows that have resulted in cultural and affective dislocations (Jay 2010). These forms of displacement are often traced in the trope of loss in Southeast Asian Anglophone literary fiction, which narrates the transnational flows of the family that significantly impact identity formation (Ho 2021b, 2023c; Ho and Amran 2021).

Given that globalisation has opposed an anachronistic nationalist frame (Jay 2010, p. 25), this chapter engages the term of transnationalism to study literary representations of a historical Malay figure. Three authors from across Southeast Asian nations have depicted this Malay man who was enslaved by the Portuguese explorer Ferdinand Magellan in 1511 in their historical fiction. Through their common Malay protagonists, these works of historical fiction highlight historical and ongoing contestations of race and religion in the region. In contextualising this research in transnational Southeast Asian studies, the Malay diaspora can be studied in relation to transnational flows of Malay cultural identity in and beyond Southeast Asia. Situating the Malay protagonist in the European colonial period of early globalisation, this present study engages the discourse of transnationalism to analyse migrant identity, culture, and homes. Aihwa Ong (1999) contends that "*trans*" highlights "moving through space or across lines, as well as changing the nature of something", which underscore the "*trans*versal, the *trans*actional, the *trans*lational, and the *trans*gressive aspects" (p. 4, original emphasis) of global capital circulating in the Asia Pacific region. Translated into the Southeast Asian region, transnationalism then offers a lens to examine the changing nature of Malay identity across different geographical spaces and national borders, which is delineated in various historical fiction with their contesting claims to the protagonist's Malay cultural identity. Notwithstanding this, the challenges of transnationalism are reflected in the way that Malayness continues to be defined restrictively by national ideologies that are not always commensurate with nor accommodate a multiple concept of Malayness.

This historical Malay figure is first fictionalised in Malaysian author Harun Aminurrashid's novel *Panglima Awang* (1958), loosely translated from the Malay language as "National Warrior". Recent literary fiction published in the new millennium do not bolster the ethnonationalist identity of the Malay/sian nation-state, as much as they employ an imaginative lens to narrate histories along exigencies of transnational mobility. Historical fictional works that exercise these artistic liberties include two subsequent novels, namely Filipina Carla M. Pacis' novel *Enrique El Negro* (2002) and Singaporean Danny Jalil's *Enrique the Black* (2021). These recent novels lay transnational claims to this Malay man by bringing up identity contestations across geopolitical boundaries and national borders, thus exemplifying political, economic, and cultural imperatives that impinge into the life of this transnational migrant slave. Significantly, such efforts at the fictional reworking of transnational travels of this Malay protagonist reflects a global reframing of the cultural origins of Malayness. These contemporary literary novels reimagine this Malay historical figure as an embodied product of flexible identities, porous boundaries, and shifting paradigms that shape transnational migrant subjects in maritime Southeast Asia. Narrative accounts of this ethnic Malay man that are produced within other non-Muslim and non-Malay-dominant nations, namely Philippines and Singapore, highlight the transnational as an operative framework for the construction and discourse of Malay identity in Southeast Asia.

Consequently, the politicisation of Malay identity is delineated in the three historical novels that have been produced within Southeast Asia, which is a region that is home to various ethnicities and religions due to its history of transcultural and transnational contact since the early European colonial period. It is evident in these novels about Enrique that his Malay identity serves a twofold purpose, that of personal identification and a cultural marker. A comparative discussion of the historical novels is briefly offered to trace his ethnoreligious identity during his transnational times, before focusing on the post-postcolonial production, imagination, and reception of the latest novel that is Danny Jalil's *Enrique the Black* (2021). As there is a scarcity of historical records on his life (Curaming 2023), the story of this Malay historical figure is predisposed to fictionalisation which engages both history and literature. Inspired by the transnational travels of this historical figure, these fiction writers have sought to imaginatively fill the gaps and silences that gloss over the life of the Malay navigator-cum-warrior. While much fame is attributed to the Portuguese circumnavigator Ferdinand Magellan, the young Malay boy that he captures has been largely occluded from historical narratives and fictional works. This occlusion perpetuates his marginal status that is reinforced through Orientalism, which is apparent through his subjection to the "violence of othering" (Ho 2020, p. 54). Collectively, the three Anglophone novels that were published the post-/postcolonial period feature Enrique as their titular character to redress the silencing of this man's story and colonial suppression of his Malay identity. These Southeast Asian fictions highlight his life to accentuate economic and cultural global flows that produce, but also untangle, an essentialist discourse of Eastern/Malay identities. Through the transnational lens of the Malay protagonist who is ascribed the Christian moniker of "Enrique" by his Portuguese master, the Malay man's cultural struggles are foregrounded in his fictional life story.

15.2 Transnational Traders in Maritime Southeast Asia: The Islamic Sultanate of Malacca

In the sixteenth century, the Malay Archipelago that includes today's Southeast Asian nations of Malaysia, Indonesia, the Philippines and Brunei Darussalam became an integral part of the East Indies, which was an important region for maritime spice trade that Europeans worked to monopolise (Vann 2014). These countries also collectively make up the Nusantara, a local term employed in the Malay world to denote the thriving transnational trade and Islamic religious expansion. The centre of the Malay world, according to the historical document *Sulalatus Salatin* or "Geneology of Kings" that was later renamed *Sejarah Melayu* or "The Malay Annals" (Leyden 1821; Winstedt 1938), was located in the Malaccan Sultanate along the Strait of Malacca which was an important trading port. Historically, Malacca is the hometown of Enrique. The cultural significance of *Sejarah Melayu* for reclaiming Malayness to mitigate colonial prescriptions of Malay identity is salient (Ho 2023a). Crucially, Malay ethnicity and its identification with Islamic religion can be traced back to these historical records, which narrates the golden age of the Malay world prior to the Portuguese invasion of Malacca in 1511. Malacca was a bustling port town where Arab, Indian and Chinese traders interacted. Notwithstanding its positionality as a Malay and Muslim stronghold, Malacca emerged as a cosmopolitan place with rich spoken languages due to transcultural contact with transnational traders. However, European explorers including the Portuguese governor Alfonso de Albuquerque posed a threat to the Islamic Sultanate of Malacca. Both Portuguese and Spanish imperial powers gained momentum during this heightened period of trade and Islamic cultural propagation in the Nusantara. Portuguese invasion of Malacca is recounted in *Enrique the Black* (2021) when the eponymous protagonist contemplates the "fall of Malacca, wondering where Sultan Mahmud Shah was now. He thought of Albuquerque" (Jalil, p. 279). Drawing on religious contestations between the Muslim Sultan of Malacca and Christian Portuguese explorer, the novels about Enrique are set in sixteenth-century Malacca to portray West–East cultural ideological conflicts amidst European navigators' strengthening efforts to monopolise trade routes and colonise indigenous people.

15.3 Economic and Cultural Globalisation: Transnational Contestations of Race and Religion

Taking into account globalisation's dual dimensions of economics and culture (Jay 2010), transnational contestations of race and religion within the European colonial enterprise in Southeast Asia are examined in the multiple fictional texts on Enrique. Coupled with a materialist approach to globalisation (Miyoshi 1998; Said 1994), cultural theories of globalisation that are articulated in global flows of people and ideologies (Appadurai 1996) and James Clifford's discussion of "diverse, interconnected histories of travels and displacement in the late twentieth century" (1992 p. 97) emphasise the deterritorialisation of cultures. On the other hand, Ullah and Ho (2020) posit that globalisation produces indigenous cultures' destruction through processes of "demise, fragmentation [and] transformation" (p. 191). As a result, ethnonationalist responses are often elicited to counteract Western cultural homogenisation in globalisation. In the Malay world, ethnonationalist ideologies are evident in the localisation of Malay and Muslim identities in Malay-Muslim dominant nations (Ho 2021a). Historically enabled through colonisation, transnational flows of culture continue in the post-postcolonial period of "late globalisation" (Turcan 2016). In a transnational world, Malay identity is not only defined but also disputed by contending cultural ideologies established in and beyond the nation (Ho 2019, 2023b; Humairah and Kamaludeen 2021; Muzhafar et al. 2015).

15.4 Shifting Homes: Transnational Migration and Mobility in and Beyond Southeast Asia

For transnational migrants, home is "an experience cultivated from the margins" (Boccagni 2022, p. 142; see also Ahmed et al 2003). The simultaneous process of homemaking and homing become significant, and their societal significance should not be overlooked. Enrique is a character who exhibits acts of homemaking in his religious prayers, while also embodying a homing process through his travels towards and memories of Malacca. Even while home is a place of physical origin, affective and social communities forged through religious rituals and practices are considered crucial avenues of identity affirmation for acquiring a sense of home. To instantiate, the practice of Islam for a transnational Malay and Muslim migrant who is dislocated from the Malay and Muslim centre of the Malay world offers him with social capital to experience home amidst shifting homes. For Enrique, his efforts at homing can be scrutinised in terms of Thomas Tweed's idea of "religion is homemaking" (2016, p. 185). As a transnational migrant, he continues to display his gravitations towards daily calls for the Muslim prayer that provide him with a semblance of his Malaccan home, and which simultaneously aids him in the process of cultural reorientation in homing. While displaced from his national origins, Enrique has nightly recollections of Malacca: "Each night as he closed his eyes he always dreamt of fire, hazy, smog-filled visions that dispersed into clarity, visions of his Malacca burning, of his village on fire" (Jalil 2021, p. 57). While situated outside of Malacca's locality, Enrique's reveries of his homeland that was established as the cultural centre of Islam in the Nusantara before European invasion signals an affective loss triggered in the vast physical destruction of the Sultanate of Malacca. This loss compels him to (re)turn to mosques when traversing the world. In employing religious acts as a recourse to homemaking and homing, Enrique as a transnational migrant relies on his physical and cultural mobilities to affirm his Malay identity.

15.5 Traversing Transnationalism: Enrique's Malay Identity in Historical Fiction

The identity of the first Malay circumnavigator of the world has been a focus of scrutiny as efforts to reterritorialise him have played out in recent historical and literary narratives from Southeast Asia (Curaming 2023). Various nations and religious cultures have attempted to lay claim to him through their fictional accounts of his life. In the most recent literary rendering in *Enrique the Black* (2021), this Malay protagonist's physical and cultural dislocations attest to his personal struggles as a colonised subject in transnational spaces. However, his enslavement also signals "forms of displacement endemic to the long history of globalisation" (Jay 2010, p. 8) that begin in colonial enterprises. Novels on this transnational migrant from various authors that document his transnational journeys during the early period of globalisation of European colonial navigation engage the concept of shifting home and homemaking in transnational spaces. While discussing literary representations of this Malay historical figure, this chapter focuses on the complexities of Malay-Muslim identity across transnational and transcultural homes.

Colonial prescriptions of Orientalism are ascribed to a historical Malay man from Southeast Asia, with historians of Malaysian heritage and nationalist authors claiming that Ferdinand Magellan's slave is from Malacca while other historians and writers have argued that he originates from either the Philippines or Indonesia (Yahya and Tawas 2014). This Malay man, who lived in maritime Southeast Asia during the early Portuguese colonial period and who was a fundamental part of the Spanish colonial expedition to Philippines, is a contested character in the historical novels produced by different authors across Southeast Asia. Historical novels about Enrique do not subscribe to the same national and religious orientation of this Malay historical figure, who played a prominent role as a translator for Ferdinand Magellan during his global expedition, from the Sultanate of Malacca to the Philippine archipelago. Mainstream historical documentation supports the view that Enrique was taken as a slave during the Portuguese invasion of Malacca in 1511, which was aimed at controlling trade in the Southeast Asian region. In Danny Jalil's narrative account of Enrique, this is delineated through the omniscient narrator's description that "Malacca was the centre of trade in this side of the world, or at least as far as the Malay empire was concerned. [Enrique] saw the Chinese and the Arabs and the Indians" (p. 34). Ferdinand Magellan's requisition of his Malay slave at this port town leads to Enrique's subordination to his Portuguese master for a full decade before sailing back to Southeast Asia to fulfil his master's ambition to reach the spice islands of Moluccas in Indonesia.

Enrique's transnational travels and reputation as the first Malay circumnavigator of the globe predisposes him to identity contestations among the three historical novels that portray his Malay race and religion. Pro-nationalist in its tone, the first historical novel about Ferdinand Magellan's slave by Harun Aminurrashid serves as a nationalist text to celebrate indelible links between Malayness and Islam that a newly politically independent Malaya was establishing. *Panglima Awang* (1958) does not complicate nor compromise on the ethnoreligious identity of this titular Malay character who is hailed as a national warrior. With its publication timed with the ethnonationalist birth of a newly independent Malaya in 1957, the novel equates his Malay identity to Muslim religion through his portrayal as a sure and steadfast defender of Islam. Even while taken as a slave for the European man's economic gains, Enrique resists the Christian coloniser's efforts at religious globalisation when upholding his Malay-Muslim identity in Harun Aminurrashid's novel. In Carla M. Pacis' subsequent historical fiction *Enrique El Negro* (2002), Islam is rejected for an animistic faith that is embraced in its Malay Filipino protagonist called Enrique. Unlike Harun Aminurrashid's portrayal, Carla M. Pacis further imagines Enrique to hail from Cebu, the Philippines, as this titular protagonist is taken by seafaring pirates instead of captured during the Portuguese invasion of Malacca. Nonetheless, the racialisation of Enrique El Negro is a similar theme that runs across the three Anglophone novels about this Malay man. Danny Jalil's *Enrique the Black* (2021) attests to Enrique's personal experiences with war and European colonisation that led to his transnational existence. From its economic and cultural dimensions, globalisation initiated through the European explorers' arrival into Southeast Asia resulted in exploitations of the Malay slave for the former's fame, financial, and cultural gains.

Set during the Portuguese and Spanish imperial periods, Danny Jalil's novel *Enrique the Black* engages the colonial discourse of a centre–periphery binary that shapes the peripheral position, marginal treatment, and orientalist perceptions of the Malay protagonist during the sixteenth century. Enrique's time was a historical period of early globalisation that was synchronously tied to European colonisation (see Robertson 1992). Contemporary fiction has portrayed the European treatment of this Malay slave as racialised, deracinated, and deculturalised in the cultural and geographical Eurocentric spheres of the world. Brought to the European locations of Portugal and Spain, this Malay man is socially ascribed a different identity in Western-defined spaces and transnational places that redefine his ethnic and religious identities along reconfigurations of his colonial master. In the most recent fictional installation of his narrative, this young man's Portuguese captors refer to him as "Garoto" (Jalil 2021, p. 61) and christens him as "Enrique" (Jalil 2021, p. 65), the eponymous protagonist of contemporary fictional works *Enrique El Negro* (2002) and *Enrique the Black* (2021). The name "Enrique" signals twofold ethnoreligious losses of Malay and Muslim orientations. In the transnational space, an occlusion of his Malay and Muslim identities is a result of colonial intent to enlighten and convert the oriental slave. This cultural globalisation is unpacked when studying the transnational histories, perspectives and contestations in Danny Jalil's fiction, which traces Enrique's enslavement, his circumnavigation of the world and his identity crisis as a Malay man.

Responding to critical theorists Arjun Appadurai (1996) and James Clifford (1992) who highlight the cultural effects of displacement, Danny Jalil has written a historical novel that is concerned with the ways in which its Malay-Muslim protagonist's transnational mobility accentuates his cultural marginalisation and physical dislocation. Due to a colonial othering of his ethnic and religious identities, he experiences racial and cultural discrimination. When heeding literary critic Caren Kaplan's "caution against critical practices that romanticise or mystify regions or figures" (1996, p. 66), *Enrique the Black* can be seen to engage the deleterious effects of global flows and disjuncture that are instigated by colonialism in exposing "the lens of the colonial discourse" (Jalil 2021, p. 66) with which European characters regard their subjugated Eastern subjects. Even after his religious conversion into Christianity by his Portuguese master, Enrique continues to be racialised and culturally othered as a "*Black*" (Jalil 2021) man. As a Malay man from the Southeast Asian region, he is frowned upon when arriving as an outsider on account of both his race and nationality in Seville. Notwithstanding his Christian conversion, this Malay slave who is "not of this [European] land" (Jalil 2021, p. 114) is subjected to descriptive labels such as "'mulatto' and 'negro', 'brown' and 'black', [spoken] in Portuguese and Spanish" (Jalil 2021, p. 114). His consciousness of the racist names and insults that are directed at him heightens not only the painful degree of cultural discrimination, but also underscores the divisive lens of regional and national differences through which he continues to be viewed in Eurocentric spaces. Pertinently, a profound vision of the incarceration of indentured slaves is evoked in his mind—"Though he was free now to roam, he felt eyes watching him. This was almost like being chained" (Jalil 2021, p. 114). His physical discomfort becomes amplified and is projected to other slave communities, in a way that recalls the transatlantic slave trade made possible through European colonial domination of African people who are relegated to the "dark continent" (Stanley 1878). Although not originating from Africa, Enrique as a Malay man is regarded as such through an orientalist perspective that racially and culturally homogenises people from the East. A poignant instance of this cultural stereotyping occurs during his encounter with a woman who sells him a red apple in a marketplace in Seville. This Spanish grocer can barely conceal her bias against him as she notices "his tan first", "his odd face" and "his larger, more rounded nose" (Jalil 2021, p. 114). Her symbolic violence is evident through her internal rhetoric of his racial difference. Her account of his physiognomy feeds the Occident's narrative of him as a binary oppositional figure, thus exacerbating his peripheral position in the marketplace and, in turn, within European sociocultural spaces and Eurocentric economic locales. Further orientalist imagery related to his lack of morality is invoked in the subliminal reference to sin in the rancid taste of the apple as he gains an acute awareness of Europeans' gaze of Orientalism—"He savoured the fruit's sweetness as he chewed, but that sweetness turned sour after he was done eating the apple" (Jalil 2021, p. 114). The omniscient narrator's simultaneous description of Enrique as a "snake trying to hide from the searing eyes of the sun" (Jalil 2021, p. 114) operates allegorically to point to a typecasting of dark-skinned figures as the racial other in the marketplace, which metonymically represents the biblical Garden of Eden. The image of the snake as a reptilian serpent emphasises his animalistic otherness through the lens of Sevillian townsfolk, who in their likeness to the sun self-assume a contrasting model of enlightenment.

In transnational studies of the interrelated histories of colonisation and globalisation, religion is a field that has also attracted much scholarly attention (Levitt 2003; Portes 2003). The Christianising mission that was intertwined with Spanish and Portuguese imperial powers who held staunchly to Catholicism, was integral to the colonial project of religious globalisation in the sixteenth century. Within Southeast Asia, a transnational expansion of Catholicism was facilitated by Spanish colonial expeditions to maritime Southeast Asia, including the Philippines where Ferdinand Magellan gets killed and, in turn, Enrique achieves his physical autonomy. Coupled with his failure to convert the Chief of Mactan, Ferdinand Magellan's physical demise during the attack on Mactan Island in 1521 signals his premature curtailment of his religious mission to convert the indigenous people whom he encounters in Philippines. As demonstrated in the literary portrayal of this navigator, economic fame and religious fervour equally motivate his colonial expansion and European expedition. Ferdinand Magellan's admiration for fellow Italian navigator Christopher Columbus "who explored the world and found strange new lands" (Jalil 2021, p. 66) is based on the latter's success of the economic expansion, but also his simultaneous acts of religious globalisation in Plymouth. Moreover, Ferdinand Magellan declares, "They [The natives of Mactan] will be begging for me to know more about the words of Christ our Lord and the Holy Spirit, my boy! I want all of them to believe!" (Jalil 2021, p. 164) While successful in converting Chief Humabon whom he christens as Rajah Charles, Ferdinand Magellan is thwarted by Chief Lapu Lapu who remains intractable and physically, as well as ideologically, challenges him. Ferdinand Magellan's pursuit of twofold economic and cultural missions of globalisation results in his death as Lapu Lapu murders him to prevent a Christian conversion of his tribe.

The transnational debates in Southeast Asia about Enrique's religious orientation are set out in the portrayals of Enrique as a Malay man with national origins that are contested in different novels. Levitt points out that "[r]eligion and nationality reinforce one another" (2003, p. 853), and this intertwining of religion and state governance is apparent not only in Malaysia but also in sixteenth-century Spain. In Danny Jalil's novel, the amalgamation of religion and nation is first signalled in Ferdinand Magellan's declaration upon his arrival in Mactan Island that "[t]hese lands I shall claim for Spain, and these souls I will claim for Church" (Jalil 2021, p. 219). Oppression via the religious organisation of the church is well-critiqued in Frantz Fanon's *The Wretched of the Earth* (1961) that discusses the epistemological system of faith that facilitates atrocities against humanity (see also Lackey 2002). On the other hand, Carla M. Pacis' *Enrique El Negro* features a Malay Filipino man who is "born of one of these many beautiful islands [in present-day Philippines]" (2002) and lives on Sugbu island (historical name of Cebu). This geographical orientation also denotes Enrique's subscription to an animistic faith. This Filipina author "has chosen to believe that Enrique was [a Malay but non-Muslim] Filipino" and her literary account of Enrique's identity lies in stark contrast with Harun Aminurrashid and Danny Jalil who assign Malaccan Malay-Muslim identity to Enrique. Both Singaporean writer, Danny Jalil, and Malaysian nationalist writer, Harun Aminurrashid, maintain in their novels that Enrique is a Muslim man from the Sultanate of Malacca. Literary debates about Enrique's identity recall transnational contestations of Malayness that abound in Southeast Asia, a region in which Malays inhabit geopolitical landscapes that define Malay identity in response to national ideologies (Curaming 2011; Ho 2019). Furthermore, Antonio Paoliello avers that there is "the common equation between conversion to Islam and embracing Malayness" (Paoliello 2019, p. 475). However, Carla M. Pacis' novel deconstructs this ethnoreligious equation as her novel explores Enrique as a (non-Muslim) Malay in the Philippines occupying an exclusionary status in the Malay world. The Malay world's definition of Malayness excludes the geographical and sociocultural realm of the Philippines, where Malays are not necessarily Muslims as illustrated in the titular Enrique El Negro who practices animism instead. In contrast, Harun Aminurrashid's earlier historical novel features a Malay and Muslim protagonist who is the epitome of Malayness in his complete defence of the Islamic religion and attendant loyalty to the Malaccan Sultanate. Enrique is aptly lauded as *Panglima Awang* who does not betray his Muslim religion and, thus, Malay cultural roots as he resists Catholic conversion in the Eurocentre. In *Enrique the Black,* Danny Jalil's representation of Enrique aligns with Harun Aminurrashid's delineation of a Malay and Muslim man from Malacca, in spite of Enrique the Black's Christian conversion and his cosmopolitan participation as a multi-linguist in the transnational spaces that he navigates.

As a transnational slave in the sixteenth century, Enrique is subjected to global flows of culture that add impetus to his ideological adaptation to new geographical environments, which is "understood in terms of existing centre-periphery models" (Appadurai 1996, p. 32). Rather than a disjunctive order of culture that disrupts this colonial binary structure, the transnational experiences of Enrique beyond Southeast Asia attest to a colonial reinforcing of an orientalist distinction between the core and periphery that enacts cultural and economic processes of globalisation. In Danny Jalil's novel, the Catholic conversion of this Malay slave reifies the Eurocentric powers that lay decisive claims to his racialised body, which is a product of the unidirectional flow of Catholicism from European centres into Eastern parts of the world. The Portuguese crown gained "access to papal funding as well as that of the Order of the Christ, a leading crusading order" (Burke 2007, p. 68) that facilitated their cultural and economic globalisation. Even while Danny Jalil maintains that Enrique is a Malay and Muslim man from Malacca, this protagonist makes compromises as a cross-border transnational slave within Eurocentric cultural and national environments. For instance, his submission to Catholic conversion in spite of his Muslim proclivities for religious communalism in mosques is evident when his Portuguese master takes him to church one Sunday. At his Christian conversion during which Ferdinand Magellan announces "You will be born anew, cleanse of all your past sins" (Jalil 2021, p. 65), Enrique draws poignant parallels between the "holy power" of the "mosque" with which he is familiar and the institution of "the church" (Jalil 2021, p. 65). This comparative point enables him to accept his religious conversion without much initial resistance as he is aware of his marginal status as a disempowered slave—"Enrique spent the rest of the week getting used to his new name and the presence of his new master", (Jalil 2021, p. 65). However, it also causes an identity crisis manifested in his internal conflict. He is torn between complying with his Christian master and his affective Muslim duty that is integral to his identity as a Malay Malaccan man. When the Portuguese invade Morocco, the omniscient narrator discloses that:

> [Enrique] heard the *adhan*, the call to prayer, a faint call in the distance, wondering if it was either in his head or in his heart, the call so great, it shifted the movement of Portuguese soldiers sideways, the way a hurricane changes the flow of the tide." (Jalil 2021, p. 81)

Enrique's cognitive dissonance opens up a chasm between his externally oriented thoughts ("head") and inner desires ("heart"), the latter signifying his attempts at diverging from colonial narratives of difference that set Muslim Moroccans and himself, as a Muslim Malaccan man, apart from the Portuguese. The mosque is central to the transnational religious space as it offers him a sense of belonging within a shared community that enables him to invoke his past home, while gesturing to a future home to which he can return. When hearing that "the call of prayer grew stronger", Enrique postulates that "he was not far away from a mosque" (Jalil 2021, p. 81) to show his cognisance of transnational Muslim men who partake of communal worship in public spaces that are both "transtemporal and translocative" (Tweed 2016, p. 185). Here, the mosque marks "an alternative cartography of belonging" (Levitt 2003, p. 861), as it facilitates religious activities that transcend time and localities to inscribe belonging in transnational settings. Considering that Enrique's migration from Malacca has corroded his sense of identity, the call to prayer and attendant Muslim communal rituals serve as acute efforts at homecoming while exemplifying attempts at forging an affective home even while away from his Malaccan homeland. Thomas Tweed avers that religion is a figurative tool of homecoming (2016, p. 185). Religious acts of homemaking that are not forsaken serve, therefore, as a symbolic means to compensate for a wanting sense of home and identity in transnational cultural contexts. Even so, the call of war that arrives immediately after this Muslim call of prayer ("the call that came after brought within him nothing but dread. Horses galloped up ahead in the distance", Jalil 2021, p. 81) accentuates (Morroccan) Muslims as adversaries of the European colonisers who position the former on the cultural and social peripheries of the European transnational centre.

In displaying a linguistic aptitude, Enrique further challenges the stereotypical image of the Malay man as a stoic savage that is used against him during cultural and religious colonisation. Cultural stereotypes are explored in Syed Alatas' *The Myth of the Lazy Native* (1977) and Edward Said's *Orientalism* (1978), which highlight the transnational project of cultural imperialism in processes of Orientalism and cultural deracination by Occidental European powers. In *Enrique the Black*, the eponymous character defies colonial ideology when portrayed as a man who becomes cognisant of racial and cultural discourses that undergird European economic and cultural globalisation. Enrique is delineated as conscientious with an "innate understanding of languages" (Jalil 2021, p. 96) and an uncanny ability to "learn [languages] via aural osmosis" (Jalil 2021, p. 34). His linguistic competence transcends national boundaries that he views as permeable while navigating the world. Given that cosmopolitanism can be understood "from the perspective of coloniality" (Mignolo 2000), Enrique's cosmopolitan participation through his linguistic proficiency in transnational and transcultural communication blurs cultural demarcations established in the myth of the unlearned or unintelligible Malay man. As a polyvalent multi-linguist with proficiency in several languages, Enrique embodies a critical kind of cosmopolitanism through his analytical insights of a common humanity accessed in transnational places and spaces. The omniscient narrator explains that "[t]he more Enrique saw the world it was the similarities, rather than the differences, that attracted him" (Jalil 2021, p. 96). His insight positions him uniquely to subvert the colonial model of binary identities. Furthermore, his acute discernment that these cultural differences have led to ideological divisions, as "borders create hatred" (Jalil 2021, p. 134), exposes the orientalist discourse that undergirds the European colonial enterprise. Even as European captors removed Enrique from his Malaccan home and enslaved him, he is adept enough to empower himself—"So they had captured him, but he was also a thief, capturing their language, stealing a word here, and a phrase there, stashing enough in the vault of his young mind to eventually withdraw whole sentences" (Jalil 2021, p. 57). Here, the metaphor of stealing is reversed or inverted to assign agency to Enrique instead, thereby, denoting this Malay man's plundering of his European captors' linguistic resources for his transnational gains. When the Spaniards invade Mactan Island in 1521, the narrator also reveals that Enrique "wondered how right it was to burn the villages, how un-Christian it was to torment the innocent" (Jalil 2021, p. 237). Enrique's awareness of the physical and symbolic forms of colonial violence demonstrates his intelligence, as he reflects on colonial hypocrisy and exposes the self-serving myth of Malay cultural stereotypes. Furthermore, his ruminations underscore the utility of Christian ideology in advancing the transnational project of colonisation for European economic and cultural gains.

Underscoring ideological contestations between East and West, *Enrique the Black* highlights Enrique's personal and cultural identity crises as he encounters transcultural contact while traversing transnational realms together with European navigators. To mitigate the orientalist discourse that dehumanises him as a "slave" (Jalil 2021, p. 67; p. 102; p. 256), Enrique affirms his unbroken links to his Malay/Malaccan family and Muslim religion, however diminished they may have become during his enslavement. Colonial authorities that coerced him into a Catholic conversion have led to his physical oppression and suppression of his Muslim identity and Malaccan origins. As the omniscient narrator explains, "In dreams, he could remember his real name but when he awoke, that name was a pyramid of sand on a shore stolen away by a Portuguese tide" (Jalil 2021, p. 91). Enrique is viscerally affected by the loss of his Malay name that becomes obscured in his Christian conversion. In the vivid metaphor of the tide to denote Portuguese navigators who have pillaged maritime Southeast Asia, Enrique's geographical, emotional and cultural displacement are also referenced in the Malaccan shores that are awash due to European colonial invasion. In further evoking the mutability of sand, Enrique's adaptability is also highlighted. Nevertheless, Enrique exemplifies an embodied psyche of the oppressed whose original identity is relegated to the realm of his "personal unconscious" (Jung cited in Hauke 2006, p. 61). He internalises an inferiority attached to Malayness that reflects his "psychoexistential complex" (Fanon 1967, p. 12) due to the process of "epidermalization" (Fanon 1967, p. 11), which results in his disavowal of his Malay/Malaccan name during his enslavement. In a cathartic moment when Enrique relinquishes himself from colonial slavery upon returning to Southeast Asia, he remembers his Malay-Muslim name: "[Enrique] called out his true name, and it echoed forward in time, into distance, into history" (Jalil 2021, p. 279). Here, he consciously embraces a geopolitically situated moniker that had been both suppressed and repressed, identifying with his cultural and religious origins in his Malaccan "homeland" (Jalil 2021, p. 279). In reclaiming his Malay name that was assigned to him upon birth by his Malay/Malaccan father, he also initiates a reversal of epidermalisation in his concerted efforts to decolonise himself from the European colonial master. Enrique disrupts and challenges colonial hegemonies when invoking "the name his real parents had given him" (Jalil 2021, p. 279), simultaneously overcoming his inferiority complex as he likewise affirms that he is "not Enrique the Black" (p. 279). Precisely because he existed as "*a rover, a wanderer, a nomad*" (Jalil 2021, p. 279, original emphasis) from 1511 to 1521, Enrique continued with his transnational travels upon his freedom. However, he does so with a renewed identity when avowing his Malay-Muslim home in which his familial and cultural identities are rooted. His continual transnational journeying signals his agency at identity transformation to restore his mind, body and spirit to be "free of bondage, of the past, free to create his own future" (Jalil 2021, p. 279). Hence, his transnational mobility not only facilitates his learning of colonial hierarchies that have led to his spiritual bondage and diminished self-worth as a Malay man, but ultimately prompts his regeneration of his Malay identity that culminates in his physical, spiritual and cultural freedom from European colonisers. Even while contemplating that he may "*never lay his eyes on his homeland ever again*" (Jalil 2021, p. 279, original emphasis), the significance of the geopolitical region of Malacca is denoted in his unlearning of European hegemony when relearning his Malay/Malaccan and Muslim birth-name. Through traversing transnational spaces while navigating the globe, Enrique gains an acute sense of his Malayness to subvert the essentialist, orientalist and Western discourse of Malay identity.

15.6 Conclusion

To conclude, this chapter has examined the literary-historical story of Enrique. Borrowing from Frassinelli et al. (2011), the title of the chapter alludes to current work on transnationalism in developing the model, relevance and application of the transnational as an operative framework within literary studies. On the one hand, the transnational mobility of Enrique underscores his experience of changing cultural landscapes and porous geopolitical boundaries that feed into the identity-making of this migrant subject in maritime Southeast Asia. On the other hand, homing and homemaking emerge as salient processes to affirm the importance of the homeland for him to gain a semblance of identity that is germane to the transnational Southeast Asian Malay subject. As illustrated in Southeast Asian historical fiction, the protagonist in Danny Jalil's novel asserts a Malay-Muslim identity that differs from Carla Pacis' flexible (non-Muslim) and Harun Aminurrashid's fervent nationalist construction of Malayness as post-postcolonial responses to historical European contact with, and Enrique's journeying from and back to, Southeast Asia. This work of reterritorialising Enrique has occurred through these diverse narrations of Enrique's Malay identity. The historical significance of his life and the pertinence of his cultural identity as grounds for regional and national contestations in post-postcolonial times are foregrounded through narrating his experiences within an increasingly transnational world. Drawing on the transnational Southeast Asian context, these fictional texts underscore cultural definitions of Malayness that depart from the orientalist discourse in early colonial times. In Danny Jalil's *Enrique the Black,* Malayness is revisited and reconstructed as the Malay historical protagonist engages in acts of homing and homemaking. Ultimately, the novel responds to global flows of culture that have shaped the transnational migrant experience.

References

Ahmed S, Castada C, Fortier AM et al (2003) Uprootings/regroundings: questions of home and migration. Taylor and Francis Group, Milton

Alatas SH (1977) The myth of the lazy native: A study of the image of the Malays, Filipinos and Javanese from the 16th to the 20th Century and its function in the ideology of colonial capitalism. Frank Cass & Co, London

Aminurrashid H (1958) Panglima awang (National warrior). Pustaka Melayu, Singapore

Appadurai A (1996) Modernity at large: Cultural dimensions of globalisation. University of Minnesota Press, Minneapolis

Boccagni P (2022) Transnational migration and homemaking. In: Yeoh BSA, Collins FL (eds) Handbook on transnationalism. Edward Elgar Publishing, Cheltenham and Camberley, pp 141–154

Burke E (2007) The sixteenth-century world war and the roots of the modern world. In: Karras A, Mitchell LJ (eds) Encounters old and new in world history. University of Hawaii Press, Hawaii

Clifford J (1992) Traveling cultures. In: Grossberg L, Nelson C, Treichler PA (eds) Cultural studies. Routledge, New York, pp 96–112

Curaming R (2011) Filipinos as Malays: historicising an identity. In: Mohamad M, Aljunied SMK (eds) Melayu: politics, poetics and paradoxes of race. National University of Singapore Press, Singapore, pp 241–274

Curaming R (2023) Historical versus practical pasts? Enrique de Malacca in Malaysia's historical imagination. História da Historiogr: Int J Theory History Historiogr 16(41)e1945:1–22

Fanon F (1967) Black skin, white masks. (trans: Markmann CL). Grove Press, New York

Frassinelli PP, Frenkel R, Watson D (eds) (2011) Traversing transnationalism: the horizons of literary and cultural studies. Brill Press, Leiden

Hauke C (2006) The unconscious: Personal and collective. In: Papadopoulos RK (ed) The handbook of Jungian psychology: theory, practice and application. Routledge, London and New York, pp 54–73

Ho HMY (2019) Women doing Malayness in Brunei Darussalam. Southeast Asian Review of English 56(2):147–165

Ho HMY (2020) The violence of othering and (non-)indigenous revival: Aammton Alias' 'the last bastion of ingei: imminent' as postcolonial speculative fiction of Brunei Darussalam. Southeast Asian Review of English 57(1):55–79

Ho HMY (2021a) Localisation of Malay Muslim identity in Brunei Darussalam: a modern nation's cultural and economic goals. In: Ho HMY, Deterding D (eds) Engaging modern Brunei: research on language, literature and culture. Springer, Singapore, pp 127–143

Ho HMY (2021b) Chinese Bruneian identity: Negotiating individual, familial and transnational selves in anglophone Bruneian literature. The Wenshan Review of Literature and Culture 14(2):1–34
Ho HMY (2023a) Rethinking 'amok': indigenous identity affirmation in Malay legends of Southeast Asia. In: Pelea I (ed) Culture-bound syndromes in popular culture. Routledge, London and New York, pp 164–180
Ho HMY (2023b) History, memory and duality: Malaysia's national amnesia in Chuah Guat Eng's Days of Change. Kritika Kultura 41:6–28
Ho HMY (2023c) Reframing loss: Chinese diaspora identity in KH Lim's Written in Black. Suvannabhumi 15(1):131–152
Ho HMY, Amran R (2021) Constructs of the modern home: negotiating identity in Anglophone Bruneian literature. In: Ho HMY, Deterding D (eds) Engaging modern Brunei: research on language, literature and culture. Springer, Singapore, pp 163–175
Humairah Z, Nasir KM (2021) The primordial modernity of Malay nationality: contemporary identity in Malaysia and Singapore. Routledge, London and New York
Jalil D (2021) Enrique the black. Penguin, Singapore
Jay P (2010) Global matters: the transnational turn in literary studies. Cornell University Press, Cornell
Kaplan C (1996) Questions of travel: postmodern discourses of displacement. Duke University Press, Durham
Lackey M (2002) Frantz Fanon on the theology of colonization. Journal of Colonialism and Colonial History 3(2). https://doi.org/10.1353/cch.2002.0045
Levitt P (2003) You know, Abraham was really the first immigrant: religion and transnational migration. The International Migration Review 37(3):847–873
Leyden J (1821) The Malay annals (trans: Leyden J). Longman, Hurst, Rees, Orme and Brown, London
Mignolo W (2000) The many faces of cosmo-polis: border thinking and critical cosmopolitanism. Public Culture 12(3):721–748
Miyoshi M (1998) Globalization, culture, and the university. In: Jameson F, Miyoshi M (eds) The cultures of globalization. Duke University Press, Durham, pp 247–269
Mohsin H (2000) Reccess. In: The Chronicle. Duke University, 18 February 2000, p 5
Muzhafar I, Hashim RS, Raihanah MM (2015) Malay cultural identities: A review. Humanities and Social Sciences Letters 3(1):1–9
Ong A (1999) Flexible citizenship: The cultural logics of transnationality. Duke University Press, Durham
Pacis CM (2002) Enrique el Negro. Cacho Publishing, Manila
Portes A (2003) Conclusion: Theoretical convergencies and empirical evidence in the study of immigrant transnationalism. The International Migration Review 37(3):874–892
Paoliello A (2019) Becoming Muslim or becoming Malay? Conversion to Islam in two Sinophone Malaysian short stories. Canadian Review of Comparative Literature 46(3):470–487
Robertson R (1992) Globalisation. Sage, London
Said E (1978) Orientalism. Pantheon, New York
Said E (1994) Culture and imperialism. Vintage, New York
Stanley HM (1878) Through the dark continent. Sampson Low, London
Turcan RV (2016) Exploring late globalization: A viewpoint. Markets, Globalization and Development Review 1(2):Article 4
Tweed TA (2016) On narratives, niches, and religion: A response to Jonathan Marks. Philosophy, Theory and the Sciences 3(2):183–187
Tyrrell I (1991) American exceptionalism in an age of international history. The American Historical Review 96(4):1031–1055
Ullah AKMA, Ho HMY (2020) Globalisation and cultures in Southeast Asia: demise, fragmentation, transformation. Global Society 35(2):191–206

Vann MG (2014) When the world came to Southeast Asia: Malacca and the global economy. Maritime Asia 19(2):21–25

Winstedt RO (1938) The Malay annals or sejarah Melayu. Journal of the Malayan Branch of the Royal Asiatic Society 16(3):1–226

Yahya H, Tawas RR (2014) Pengeliling bumi pertama adalah orang Indonesia: Enrique Maluku. (The first circumnavigator is an Indonesian: Enrique Maluku). PT Ufuk Publishing House, Jakarta

Hannah Ming Yit Ho is an Assistant Professor at the Faculty of Arts and Social Sciences at Universiti Brunei Darussalam. She is also a research associate at the Asia Research Institute, National University of Singapore. Formerly, she was a postdoctoral fellow at King's College London and University of California, Berkeley. She is an associate editor of *Southeast Asia: A Multidisciplinary Journal*. Her books include the edited volume entitled *Engaging Modern Brunei: Research on Language, Literature and Culture* (with David Deterding).

Chapter 16
The Portuguese in Southeast Asia: Cultural Transfers and Transnational Identities in Malacca

Rui Oliveira Lopes

Abstract With the capture of Malacca in 1511, the Portuguese became the first European power to engage in maritime Southeast Asian commerce and trade. Malacca was a strategically vital location facilitating trade across the South China Sea, leading to valuable camphor and other spices from Borneo and then to the trade expeditions in the Banda and Maluku Islands. The Portuguese rapidly understood the value of textiles within Southeast Asia maritime trade and among Malay communities, bringing textiles from Cambay and circulating Javanese batik and Sundanese ikat textiles within several port cities across the region. Rather than a colonial settlement, Portuguese Malacca was a commercial entrepôt linking a highly profitable maritime trade route from Goa, Cochin, Ceylon (Sri Lanka), the Coromandel Coast, Malacca, Macao, and Japan. Due to Portuguese diplomacy and trade, significant socio-cultural transformations occurred across the Malay world in Southeast Asia. At the end of the sixteenth-century, Malacca was gradually transformed into a Luso-Malay society. By studying historical documentation, material culture, and a discussion on expressions of tangible and intangible cultural heritage among Luso-Asian communities, this chapter examines the formation of transnational identities and cultural transfers resulting from Portuguese presence in Southeast Asia between 1511 and 1641.

Keywords Transcultural identities · Connected history · Transnational identities · Portuguese Malacca

16.1 Introduction

The discussion of transcultural and transnational identities emerging in territories controlled or partially controlled by the Portuguese throughout Asia requires a postcolonial approach to the history of the socio-cultural fabric woven within the context of commercial, religious, and political activities between port merchants and port

R. O. Lopes (✉)
Museum of Convergences, Porto, Portugal
e-mail: rui.o.lopes@gmail.com

H. M. Y. Ho and Y. Chan (eds.), *Transnational Southeast Asia*, Asia in Transition 29,
https://doi.org/10.1007/978-981-96-3608-2_16

communities (Subrahmanyam 1993, 2016). It is also crucial for a critical analysis of both European and indigenous discourses of that history and to connect them to expressions of tangible and intangible heritage. However, rather than examining the cultural, political, and sociological legacy of imperialism or the exploitation of the colonialist over the colonised, in this chapter, we assume that a shared history emerged from civilisational encounters and that transcultural and transnational identities developed in the context of historical phenomena against the idea that singular traditional cultures were confined within the boundaries of nation-states and regimented cultural conventions (Subrahmanyam 2022).

The history of civilisational encounters and cultural exchanges led to a transformative process informed by notions and perceptions of circulation and connections between societies (North 2010; Swope and Andrade 2018). The cultural dynamics which occurred in a specific time and location, such as the Portuguese rule in Malacca from 1511 to 1641, had a profound effect on the physical landscape of the port city, through the construction of the fortress and other administrative buildings. It also transformed the social fabric of its inhabitants, both the indigenous Malay Muslim communities and all the migrants from Japan, Korea, Macau, India, and Africa who came along with the Portuguese who settled and became residents in Malacca and blended in the multicultural and transnational landscape through the cultural and social interactions with other port traders from other parts of Southeast Asia. From a wider perspective, Portuguese Malacca was linked to several maritime trading routes and mercantile communities from Borneo, Indonesia, and the larger trade networks across the Java Sea, the Banda Sea, and the Celebes. All these connections and circulations of commodities, ideas, and identities had a ripple effect and were equally transformative to the European civilisation and culture (Lach 1968). The history of cultural encounters cannot be compartmentalised into binary narratives of one and the other, coloniser and colonised, here and there, oppressors and oppressed as there is a global interdependency in a connected history (Douki and Minard 2007). This Braudelian theoretical framework towards historical analysis is particularly applicable to the Portuguese seaborne empire as a mercantile network of port cities that were interdependent (Braudel 1949). The control of the trade of cotton textiles in Cambay, Bengal and in the Coromandel Coast was linked to the spice trade in Southeast Asia, exchanging these products with clove from the Maluku Islands, nutmeg from Banda, sandalwood from Timor, camphor from the Sultanate of Brunei, and pepper from Sumatra and Java. These commodities were then exported to China, India, and Europe. It was not just food products that were circulating and changing culinary traditions all around the world, but a profound knowledge of the things of the world awakened to a wider consciousness of the world, in its profound natural and cultural diversity.[1] These global commercial and cultural interactions contributed to the making of the European Renaissance and prompted the Age of Enlightenment in the seventeenth and eighteenth centuries, and perhaps to our contemporary concerns on migration, dislocation, racial discrimination, and how museums and other educational and cultural institutions are more concerned with cultural decolonisation and inclusive histories.

16.2 Understanding the Socio-Cultural Fabric of the Portuguese Seaborne Empire (1415–1825)

Vasco da Gama's arrival to India in 1498 opened a new chapter in the history of Portuguese Seaborne Empire, paving the way to unprecedented global cultural dynamics and prompting the emergence of transnational cultural identities throughout the world. It is important to bear in mind that the Portuguese presence in the Indian Ocean and, subsequently extending its influence to the Malay world all across mainland and maritime Southeast Asia, and as far as to the China Sea and the shores of Japan, is an amplification of the Portuguese commercial, missionary, diplomatic and cultural exchange experience developed across the Atlantic coast of Africa for nearly 100 years, since the conquest of Ceuta in 1415.

As referred by Charles Boxer (1969), what started as a feverous crusade against the Moorish infidels, soon became a strategic move to reach the Trans-Saharan caravan trade routes through the port cities of the West Coast of Africa, particularly in Arguin (today's Mauritania), Sierra Leone, Lagos, Benin and across the whole coast of Congo, mostly looking for gold and slaves. Rapidly, the Portuguese acknowledged the high value of luxury and rare commodities of African origins among European courts and an emerging bourgeoisie of wealthy merchants and noblemen, who avidly sought exquisite objects that were then collected and displayed in princely *Kunst- und Wunderkammern* (Cabinets of Wonders and Curiosities) as expressions of diverse cultural identities and the cultural dialogues between European and other cultures across the world (Gschwend 1995). Gradually, the Portuguese motivations to proceed south along the coast of West Africa and eventually to cross the Cape of Good Hope towards the Indian Ocean were to find the kingdom of Prester John, a legendary Christian patriarch who, according to popular tradition, had been ruling a Christian nation among the pagans and Muslims in the Orient. In addition, as it is widely known, the Portuguese were aiming to take control of the highly profitable spice trade that was mostly controlled by the Arabs through the Red Sea or the Persian Gulf and the Levant, motivating the Portuguese to take the ports of Ormuz (Oman) and Aden, today's Yemen (Strootman et al. 2019; Subrahmanyam 1993). Although the details of the true reasons for the Portuguese global enterprise are still debatable, scholars generally agree that the Portuguese empire was mostly determined by commercial, religious, diplomatic, and strategic factors in a balance of proportions that would vary according to local, regional, and global circumstances (Thomaz 1992; Costa 2022). There was a set of conveniences and needs, in which one interchangeably justifies the other, usually supported and legitimised through public decrees issued by a Pope.

Therefore, rather than a colonial empire in its true sense of a territorial domain and overpowering control of other nations, the Portuguese empire was essentially characterised as an encompassing mercantile operation based on the possession of fortifications and posts strategically situated for trade, secured by political and military control under the governance of the *Estado da Índia* (State of India), which was based in Goa. In this context, the Portuguese presence was rather scattered and limited to port cities operating as multicultural and transcultural emporiums where the boundaries of cultural identities, nationality, ethnicity, customs and traditions, language, and even religious identity were often not quite clear. It is important to mention that, as well it will be discussed later in this chapter, these port cities and their communities were already established and were already part of a wide maritime trade network prior to the arrival of the Portuguese.

The Portuguese *Estado da Índia* served as a governing centre responsible for administrating all matters under the rule of the King of Portugal in a wide string of trade ports and military forts controlled by the Portuguese across Asia, connecting Goa on the west coast of India to Malacca on the Malay peninsula, Macau, and Nagasaki in the Far East. Through Nagasaki, the Portuguese-controlled ports were then connected to a wider commercial network outside the direct control of the Portuguese, which extended to Manila, Acapulco, and across the Pacific (Lieberman 2009).

Through the maritime trade routes controlled by the Portuguese connecting several Asian ports, the circulation of individuals at all levels of society and the flow of commodities gave rise to transcultural and transnational port communities. The notions of transcultural and transnational identities in Portuguese port communities were mostly expressed in tangible heritage, through architecture, artefacts and valuable objects, and intangible heritage through religious beliefs, culinary traditions, language, customs, and social practices.

The political, administrative, and commercial success of the Portuguese enterprise in Asia relied significantly on cultural and social interactions, namely on Afonso de Albuquerque's policy of intermarriage which encouraged Portuguese men (*casados*) to marry Asian women who would give birth to *mestiços* (mixed) descendants (Boxer 1963; Chaturvedula 2010; Thomaz 1992; Xavier 2008). Although this policy contributed to a broad notion of a Luso-Asian transnational identity expressed through the practice of syncretic religious rites, the hybridisation of social customs, and the dissemination of Portuguese and Malay languages as a *lingua franca* across Asia, towards the seventeenth century there was an increasing preoccupation among Portuguese secular and religious authorities over this cultural degeneration and contamination related to physical and spiritual purity of the Portuguese as markers of their social, racial, and intellectual superiority (Chaturvedula 2010). It is undeniable that the Portuguese authorities imposed severe restrictions on religious practices, rites and ceremonies, the destruction of temples, and prevented other expressions of social practices that were considered acts of idolatry and offence to God (Chaturvedula 2010).

Throughout the seventeenth century, Albuquerque's policy of tolerance was replaced by an increasing religious imperialism, mostly expressed in Goa as demonstrated by cases judged by the Goan Inquisition (Cunha 1995; Dellon 1688). However, historical sources documented that Muslims and Hindus in Diu, Daman, Malacca, and the Buddhists in Macau were allowed to keep their religious practices and preserve their social and cultural customs (Thomaz 1992). Ironically, it was through the missionary efforts of the Jesuits, after their establishment in Asia in 1542, that a profound socio-cultural transformation occurred within Asian societies. The Jesuit method of accommodation encouraged cultural adaptation to local societies, leading many missionaries to study indigenous traditions, beliefs, ritual practices, languages, and material cultures to establish a bridge of mutual understanding towards a culturally and socially inclusive society under the guidance of Christian values. As Goa was under political pressure to stand as the "Rome of the East", an epitome for Goa as the capital of the Roman Catholic Church in the Eastern world, the Jesuit approach of accommodation enjoyed more freedom in Southeast Asia, Macau and, particularly in Japan, although their missionary methods were occasionally condemned by religious authorities in Goa.

As a result of a scattered and strategical presence in four continents (Europe, Africa, Asia, and America) and three oceans (Atlantic, Indian, and the Pacific), the perception other people had of the Portuguese is generally that of a transcultural and transnational identity. Around 1600 a British traveller described the Portuguese in Guinea where some were white, others were *mulatto* (mixed Portuguese and African), but most of them were black (Hespanha 2019). The identification of the Portuguese was tied to specific cultural practices and appearance such as their dress, objects of daily use, religion, language, and expressions of behaviour, not necessarily associated with skin colour or facial features, and even less so with the country of origin or descent, or political subjection. Portuguese identity was sometimes appropriated and reclaimed by Eurasians or other peoples who sought for concessions, privileges, or simply to cause an impression. On other occasions it was imposed on other individuals. Although the Jesuit order was a religious organisation composed of men of several nationalities serving under the auspices of the Portuguese *Padroado*,[2] the missionaries were frequently identified or perceived as being Portuguese, just because they travelled in Portuguese trade ships and spoke Portuguese language.

The Japanese lacquer folding screens and other material culture produced by the Japanese during the sixteenth century recurrently documented the arrival of the Portuguese to Japan and the festive atmosphere around the unloading and loading of the black ship (*kurofune*) in the port of Nagasaki (Curvelo 2010; Yamafune 2012). The folding screens also represent the parade of Portuguese merchants, Jesuit priests and a multitude of crew members harnessing dogs, camels, horses, monkeys, peacocks, and other caged exotic animals, while Indian mahouts control enormous elephants. Among the individuals aboard the Portuguese black ship who were dressed in typical Portuguese clothes (balloon-like trousers, wide or high stiff collars, and the use of buttons), Europeans, Asians, and Africans can be easily identified. According to the *Teppô-Ki* (Chronicle of the Firearms), a document written by a Zen monk Bunshi Genshô in 1606 describing the arrival of the Portuguese at Tanegashima in 1543, confirms that individuals of Chinese, Malay, Indian, and Japanese origins on board of the Portuguese ships were perceived as *namban-jin* (southern barbarians), a term used as a reference to the Portuguese (Kouamé 2020).

The notion of this expanded Portuguese transnational identity is the backbone of the Portuguese empire in Africa, America, and Asia, connecting two distinctive worlds: one of the Portuguese-speaking European and Christian, and another non-European who is connected, through his or her mother, to another religion, culture, language, and social practices (Costa 2022). These Luso-Asians circulated in both worlds aware of the cultural codes and traditions, from one side and the other, becoming grassroots protagonists for the long-lasting Portuguese presence in Asia, which is particularly noticed in the cultural pride of the *Kristang People* in Malacca and other Eurasian[3] Portuguese diasporic communities in Asia, which will be discussed later in this chapter.

16.3 Global Identities and Transnational Communities in Portuguese Malacca (1511–1641)

Historical sources from the early sixteenth century, such as *Suma Oriental*, written by Tomé Pires, a Portuguese apothecary who lived in Malacca from 1512 to 1515, as well as several other archival documents such as letters, reports, and petitions exchanged between private or crown merchants and the king of Portugal or his delegates, demonstrate that Malacca was mostly administrated as a commercial emporium, connecting the Indian Ocean and the maritime trade routes to Southeast Asia and East Asia (Leão 1998; Pires 1944).

Although historians often refer to the "conquest of Malacca" by the Portuguese in 1511, it is generally accepted that the Portuguese never overthrew the Sultanate of Malacca, but simply took the control of the port and the maritime trade (Thomaz 1992; Borschberg 2019). Arguably, the military siege and capture of the port of Malacca and the subsequent expulsion of the Sultan of Malacca, Mahmud Shah, from his capital city, who continued ruling the Sultanate of Malacca from the Riau Islands for another two decades until his death in 1528 (Borschberg 2019), were a response to a plot arranged by Muslim Gujaratis and Javanese merchants who convinced Sultan Mahmud and the *bendahara* (a kind of vizier in classical Malay administration) that the Portuguese were a threat to the rule of the Sultan. In 1509, the Portuguese ships were attacked by the Malayan fleet, and the Portuguese sea Captain, Diogo Lopes de Sequeira, successfully repelled all boarding attempts but was unable to rescue the Portuguese traders left at the factory (trading post), who were then made captives (Rodrigues and Costa 2012). Afonso de Albuquerque, who was the Governor of Portuguese India at the time received messages from the captives through the Portuguese factor in Malacca. Albuquerque sent a powerful Hindu merchant, Nina Chatu, who travelled to Malacca to intercede for the Portuguese, unsuccessfully (Thomaz 1991). After securing Goa in 1510, Albuquerque decided to sail to Malacca in the following year, which would become pivotal to the establishment of a Eurasian community in Southeast Asia.

Rather than a territorial empire, the Sultanate of Malacca was itself an established mercantile power in the Malay peninsula, extending its commercial influence and its aspired reputation to parts of central and eastern Sumatra, Borneo, and the Riau Archipelago. With the capture of the port of Malacca, the Portuguese merchants inherited and sought to maintain not just the expanded trade network across maritime Southeast Asia—Aceh, Java, Borneo, Banda and Maluku Islands (*negeri diatas angin*), Peninsular Southeast Asia—Pegu (Myanmar) and Siam (Thailand)—and the Far East—Macau and China (*negeri dibawah angin*). In addition, the Portuguese recognised the importance of maintaining the traditional trade system known as *beli-belian* (reciprocal purchase), the role of the Malay merchant-officials (i.e. the *bendahara*, the *temenggong*, the *laksamana*, and the *shahbandar*) and certain aspects of the *adat* (Malay customary law) (Borschberg 2019; Thomaz 1992). In fact, it would be more accurate to say that the success of the Portuguese in Malacca was determined by their ability to adapt to the local circumstances rather than introducing a new system of governance and trade. Native merchants were integrated and used as commercial advisors, while most of the crew on board Portuguese ships were from Pegu and Java, allowing the Portuguese traders to adapt to local Malay customs and trade system.

According to Malay and Portuguese sources, the port communities in Malacca before the arrival of the Portuguese consisted mostly of Gujarati, Tamil, Bengali, Persian, Arabs, and Javanese (Fang 1976; Pires 1944). These communities were generally Muslims, except for the Tamils from Coromandel Coast, who were Hindus established in *Kampong Keling*, but equally influential in the trade system and within Malaccan society. The Javanese were by far the largest community, but they were mostly focused on the small trade of food products and services, as many Javanese had occupations as peddlers, hawkers, craftsmen, sailors, fishermen, and metalsmiths, among others. The Javanese were frequently hired by the Portuguese to serve on ships connecting to Siam and Pegu as crew members or as interpreters (Thomaz 1991, 1993).

The Portuguese chronicler Tomé Pires in his *Suma Oriental* also mentions the community from the Luzon Islands, established in a small port northeast of Malacca, between Beruas and Klang, occupied with local commerce and serving in the fleets of important Luzon merchants who kept profitable connections to Siam, Brunei, the Sunda Islands, and China (Pires 1944). The Portuguese sources identify them as Muslims, which allow us to assume that it is possible that the Luzon community from Malacca may have contributed to the Islamisation of Luzon.

In its turn, the Chinese merchants have been in Malacca since the beginning of the Sultanate of Malacca in the early fifteenth century. The Chinese community occupied *Kampong China*, located at the heart of the trade centre, upstream of *Kampong Keling*. Chinese Junks established important trade connections to Champa and Tonkin (Chăm Pa and Đông Kinh, Vietnam), to Fujian (China), and Japan, through the Ryukyu Islands. Interestingly, sixteenth-century Portuguese documents consistently refer to the Chinese in two ways, perhaps to distinguish two types of communities, the "*chins*" *or* "*chinas*" and the "*chincheus*" (Thomaz 1992). These two terms were used to distinguish ethnic identity (Cantonese and Hokkien), the use of different dialects, or perhaps to distinguish Chinese Muslims from other Chinese *gentiles*. Following the Portuguese presence in Malacca and the establishment of the Portuguese in Macau in the first decades of the sixteenth century, the Chinese community in Malacca increased significantly as they were given privileges, in a strategy to counterbalance and control the influence of the Muslim communities in Malacca.

There were other numerous, but not influent, communities in Malacca during this transition from the Sultanate to the Portuguese rule of the Port of Malacca. Among these were the Bengalis, who were mostly working as tailors, fishermen, and servants for heavy labour and menial tasks, and the Burmese from Pegu who served almost exclusively as pilots and merchants. As an international commercial hub, Malacca was also the home of other transient foreign communities from Brunei and other parts of Borneo, Palembang, Aceh, and other port cities in Sumatra, the Sunda Islands, and Maluku.

The official policy, during the Sultanate and Portuguese rule, was to encourage the settlement of foreign traders in Malacca, giving them privileges such as reduction of import duties and being allowed to participate in the Portuguese trade with China. Additionally, Muslim, and Hindu merchants only paid six per cent in customs duties, while Christians paid ten per cent (Thomaz 1993). Although this may sound contradictory, the reasoning was most likely to keep the Portuguese soldiers from abandoning their posts for the lucrative trade. The Portuguese clearly recognised the advantages of keeping the ethnic, religious, and neighbouring solidarities of the Malaccan port communities.

As a pluralistic and cosmopolitan society, Malacca became an oasis for religious tolerance despite its reputation as a Muslim stronghold. Although Islam was established as the main religion in Malacca since its inception, the observation of Islamic teachings and precepts was open and permissive, generating aversive reactions from Arab traders who criticised local and transient regional Muslims for eating non-halal food, drinking wine, and taking pagans as wives (Tibbetts 1979). These practices are corroborated by Portuguese sources (Pires 1944). Islamic law (*fiqh*) was more subsidiary and prevailed over Malay customary law (*adat*) only when more strict action was needed, such a theft and adultery (Fang 1976).

The transcultural and transnational characteristics of Malacca's society predate the Sultanate itself, as the region has been an increasing cluster for maritime trade and a centre for the dissemination of Hindu, Buddhist, and Islamic social groups since the seventh century. However, the arrival of the Portuguese and other Europeans added another layer to the social and cultural fabric of Malacca and expanded it into a larger and more intricately interconnected human geographies, sharing tangible and intangible cultural expressions, which will be discussed in the following sections of this chapter.

16.4 Portuguese Tangible and Intangible Heritage in Malacca

Architectural heritage is one of the most impactful expressions of cultural identity and tangible heritage, not just for the imposing structure and scale of a public construction, but mostly for standing as a carrier of memory, power, and its imposing strong ideological statements. Portuguese-built heritage sites in Malacca consist of military architecture (the Portuguese Fortress), urban development (Portuguese settlement, *Pelourinho* (pillory) square), and religious architecture (St. Peter's Church, Saint Paul's Church, Church of Saint Lawrence), which was pivotal to inscribe Malacca in the UNESCO World Heritage listing in 2008.

Vernacular architecture in Malacca consisted mostly of ephemeral structures made of timber, palm leaves, and other materials rapidly available in the surrounding environment. Masonry buildings were generally limited to mosques, Hindu temples, and other few urban structures. The built environment in Malacca was already a result of cultural and religious hybridisation, denoting a confluence of Hindu-Buddhist styles and architectural elements infused in traditional Austronesian building materials and techniques which integrated adaptation to local climate, belief systems, and Malay traditions. The urban landscape of Malacca changed significantly under the rule of the Portuguese, through the construction of the fortress with bastions built at angles along the compact city walls, repurposing the materials of destroyed mosques and palaces. The fortress housed all urban institutions typically found in most Portuguese fortresses, which included the town hall, Catholic churches, the residence of the bishop, the Jesuit college, three monasteries, and the *Misericórdia*, a charitable institution to take care of widows, orphans, indigents, which also served as a hospital and handled the administrative matters related to the estate of Portuguese deceased.

Coeval drawings of Portuguese fortifications remain an important source of information about the urban setting, which allowed on some occasions for an understanding of the social fabric within and on the outskirts of the fortresses. These drawings demonstrate that in most port cities, the Portuguese settlement or territorial occupation was limited to the fortresses, the ports, and the markets. On the outskirts of the fortresses, we often see the representation of mosques, temples, pagodas, and other expressions of local socio-cultural practices, along with a few scattered churches, crosses, and the pillory documenting the coexistence of local and European cultural realities. This also shows that it was mostly the catholic missionaries who ventured more within the heart of the local and other foreign non-Portuguese communities, leading to a transformative cultural and identity transformation, which will be discussed later in this chapter.

According to a panoramic sketch documenting the siege of Malacca led by the Sultan of Aceh Alauddin in 1568, most of the structures inside the walls were made of masonry, but some of the top storeys were made of wood. They had lofty, pitched roofs that were probably tiled in part to prevent fires. Some of the buildings in the drawing are strikingly similar to a few sixteenth-century structures that are still standing in India, namely in Goa and Cochin. The fact that many of the structures that are still standing in India feature exterior steps to the top floor suggests that the lower level was intended for individual traders' storage. Many of the buildings were contiguous to one another, forming terraces and streets, although they have relatively vast gardened areas behind them. This type of urban layout was also typical in Portuguese Goa, as can be confirmed in the illustrations by Jan Huyghen van Linschoten, a Dutch merchant and traveller who lived in Goa during the second part of the sixteenth century, known for his *Itinerario*, an illustrated travel journal documenting his journey through the East Indies describing the peoples, their habits, religious practices, cultural expressions, fauna, and flora.

Other drawings and visual representations of Portuguese Malacca clearly highlight the magnificence of the 60 m tall and four-storey keep as well as the main churches, demonstrating that most of the Portuguese settlers, soldiers, along with the *casados* and their native wives, their Luso-Asian descendants, and their personal slaves lived within the walls of the fortress. On the other hand, the diverse populations of non-Portuguese peoples and the slaves[4] working for the *Estado da India* normally lived beyond the walls of the fortress. In the surroundings of the fortress, Portuguese descendants *mestiços*, often born from illicit unions between a slave mother and a Portuguese settler, grouped with other communities of freed slaves, congregating in their own Catholic churches, and socially separated from the communities of *casados*, were considered the Creole elite of Portuguese Malacca.

The *casados* of Malacca, the Portuguese *fidalgos* (noblemen), soldiers who deserted, the private merchants, and a larger number of foreign slaves and free men, originally from Africa and other parts of Asia, engaged with the original trading communities and local Malays, leading to a diverse transnational and transcultural identity that had mostly Catholicism, Portuguese language, and other Portuguese/European traits as an identitarian binder within this heterogenous society. As the Portuguese and other free men took local women as concubines or wives, many of the women converted to Christianity, and their children were raised as Christians, who also became known as *Cristãos da Terra* (Christians of the Land).

In Malacca, the Portuguese Eurasians became known as the *Kristang* community due to their upholding of the Catholic Christian values and Portuguese cultural origins. Nevertheless, the *Kristang* identity is far more complex than just a blend between Malay and Portuguese identities and traditions. As already discussed, the Portuguese identity in Asia was a fluid one including individuals from a variety of races, ethnicities, languages, and social practices who originated from Portugal or were born in any part of Asia, Africa, or South America. In its turn, the *Kristang* community, although it holds a strong heritage and culture of Portuguese origin, it also adopted layers of cultural eclecticism from the Dutch, the British, the Indian, and the Chinese.

The Portuguese Eurasian community of Malacca formed a miscegenic transnational identity, shaping not only the socio-cultural fabric of Malacca but also that of Macau. This is a result of several Portuguese, Malaccan and *mestiço* individuals who ventured to China and settled there. Several scholars have been studying the Malay legacy on Cantonese culture, particularly in the culinary traditions and linguistics (Jackson 2020; Moody 2021). For example, the *belacan*, a fermented shrimp or krill paste mixed into a chili sauce commonly used in Malay cuisine, was adapted into Macanese cuisine and from there to Goa where it is known as *balichão* or *balichow* and used as a side condiment. In Macanese cuisine, *Balichão* is one of the traditional sauces of Portuguese/Malaccan origin, used to cook many traditional dishes. The main ingredients are *belacan* (fermented shrimp or krill paste) mixed with bay leaves, salt, chili, cloves, lime, and peppercorns laced with fortified wine, a Portuguese distilled spirit known as *aguardente* (Silva 2015). The words *balichão* and *balichow* are, respectively, Portuguese and Macanese adaptations from the Malay *belacan*. The well-known Macanese cuisine is celebrated for its fusion character blending the use of spices from Goa, Malacca, Brazil, and East Africa. Likewise, the wide use of bird's eye chili[5] in Asian cuisine, namely in India, Thailand, and Malaysia was introduced by the Portuguese in the sixteenth century, who brought it from Africa and South America (DeWitt 2020). In fact, the most typical type of chili peppers used in Southeast Asia today is bird's eye chili, which is commonly known as *cili padi* in Malay cuisine. Some of the Malaccan *Kristang* culinary traditions also denote this Portuguese/Macanese cuisine traditions, such as the baked stuffed crab known as *inchimintu karangezu* (an adaptation from the Portuguese *carangueijo* word for crab), very similar with the Macanese *casquinhas*. The unique *sambal binagre pesi* (adapted from the Portuguese *vinagre* word for vinegar and *peixe* word for fish) or the Malaccan-Portuguese *kari debal* reveals the preference of the Portuguese in seasoning food with vinegar, which is still very present in Portuguese dishes. The use of vinegar is also noticed in several Goan/Portuguese dishes such as lamb *vindalho* (from the Portuguese *vinha de alho*, a seasoning with wine and garlic to marinate meat or fish), which is the origin of the famous *vindallo* sauce in South Indian cuisine. Due to the high value of wine and its scarcity in India, this ingredient was replaced by palm vinegar. There is a myriad of foods and culinary traditions in Southeast Asia that resulted directly from the Portuguese presence in Malacca or from the larger maritime trade routes that were controlled by the Portuguese. This contributed not only to the introduction and integration of plants, fruits, vegetables, and spices from other parts of the world into Malay cuisine, but it consolidated the resilience of *Kristang* creole, a Malay-Portuguese language, that is still spoken today around 800 people (Baxter 2005).

Kristang or *Papia Kristang* is a Malaccan-Portuguese creole language that has been spoken by the *Kristang* communities, mainly in Malacca, but also in Singapore since the sixteenth century. The language is severely endangered, and the number of speakers has been decreasing significantly in the last few decades. According to a UNESCO report, in 2007 there were approximately 2200 speakers. The number has dropped to more than half in less than ten years, mostly due to rapid socio-economic change, migration, and other sociological factors (Baxter 2013).

This language emerged within the multicultural and transnational context of Portuguese Malacca and the use of Portuguese language as a *lingua franca*, which persisted during the Dutch rule of Malacca, from 1641 to 1825. The extended Portuguese communities adapted Portuguese foreigner talk and pidgin Portuguese models that were also formed in India and other regions across Asia. When the Portuguese lost Malacca to the Dutch in 1641, several groups of people within the extended Portuguese community migrated to other regions, spreading the influence of Portuguese creoles and the growth of other diasporic Luso-Asian communities (Fernandis 2000). While high-level officials settled in Goa, others migrated to Java and Sumatra, some migrated to Macau, where Macanese *Patuá* language is currently critically endangered or nearly extinct. Others still migrated to Makassar, where the Portuguese community was known as *bandel* (Loureiro 2016). There was a group of Portuguese and Portuguese descendants who were made captives and sent to Batavia, becoming part of the *Mardijker* community in the Dutch East Indies. The *Mardijker* were freed Christians slaves of various ethnicities and nationalities under the ample umbrella of Portuguese identity as they spoke a Portuguese-based creole, mostly migrants from Portuguese-occupied regions of India and Bengal. The term *Mardijker* is a corruption of the Malay *Merdeka*, which means "independent" as a form of categorising the free social status of these individuals (Choudhury 2014). Arguably, the *Mardijker* gained their freedom because of their ability to speak both Portuguese and Malay, which were the predominant languages spoken in Southeast Asia, serving as interpreters and cultural mediators as they were knowledgeable about language and customary law. The *Mardijker* language in Jakarta and other Portuguese-based creoles in Southeast Asia, such as the *Papiá Tugu* in Kampung Tugu, Jakarta, the *Portugis* in the Ambon, Ternate Islands and Minahasa, and the *Bidau Portuguese* in the Bidau area of Dili, East Timor, became all extinct since around 1950s (Cardoso 2020; Cardoso et al. 2015; Ferraz 1987).

The resilience of language is related to the importance of language in other social practices such as religious rituals, festive events, economic activities, oral traditions, performing arts, and other intangible cultural expressions as an important identitarian element for the *Kristang* who still consider themselves as a Portuguese community in the present.

One of the aspects that contributed to the establishment of Portuguese creoles was the accommodation approach of the Jesuits and other Christian missionaries as they tried to harmonise local cultural elements with Christian teachings, values, and worldviews. Missionaries often studied local languages, religious systems and developed fruitful approaches to inculturating Christian beliefs, practices, and values into Asian cultures (Lopes 2021). In Malacca, as in other parts controlled by the Portuguese, the missionaries utilised music and other religious festive events as a form of cultural and religious mediation. The missionaries often adapted indigenous musical instruments and melodies to introduce and infuse Christian chants, accompanied with folklore performances (Costa 2022). The *Kristang* community in Malacca and Singapore is known for its socio-cultural performance dance called *branyo*, a popular dance performance that has been an integral part of *Intrudo*. This festival is known in the Catholic Christian tradition as Shrove Sunday or *Quinquagesima*, which is the Sunday just before Ash Wednesday, marking the beginning of Lent (Md Nor 2020). *Branyo* performances are also played during other Christian religious ceremonies such as wedding and the *Festa Senjuan* (Saint John's Festival) and *Festa San Pedro* (Saint Peter's Festival) at the churches consecrated to these Catholic saints. Similar dance and singing performances, also inspired in the Portuguese folk dance called *Corridinho*, are found in several Luso-Asian diasporic communities in Goa, Daman, Diu, Sri Lanka, and Macau (Sarkissian 2000).

16.5 Conclusion

Based on this discussion, several conclusions can be drawn. One of the key points of contention is the theory that the phenomenon of cultural globalisation, defined as a process marked by the transmission and assimilation of ideas, meanings, and values imposed from a dominant culture over a non-dominant culture, has led to cultural homogenisation through the exchange of commodities and colonisation (Ullah and Ho 2020). However, in a contrasting thesis, as shown in this chapter, the intensification of social and cultural relations, despite power dynamics and hierarchies of socio-anthropological and religious factors, demonstrates increasing opportunities of cultural diversification and the emergence of new cultural identities as a result of a transcultural exchange (Wang 2007). Historically, port cities and port communities have served as a booster towards the emergence of multicultural and transnational identities (Andaya and Andaya 2016). From a historical perspective, framing cultural identity has often been related to distinct forms of cultural expressions based on collective and social memories. The sense of belonging in which cultural formations develop is a biocultural phenomenon emerging from an ecological and sociological context. Socialisation and customs within a specific ecological setting result in the formation of a collective cultural memory, which defines the cultural identity of social groups. Cultural memory is a collective concept for tangible and intangible knowledge that determines behaviour and experience in social interactions transmitted through generations in repeated societal. Collective memory includes expressions of cultural identity, such as language, oral traditions, religious beliefs, customs, and social practices, as well as physical objects and artistic expressions as belonging to specific biocultural structures.

The apparent problem posed by globalisation lies in the migration, relocation, and disintegration of the biocultural setting of individuals as a result of the processes of colonialism and postcolonialism and other circumstances provoked by a global economy (Appadurai 1991). However, cultural mobility is suggested to increase cultural dynamics and a fluidity that generates blended cultural expressions, cultural appropriation, and the reinvention of cultural traditions (Appadurai 1990).

Another interesting conclusion that can be drawn is related to cultural resilience and cultural pride of the Portuguese Eurasian communities, particularly in the regions where the Portuguese lost their political influence a long time ago. Between 1580 and 1640, Portugal was subject to the power of a dual monarchy under the rule of the Spanish Crown. During this period, the Portuguese settlements across Asia were threatened by the Dutch, especially in Southeast Asia, where the Dutch East India Company was dominant. The Portuguese Eurasian communities lost contact with their Portuguese origins and, as Catholic Christians, were often persecuted or had sanctions imposed on them. A similar situation happened in Japan at a much larger scale, when European missionaries and Christian-converted Japanese were persecuted and exiled from Japan by the *shogun* Tokugawa Ieyasu between 1613 and 1642. The Japanese Christians who remained in Japan resiliently continued their Christian religious practices and Christian identity in secret, becoming known as *kakure kirishitan* (hidden Christians) despite the violent oppression imposed by the shogunate. Drawing from the concept of historical oppression as a result of power dynamics that tend to impose inequality and oppression based on religious values, ethnic and racial discrimination, socio-cultural marginalisation, or economic power, it seems to lead either to cultural withdrawal or to transcendence (Burnette and Figley 2017). The processes of transcendence emerge as a resilient response to oppression and to seek liberation and emancipation, which is rooted in a strong foundation of collective consciousness and elements of collective identity, such as religious beliefs, strength of spirit, upholding cultural codes, congregational or community unity, in which language, and particularly creoles, emerge as a facilitator for cultural connection.

In most cases, one cannot say that the Portuguese Eurasians were severely or violently persecuted but they were mostly segregated and marginalised, allowing them to continue their shared identity and develop a sentiment of longing and pride over their Portuguese origins and historical heritage. A recent statement from a member of the *Kristang* community in Malacca gives a very clear idea of this cultural pride.

> "I would like to take a Malacca group to Portugal – dancers, musicians, cooks, politicians, say 50 people – to share our culture with the Portuguese.". Joseph Sta. Maria (pers. comm., 9 August 2003) (cited from Sarkissian 2005).

To conclude, the understanding of historical transnational identities allows for a nuanced analysis of contemporary cultural and political power dynamics. Many societies today are characterised by increasing diversity resulting from centuries of migration, trade, and cultural exchange. By tracing the historical trajectories of transnational identities, we can better understand the interplay of cultures, languages, and traditions in shaping modern identities. This understanding is indispensable for fostering social cohesion, mitigating intergroup tensions, and promoting inclusive practices in multicultural societies.

Notes

1. Here, this chapter refers to the circulation of scientific, anthropologic, and encyclopaedic knowledge that was compiled by botanists, geographers, travellers, physicians who published detailed information about medicinal plants, fruits, fauna, flora, and the vast anthropological landscape of the peoples and their customs, traditions, beliefs, and social practices around the world (Lopes 2020; Mathew 1997; Spudich 2008).
2. A privilege delegated by the Pope to the king of Portugal to administrate the Catholic missions in all territories controlled by the Portuguese (Rego 2008).
3. Again, the term "Eurasian" used in this chapter refers to this expanded Portuguese identity, which included individuals of other European nationalities as well as peoples from Asian and African origins.

4. It is important to bear in mind that the term "slaves", particularly in Malacca is an elastic one that encompassed different degrees of bondage. Both Portuguese and Malay sources clearly demonstrate that normally slaves were paid and had the right to own personal commodities (Borschberg 2019; Thomaz 1994).
5. A variety from the species *capsicum annuum*, native to South America and Africa.

References

Andaya BW, Andaya LY (2016) Divergence, convergence, integration. Port cities and the dynamics of multiple networks, 1500–1900. In: Lee P, Andaya LY, Andaya BW et al (eds) Port cities. multicultural emporiums of Asia, 1500–1900. Asian Civilisations Museum, Singapore, pp 10–27
Appadurai A (1990) Disjuncture and difference in the global cultural economy. Theory, culture & society 7:295–310
Appadurai A (1991) Global ethnoscapes: notes and queries for a transnational anthropology. In: Fox RG (ed) Interventions: anthropologies of the present. School of American Research, Santa Fe, pp 191–210
Aranas U (2014) Melaka Malaysia A-Famosa-01.jpg. Malacca. https://commons.wikimedia.org/wiki/File:Melaka_Malaysia_A-Famosa-01.jpg. Accessed 9 August 2024
Baxter A (2013) Papiá Kristang. In: Michaelis SM, Maurer P, Haspelmath M et al (eds) The survey of pidgin and creole languages, vol 2. Oxford University Press, Oxford
Baxter AN (2005) Kristang (Malacca Creole Portuguese) – a long-time survivor seriously endangered. Sociolinguistic Studies:1–37
Beeckman A (1658–1664) Mardijker, watercolour, 41 x 26 cm, assession number: EST-389, nr. 7, Album de Paulmy, département Arsenal, Bibliothèque nationale de France, Paris (Paulmy's album, Arsenal department, National Library of France, Paris)
Bocarro A (1635) Livro das plantas de todas as fortalezas, cidades e povoaçoens do Estado da India Oriental (Book of plans of all forts, cities and towns in the State of east India). Biblioteca Pública de Évora, Évora
Borschberg P (2019) The Melaka Empire, c. 1400–1528. In: Strootman R, van den Eijnde F, van Wijk R (eds) Empires of the sea. BRILL, Leiden, pp 263–293
Boxer C (1963) Race relations in the Portuguese colonial empire, 1415–1825. Clarendon Press, Oxford
Boxer C (1969) The Portuguese seaborne empire, 1415–1825. Hutchinson and Co, London
Braudel F (1949) La Méditerranée et le Monde méditerranéen à l'époque de Philippe II (The Mediterranean and the Mediterranean world at the time of Philippe II). Armand Colin, Paris
Burnette CE, Figley CR (2017) Historical oppression, resilience, and transcendence: can a holistic framework help explain violence experienced by indigenous people? Social Work 62(1):37–44
Cardoso HC (2020) Contact and Portuguese-lexified creoles. In: Hickey R (ed) The handbook of language contact. Wiley Blackwell, New Jersey, pp 469–488
Cardoso HC, Hagemeijer T, Alexandre N (2015) Crioulos de base lexical portuguesa (Creoles with a Portuguese lexical base). In: Iliescu M, Roegiest E (eds) Manuel des anthologies, corpus et textes romans. De Gruyter, Berlin, pp 670–692
Chaturvedula N (2010) Preserving purity: cultural exchange and contamination in late seventeenth century Portuguese India. Ler História 58:99–112
Choudhury M (2014) The Mardijerks of Batavia: construction of a colonial identity (1619–1650). Proceedings of the Indian History Congress, 75:901–910
Collection of the National Library of Brazil (n.d.) Planta ms., desenho a nanquim (ink drawing), 44 x 81.5 cm, Collection of the National Library of Brazil. http://objdigital.bn.br/objdigital2/acervo_digital/div_cartografia/cart531967/cart531967.html. Accessed 20 March 2024
Correia G (1858) Lendas da India (Legends of India). Edited by Felner RJL. Typographia da Academia Real das Sciencias, Lisboa

Costa JPO (2022) Portugal na história: uma identidade (Portugal in history: an identity). Círculo de Leitores, Lisbon
Cunha AC (1995) A Inquisição no Estado da Índia: origens (1539–1560) (The inquisition in the state of India: origins (1539–1560)). Arquivo National da Torre do Tombo, Lisbon
Curvelo A (2010) Namban folding screens: between knowledge and power. In: Couto D, François L (eds) Empires éloignés: L'Europe et le Japon XVI-XIX siècle. École Francaise d'Extrême-Orient, Paris, pp 201–215
Dellon C (1688) Relation de l'Inquisition de Goa (Relation of the Goa inquisition). Chez Daniel Horthemels, Paris
DeWitt D (2020) Chile peppers: a global history. University of New Mexico Press, New Mexico
Douki C, Minard P (2007) Global history, connected histories: a shift of historiographical scale?. Revue d'histoire moderne & contemporaine 54–4(4):7–21
Fang LY (1976) Undang-undang Melaka (The laws of Melaka). Martinus Nijhoff, The Hague
Fernandis G (2000) The Portuguese Eurasians in Malaysia: bumiquest, a search for self identity. Lusotopie 7:261–268
Ferraz LI (1987) Portuguese creoles of West Africa and Asia. In: Gilbert GG (ed) Pidgin and Creole languages: essays in memory of John E.~Reinecke. University of Hawaii Press, Honolulu, pp 337–360
Gschwend AJ (1995) In the tradition of princely collections: curiosities and exotica in the Kunstkammer of Catherine of Austria. Bulletin of the Society of Renaissance Studies XIII(1):1–9
Hespanha AM (2019) Filhos da Terra: identidades mestiças nos confins da expansão portuguesa (Children of the earth: mixed-race identities in the confines of Portuguese expansion). Almedina, Coimbra
Jackson A (2020) The making of Macau's fusion cuisine: from family table to world Stage. Hong Kong University Press, Hong Kong
Kouamé N (2020) Chronique des armes à feu (La) (Teppô ki), 1606 (Chronicle of firearms (The) (Teppô ki), 1606). In: Kouamé N, Meyer ÉP, Viguier A (eds) Encyclopédie des historiographies: Afriques, Amériques, Asies, vol 1. sources et genres historiques (Tome 1 et Tome 2), vol I. Presses de l'Inalco, Paris, pp 267–272
Lach DF (1968) Southeast Asia in the eyes of Europe: the sixteenth century. University of Chicago Press, Chicago
Leão FC (ed) (1998) O Índico na biblioteca da Ajuda: catálogo da documentação manuscrita referente a Moçambique, Pérsia, Índia, Malaca, Molucas e Timor (The Indian ocean in the Ajuda library: catalog of manuscript documentation regarding Mozambique, Persia, India, Malacca, Moluccas and Timor). Comissão Nacional para as Comemorações dos Descobrimentos Portugueses, Lisbon
Lieberman V (2009) Strange parallels: Southeast Asia in global context, c. 800–1830, vol 2: mainland mirrors: Europe, Japan, China, South Asia, and the Islands. Cambridge University Press, Cambridge and London
Linschoten JH van (1598) Jan Huighen van Linschoten his Discours of voyages into ye Easte & West Indies: deuided into foure bookes. Iohn Wolfe, London
Lopes RO (2020) A 'Flora Sinensis' de Michał Piotr Boym (1654): Animais, plantas e frutos medicinais da China e do Sudeste Asiático (The 'Flora Sinensis' by Michal Piotr Boym (1654): animals, plants and medicinal fruits of China and Southeast Asia). Revista Oriente 28:92–101
Lopes RO (2021) Art and religion in the context of the Portuguese expansion. In: Silva NV (ed) Stories of an empire. The Távora Sequeira Pinto Collection. Museu do Oriente, Lisboa, pp 51–77
Loureiro RM (2016) Uma cidade cercada: Malaca durante a União Ibérica (1580–1640) (A city under siege: Malacca during the Iberian Union (1580–1640)). e-Spania (25)
Mathew KS (1997) The Portuguese and the study of medicinal plants in India in the sixteenth century. Indian Journal of History of Science 32(4):369–376
Md Nor MA (2020) Eclecticism and syncretic traditions: the making of Malay folk dance. In: Md Nor MA, Burridge S (eds) Sharing identities: celebrating dance in Malaysia. Routledge, New Delhi and London, pp 37–55

Moody A (2021) Macau's languages in society and education: planning in a multilingual ecology. Springer, Singapore
North M (ed) (2010) Artistic and cultural exchanges between Europe and Asia, 1400–1900. Routledge, London and New York
Pires T (1944) The suma oriental of Tomé Pires: an account of the east, from the Red Sea to Japan, written in Malacca and India in 1512-1515; and the Book of Francisco Rodrigues rutter of a voyage in the Red Sea, nautical rules, almanac and maps, written and drawn in the east before 1515. The Hakluyt Society, London
Rego S (2008) O Padroado português do Oriente visto da Índia instrumentalização política da religião (The Portuguese patronage of the East seen from India political instrumentalization of religion). In: Revista Lusófona de Ciência das Religiões – Ano VII, n. 13/14, pp 413–430
Rodrigues VLG, Costa JPO (2012) Conquista de Malaca 1511: Campanhas de Afonso de Albuquerque (Conquest of Malacca 1511: Afonso de Albuquerque's campaigns). Tribuna da História, Lisboa
Sarkissian M (2000) D'Albuquerque's children: performing tradition in Malaysia's Portuguese. Chicago University Press, Chicago
Sarkissian M (2005) Being Portuguese in Malacca: the politics of folk culture in Malaysia. Etnografica 9(1):149–170
Silva AMJ da (2015) Macaenses: the Portuguese in China. International Institute of Macao, Macau
Sousa MF (1666). Asia Portuguesa, vol 1–3. en la Officina de Henrique Valente de Oliveira Impressor del Rey N.S, Lisboa
Spudich A (2008) Such treasure and rich merchandize: Indian botanical knowledge in 16th and 17th century European books. National Centre for Biological Sciences / Tata Institute of Fundamental Research, Bangalore
Strootman R, van den Eijnde F, van Wijk R (2019) The Portuguese maritime empire: global nodes and transnational networks. Empires of the Sea. BRILL, Leiden, pp 294–311
Subrahmanyam S (1993) The Portuguese Empire in Asia, 1500–1700. Wiley Blackwell, Chichester
Subrahmanyam S (2016) Merchant networks in the early modern world, 1450–1800. Routledge, London
Subrahmanyam S (2022) Connected history essays and arguments. Verso, Brooklyn, New York
Swope KM, Andrade T (eds) (2018) Early modern East Asia: war, commerce, and cultural exchange. Routledge, London and New York
Thomaz LF (1991) Nina Chatu and the Portuguese trade in Malacca. Luso-Malaysian Books, Melaka
Thomaz LF (1992) De Ceuta a Timor (From Ceuta to Timor). Difel, Lisbon
Thomaz LF (1993) The Malay sultanate of Malacca. In: Reid AJS (ed) Southeast Asia in the early modern era trade, power, and belief. Cornell University Press, Ithaca and New York, pp 69–90
Thomaz LF (1994) A Escravatura em Malaca no Século XVI (Slavery in Malacca in the 16th century). STVDIA, Lisbon
Tibbetts G (1979) Study of the Arabic texts containing material on South-East Asia. BRILL, Leiden
Ullah AKMA, Ho HMY (2020) Globalisation and cultures in Southeast Asia: demise, fragmentation, transformation. Global Society 35(2):191–206
Wang Y (2007) Globalization enhances cultural identity. Intercultural Communication Studies 16(1):83–86
Xavier ÂB (2008) A invenção de Goa (The invention of Goa). Imprensa das Ciências Sociais, Lisbon
Yamafune K (2012) Portuguese ships on Japanese namban screens. Dissertation, Texas A&M University, Texas

Rui Oliveira Lopes is the Director of the Museum of Convergences in Porto, Portugal since July 2024. He was a Senior Assistant Professor at the Faculty of Arts and Social Sciences, Universiti Brunei Darussalam from 2015 to May 2024. He received his Ph.D. in the History of Art in

2011 from the Faculty of Fine Arts, University of Lisbon, completing a dissertation on the confluence of Christian art in India, China, and Japan between the 16th and the eighteenth centuries. His research is focused on the visual aspects of the sacred narratives in the arts of the Asian civilizations, the artistic and cultural interactions between Asia and Europe, the museum as a place for intercultural dialogue, the role of museums in promoting and protecting Asian (in)tangible cultural heritage.

Chapter 17
Investigating Mental Health as a Transnational Health Issue: The Reporting of Mental Health in Brunei Darussalam and Malaysia

Sharifah Nurul Huda Alkaff

Abstract Mental health has been recognised as a significant public health issue by governments all over the world. In the midst of this transnational health crisis, the media are widely regarded as playing a crucial role in mitigating this serious issue. However, studies have shown that negative views regarding mental health issues were often overrepresented in the news media. More recent studies have reported a more positive trend that attempt to destigmatise mental illness. Most of these studies are based in the West with fewer studies conducted in Asian and Muslim societies. In this chapter, English and Malay language news reports from Brunei and Malaysia will be analysed in order to investigate how mental health issues are reported in these societies. This chapter will also examine to what extent newspapers in these two countries attempt to destigmatise mental health problems. The media texts will be analysed using content analysis and discourse analysis. The findings reveal that the authority-order journalistic norm of reporting adopted by the newspapers investigated has produced news articles on mental health that are generally positive in tone. However, the quality of mental health reporting can be improved further by greater adherence to international guidelines on mental health reporting.

Keywords Media reporting · Mental health · Transnational · Muslim societies · Southeast Asia · Malaysia · Brunei

17.1 Introduction

The Covid-19 pandemic has thrust mental health issues (henceforth MHIs) in the limelight globally due to the impact the pandemic has wrought on the mental health of people. However, even before the Covid-19 viral outbreak in 2020, mental health has been recognised as a significant transnational health issue by the World Health

S. N. H. Alkaff (✉)
Faculty of Arts and Social Sciences, Universiti Brunei Darussalam, Gadong, Brunei Darussalam
e-mail: sharifah.alkaff@ubd.edu.bn

H. M. Y. Ho and Y. Chan (eds.), *Transnational Southeast Asia*, Asia in Transition 29,
https://doi.org/10.1007/978-981-96-3608-2_17

Organisation (henceforth WHO). It has caused serious damage to the well-being of individuals and families and has also cast a toll on the economies of many countries. For example, mental health conditions in the workplace are estimated to cost the Malaysian economy RM14.46 billion in 2018 (Chua 2020). However, Tan et al. (2023) state that "Mental health has not been given high priority in Southeast Asian countries. Looking back, in some Southeast Asian countries, the healthcare sector has primarily focused on infectious and tropical diseases, and other emerging public health concerns. Mental health, which falls under the domain of non-communicable diseases, has received comparatively less attention" (p. 1).

In the midst of this public health crisis, there is a general recognition that the media can play a crucial role in combatting the global problem. According to de Vries (2014), there are three components to mental health promotion (mental health literacy, destigmisation, and prevention). He further states that the media can play an effective role in promoting all three aspects cited above as it "has become one of the cornerstones of modern life and has evolved into a potential ally for improving the mental health of a population" due to "its reach and interactive aspects" (pp. 358–59). Effective use of media can increase well-being and mental health in communities. However, Wahl (2003) reports that negative views were often overrepresented in the news media when mental illness is reported. Patients or characters with psychiatric disorders were significantly more frequently associated with dangerousness, criminal activities, unpredictability, and dysfunction. More recent studies have, however, reported a more positive trend. Rhydderch et al. (2016) state that newspaper coverage of mental illness in the United Kingdom from 2008 to 2014 suggests that there had been a small rise in articles that attempt to destigmatise mental illness along with a proportional decrease in portraying persons with mental illness as dangerous to others. They credited the general improvement of the public's attitudes towards mental illness as part of the UK's "Time to Change" campaign which had solicited newspaper journalists and editors with advice and training. The WHO (2022) has also established guidelines for media professionals when reporting on suicide or suicidal behaviours. The guidelines were established after reviewing scientific evidence on the impact of media reporting on mental health. Amongst the advice given is to avoid providing information about the method of suicide or the location it took place as well refraining from using sensational language, especially in the headlines which attract the most attention from readers. Similarly, an American style guide on reporting mental health (Mental Health, n.d.) places great emphasis on the use of language to avoid any prejudice or stereotyping when reporting MHIs. For example, the phrase "a person with schizophrenia" is preferred over "a schizophrenic person" as this will minimise the risks of identifying a person with their medical diagnosis.

A cursory review of the literature in the reporting of MHIs in the media has also revealed an obvious lack of scholarly research of the topic in Asian societies. As stated by Tan et al. (2023), "Despite the fact that Southeast Asia comprises 8.58% of the global population, authors from this region have not received proportional attention or inclusion in the body of literature addressing mental health" (p. 1). Similarly, Chompoosri et al. (2023) reported that although suicide is a major issue in Thailand, there are few studies on media reporting of suicide in Thailand. This can perhaps be attributed to the overwhelming stigma of mental illness in Asian cultures (Ng 1997). Ng posits that Asian societies are typically defined as family-oriented entities "which place high value on group harmony in place of autonomy and individualism... As a result, stigma may be more severe because it is attached to the family as a whole" (p. 384). More recent studies also reveal that stigma towards mental health issues is still prevalent in Southeast Asia. Subramaniam et al. (2017) find significant stigma towards mental illness amongst Singaporeans of different age-groups and ethnic backgrounds with older people, males, and those with less education having higher stigmatising attitudes. In contrast, a 2018 poll on Americans adults showed a generally positive attitude towards the issue with the vast majority (87%) agreeing that having a mental health disorder is nothing to be ashamed of (American Psychological Association 2019).

In Muslim societies, the void in research on MHIs appears to be even greater. Amongst the few studies on this topic in Muslim majority societies in Southeast Asia was conducted by Nurul Hartini et al. in 2018. They reveal that there exists significant stigma and discrimination amongst the Indonesian public towards individuals with health disorders. According to them, this negative attitude "has been found not only limited to lay population but also to occur amongst health professionals like psychiatrists, psychologists, and nurses who work with people with mental disorder" (p. 535). The researchers also conclude that in view of the dire situation regarding stigmatising attitudes towards MHIs found in their study, there should be more literature on the issue in societies like Indonesia. This is affirmed by Ciftci et al. (2013) who view that the lack of Muslim mental health literature has become increasingly glaring in view of the percentage of people of the Muslim faith in the world. It is imperative that more research in this subject needs to be conducted as it is a vastly unexplored area. Thus, this study aims to investigate how mental health is reported in mainstream newspapers in Brunei and Malaysia, two Muslim majority societies in Southeast Asia. In addition, this study also seeks to examine to what extent do newspapers in Brunei and Malaysia attempt to destigmatise mental health problems in view of current global trends supporting this initiative.

In this chapter, mental health is referred to as a transnational issue as this study utilises a broad definition of transnationalism that refers to exchanges, connections, and practises of individuals, communities, and institutions across national borders. Levitt and Schiller (2004), for example, provided a broad definition of transnationalism that included constructed identities, social groups, social expectations, and cultural values, amongst others. Other academics like Vertovec (1999) and Basch et al. (1994) focus more on migrants and diaspora groups and the linkages they forge across national boundaries. Although Brunei and Malaysia are two independent countries with different political institutions and economic structures, they are relatively similar in terms of cultural and social values and norms as both are part of the *Nusantara* or Malay World and both were governed by the British prior to independence in the twentieth Century. Both are Muslim majority countries in Southeast Asia, as stated earlier, thus, it is perceived that the majority of their population may share similar sentiments on an issue that transcends national boundaries. Clifford (1994), for example, refers to "ideas" and "feelings of solidarity" across borders when discussing his views on the concept of transnationalism. It is with these notions in mind that this study attempts to locate itself within the broad field of transnationalism.

17.2 Methodology

Data was collected from online mainstream Malay language and English language newspapers in Brunei and Malaysia. The search terms used were phrases typically associated with MHIs such as "depression", "mental anxiety", "mental health", "suicide", "psychiatrist", "schizophrenia" and "bipolar". Using these search terms, a total of 120 news articles were collected. 60 articles were collected from *Media Permata* and the *Borneo Bulletin* in Brunei (30 articles from each paper). Another 60 articles were collected from *Berita Harian* and the *New Straits Tim*es in Malaysia (30 articles respectively from each newspaper). The newspapers in both countries are mainstream newspapers published by well-established news organisations, that is, the Brunei Press in Brunei and the New Straits Times Press (NSTP) in Malaysia.

The period of reporting was from July 2022 to September 2023. This time frame is acceptable as it is not expected that there will be significant changes in the reporting of MHIs within this relatively short time period. The number of articles collected were also deemed sufficient as according to Merriam (2009), "in discourse analysis, the researcher is the primary instrument for data collection and analysis and therefore has to decide on the number or size of data to be gathered as a corpus for the study. The guiding principle should be the concept of saturation which is the point at which there are fewer surprises and there are no more emergent patterns in the data" (p. 15). Both content analysis and discourse analysis were used to analyse the data. The former was utilised to examine the content of these news articles, that is, the existence and frequency of terms related to mental health while the latter was adopted to investigate how MHIs are being talked about in these papers. In a review of mental health studies in Southeast Asia, Tan et al. (2023) revealed that out of 12 studies analysed, 10 of them adopted a quantitative approach. There was only one qualitative study and one which combined a mixed-methods approach. This suggests that there is a need to look beyond quantitative approaches when investigating a multi-layered issue such as mental health.

17.3 Findings

17.3.1 Media Permata and the Borneo Bulletin (Brunei)

In the Malay language *Media Permata* (henceforth MP), the content of almost all the articles were about various workshops and seminars organised by the authorities in explaining about the causes and effects of mental health issues in the country.

Many of these articles were related to activities held to mark World Mental Health Day which normally falls in October of every year. Some of the articles contained statistics about the prevalence of MHIs in Brunei and globally while others reported in great detail the various causes and solutions proposed by authority figures such as Ministry of Health officials. Several articles reported talks given by the Ministry of Religious Affairs officials in relation to the promotion of spiritual guidance to combat MHIs. The term "*vaksin ketuhanan*" (divine vaccine) was used to refer to the call made by His Majesty the Sultan of Brunei in his speech to the nation in July 2022 in which he stressed the importance of prayers in combatting the increasing rate of MHIs in the country. None of the articles in the data refers to an actual case of a person suffering MHIs.

Similar to the Malay language MP, the vast majority of the articles reporting on MHIs in the *Borneo Bulletin* (henceforth BB), contained information about various seminars and workshops on MHIs by government authorities. Except for two articles which are classified as opinion pieces by readers, the rest of the news items are hard news pertaining to the various events held to raise awareness of MHIs. Most of these articles are lengthy as often the speeches of various officials are reported verbatim. They are often accompanied by several photos of the dignitaries and participants present. The hard news articles on MHIs in the Bruneian newspapers do not adhere to the typical features of this genre as they are more akin to feature articles which are often longer. One of the reasons for this is because the articles often contain substantial information on current statistics and trends in MHIs including possible causes and solutions, either in Brunei or overseas. The language employed is professional and concise. There is no evidence of any element of sensationalism in these news articles. With regard to suicide, which is, perhaps, one of the most tragic manifestations of severe mental health breakdown, no reports are found in both Bruneian newspapers. Only suicide cases that occur outside of the country are reported.

17.3.2 Berita Harian and the News Straits Times (Malaysia)

The majority of the news articles reporting on MHIs in the *Berita Harian* (henceforth BH) are from press statements and speeches of various authority figures such as politicians, civic leaders, and government officials. This is similar to the findings from the Bruneian newspapers except that in the case of Brunei, only government officials are quoted. However, unlike in Brunei, the contexts surrounding these statements are rather diverse, ranging from cyber-bullying to sexual harassment cases, amongst others. MHIs is also associated with a blasphemy case which was reported in July 2022. Blasphemy is a crime in the Malaysian Penal Code, subjected to a prison term or a fine. In this case, various officials including a Government Minister stated that the accused individual must be "*under the influence of drugs or mental stress*" as "*a person of sound mind would not commit such a despicable act*". The association with MHIs and crime is also noted in another article that was published in August 2022 that reported the case of a teenaged girl who kidnapped an infant. The BH quoted a police official who stated that their investigations showed that the accused was found not to have any MHIs. Instead, her actions were probably prompted by the fact that she came from a broken family. Based on the data, it appears that there is a tendency on the part of this newspaper to associate all forms of social ills including crime to MHIs. However, the language used in the reporting is overall neutral and does not lead to sensationalism. There is one exception in our data in which inappropriate language is used. In an article published in October 2022 on a call by a psychiatrist to destigmatise MHIs, the word "*gila*" (insane) was used in both the headline and the text itself as well as the word "*sewel*" (crazy). Additionally, in an article in July 2022 that reported the murder of an elderly woman by her son who had a previous mental health history, the article depicted at length the actions and behaviour of the accused. In another article reporting an attempted suicide in June 2023, details such as the method of suicide as well as the location it took place were provided. All this information does not conform to the WHO guidelines on the reporting of suicide, as cited earlier.

With regard to the *New Straits Times* (henceforth NST) an English language newspaper which is owned by the same media organisation as the Malay language BH, there is a marked difference in the reporting of MHIs compared to the other three papers investigated in this study. Although the NST included many articles that report statements and announcements regarding MHIs by politicians, civic leaders, and medical professionals, its style of reporting is noticeably different. For instance, their articles often start by featuring personal narratives of people suffering from MHIs. This is then followed by statements from an authority figure in relation to a particular MHI. Most of the photographs that they provide also show illustrations or images of a person suffering from MHIs or crowds of people in busy urban settings. This is unlike the other three papers where there is a preference to show photos of officials and participants of various events pertaining to MHIs. Thus, there appears to be a deliberate attempt by the NST to personalise the reporting of MHIs rather than emphasise the official response or engagement with this issue. In addition, whenever this paper reports a suicide case (there are several such articles in the data), the article always includes information about where to seek help for anyone with suicidal tendencies. This information is not included in the BH although there are also a few articles pertaining to suicide in this paper. However, there are instances when the NST too fails to adhere to the WHO guidelines. For example, a July 2023 article had a sensational headline ("*Man with psychiatric problems attempts suicide by slitting throat, stomach*"). This headline associates MHIs with the violent behaviour of individuals. In another suicide reporting in January 2023 which had the following headline, "Girl leaves 'suicide' note before allegedly leaping off her flat", not only is the headline considered sensational but information about the method, location, and reference to suicide note are all against the WHO guidelines.

17.4 Discussion

The findings above indicate some similarities and differences in how MHIs are presented in the newspapers from the two contexts investigated. With regard to the Bruneian newspapers, one of the most outstanding feature surrounding these topics is the discourse of officialdom that permeates throughout all the news articles. This is the case for both the Malay language and English language newspapers in Brunei. The news articles on MHIs contain numerous direct quotes ascribed to Government officials, as stated in the previous section. The fact-heavy nature of these news articles and the rather sterile reporting style often result in the reporting of MHIs in Brunei as belonging to the realm of academia rather than the mass media. Similar observations were made by Alkaff et al. (2022) in their study on the reporting of environmental news in Brunei. They state that the "authority-order" journalistic norm is the most common form of reporting in Bruneian newspapers. This journalistic practice is defined as one in which there is a heavy reliance on authority figures, who are mainly government officials, for information. In relation to this, in most news articles in Brunei, "the list of dignitaries present and the accompanying official photographs often take centre stage. The speeches made at these events are often reproduced verbatim in the paper without much analysis and any commentary made" (p. 219). This style of reporting, which is devoid of any element of personalisation, can be argued to be aligned with the objective of destigmisation of MHIs in the media as all elements of sensationalism are removed. However, there are concerns on how effective these news articles can be in raising awareness and ultimately in reducing the stigma of MHIs as there is a possibility that most readers would just scan these news articles without reading them. In addition, the complete absence of personal elements or reporting of actual cases of MHIs in the society can increase rather than decrease the stigmisation of these issues. For example, no local cases of suicide are reported in Bruneian newspapers despite the fact that suicide cases do occur in Bruneian society although they are relatively rare. One of the articles analysed in this study quoted a Government official announcing that there were 14 cases of suicide reported in Brunei in 2020. About half of this figure involved locals. However, these cases are not reported in any Bruneian newspapers, including the two newspapers in this study. A case that occurred in October 2022 comes to mind. News about a young local girl who committed suicide by jumping off the newly built Temburong Bridge in the country was widely shared amongst the community through WhatsApp. There was also a thread regarding this case on reddit Brunei. This case was not reported in the local papers until a week later when a one-sentence article was published about a dead body that was found by members of the public. No other details were given. The complete secrecy surrounding suspected suicide cases in the media can hinder any meaningful conversations that can reduce stigma. It has also allowed rumours and speculations that are unverified to circulate on social media. As Brunei is a relatively small country with a high degree of control over its media, WhatsApp has often become a source of news for many Bruneians.

In contrast, the English language NST in Malaysia has a very different approach towards reporting MHIs. As stated in the previous section, this paper often uses personal narratives rather than relying solely on statements from authority figures. When reporting on suicide cases, information on helpline numbers and counselling services are always included at the end of the article. As Malaysia is a much larger country with multiple media platforms competing for an audience, it is expected that there will be a substantial difference in how this issue is being presented in their newspapers compared to Brunei. However, the findings from the Malay language BH in Malaysia are quite similar to the findings from the Bruneian newspapers as there is a tendency to employ the authority-order journalistic norm in which there is a heavy reliance on authority figures for information when reporting on MHIs. These articles are often produced without any commentary or alternative views being presented. A possible reason for this stark contrast in the reporting style of MHIs by these newspapers can be attributed to their differing world views. According to Sapir-Whorf's theories of cultural and linguistic relativism, one's worldview can depend on features of one's language. For example, the Malay language is generally governed by ideals of decorum and respect as stated by scholars such as Lim (2003). These beliefs often result in Malay language newspaper texts containing more omissions than parallel English texts when reporting on topics that are perceived as sensitive, as revealed by Nurkhalisah (2013). MHIs is regarded as a sensitive topic by most Malay Muslims as there is a perception that people suffering from MHIs lack a strong faith in God. Suicide is also considered a grave sin in Islam. These beliefs which are widely held by the community can result in some reticence in reporting MHIs through a personal perspective. This is certainly true in the case of the Bruneian newspapers, even the English language BB. One reason why the news articles on MHIs appear similar despite the difference in language is attributed to what has been observed by Alkaff et al. (2020) regarding newspapers in Brunei. They state their belief that "most of the BB journalists writing in English share the same linguistic and cultural background as the MP journalists writing in Malay. That is, both sets of journalists come from a predominantly Brunei Malay socio-cultural milieu" (p. 25). A study by Raj et al. (2021) on the portrayal of MHIs by Hindi and English language newspapers in India also found differences in the reporting by media of these two languages. According to the researchers, the Hindi language media had significantly higher positive tone and content compared to English language media. They did not attribute these differences to any linguistic cultural thought patterns. Instead, they claim that "the greater involvement of the third party (patients, families, and mental health experts) in Hindi media reports" due to "the convenience of conducting an interview in Hindi with a person with mental illness, their family members, and even local mental healthcare experts" have resulted in the differences noted by the media of both languages (p. 582). The authors believe that the inclusion of persons with mental illness and their families in news reports have been shown "to improve the unity and acceptance and reduce discrimination" (p. 581).

It is worth referring again to de Vries's (2014) three components to mental health promotion (mental health literacy, destigmisation, and prevention) as stated in the previous section. A fair assessment of the four newspapers investigated would indicate that only the English language NST in Malaysia attempts to use their influence as a media platform to achieve all three objectives when reporting on MHIs. However, by focusing on the reporting of individual cases and not just on issue-based reporting of MHIs which provides more emphasis on mental health literacy rather than the other two aspects, this newspaper also runs the risk of demonstrating a more negative tone and stigmatising content. As revealed in the previous section, there were at least two news reports from this newspaper that did not adhere to the WHO guidelines on the reporting of suicide cases. Thus, it can be deduced that trying to achieve all three aspects of mental health promotion would require a newspaper to be adept in navigating a balancing act between producing news on MHIs that are of interest to readers while at the same time promoting a positive and destigmatising content.

17.5 Conclusion

The post-Covid-19 global scenario has witnessed a cost of living crisis, greater social inequality, and rising unemployment, especially amongst youths in the Southeast Asian region. These social and economic crises will inevitably result in an exponential increase in MHIs, even more than what the world is currently experiencing. In this sense, mental health should be viewed as a transnational public health issue as the causes and impacts often extend beyond national borders. Thus, it is imperative that societies realise the importance of the media in reporting MHIs without discrimination especially in communities where mental health care is relatively inadequate which is the case for most of the countries in the region. To summarise, a better quality of mental health media reporting is urgently needed in the two societies investigated in this study. This should inevitably involve media and health practitioners engaging in long-term and continuous dialogues in a concerted attempt to improve mental health reporting in both countries. As implored by Tan et al. (2023, p. 3), "the inclusion of culturally diverse studies conducted in various regions within Southeast Asia... will bring valuable insights and perspectives vis-à-vis prevention, assessment, evaluation, and management of mental health issues within the region". Further studies on this issue should also focus on the role of social media in reporting MHIs as information is increasingly being produced and consumed by this platform rather than by traditional mainstream media. Additionally, a large-scale study on the perceptions of the people in these two countries on the issue of MHIs should be undertaken as it would provide much needed information on how the media that they consume have made an impact on their beliefs and perceptions towards this transnational health issue.

References

Alkaff SNH, Khairunnisa I, Najib N (2022) The reporting of environmental news in an English language newspaper in Brunei Darussalam. In: Telles JP, Ryan JC, Dreisbach JL (eds) Environment, media, and popular culture in Southeast Asia. Springer, Singapore, pp 207–226

Alkaff SNH, McLellan J, Najib N (2020) A comparative media discourse investigation of hard news texts in English and Malay language newspapers in Brunei: the role of language and culture. In: Deterding D, Ho H (eds) Engaging modern Brunei: research on language, literature, and culture. Springer, Singapore, pp 21–34

American Psychological Association (2019) Americans becoming more open about mental health. https://www.apa.org/news/press/releases/apa-mental-health-report.pdf. Accessed 1 October 2023

Basch L, Schiller NG, Blanc-Szanton C (1994) Nations unbound: transnational projects, postcolonial predicaments, and deterritorialized nation-states. Routledge, New York

Chompoosri P, Suthkam N, Arafat SY (2023) Suicide news coverage in Thailand in the light of the World Health Organization guideline for media professionals. Journal of the Psychiatric Association of Thailand 68(2):163–172

Chua SN (2020) The economic cost of mental disorders in Malaysia. The Lancet Psychiatry 7(4):e23. https://doi.org/10.1016/S2215-0366(20)30091-2

Ciftci A, Jones N, Corrigan PW (2013) Mental health stigma in the Muslim community. Journal of Muslim Mental Health. https://doi.org/10.3998/jmmh.10381607.0007.102

Clifford J (1994) Diasporas. Cultural Anthropology, 9(3):302-338

de Vries MW (2014) Mental health, mass communication & media. In: Okpaku SO (ed) Essentials of global mental health. CUP, Cambridge, pp 356–366

Lim KH (2003) Budi as the Malay mind: a philosophical study of Malay ways of reasoning and emotion in peribahasa. Dissertation, University of Hamburg

Levitt P, Schiller NG (2004) Conceptualizing simultaneity: a transnational social field perspective on society. International Migration Review, 38(3):1002–1039

Mental Health (n.d.) Mental health in media - style guide. https://www.apmdistribution.org/about-apm-distribution/docs/apm-call-to-mind-style-guide.pdf. Accessed 1 October 2023

Merriam SB (2009) Qualitative research: a guide to design and implementation, 2nd edn. Wiley Imprint, San Francisco

Ng CH (1997) The stigma of mental illness in Asian cultures. The Australian and New Zealand Journal of Psychiatry 31(3):382–390. https://doi.org/10.3109/00048679709073848

Nurkhalisah M (2013) Issues in translation between Malay and English. Southeast Asia: A Multidisciplinary Journal 13:27–34

Nurul H, Nur Ainy F, Atika DA et al (2018) Stigma toward people with mental health problems in indonesia. Psychology Research and Behavior Management:535–541. https://doi.org/10.2147/PRBM.S175251

Raj S, Ghosh A, Sharma B et al (2021) Portrayal of mental illness in Indian newspapers: a cross-sectional analysis of online media reports. Indian Journal of Psychiatry 63:575–583

Rhydderch D, Krooupa AM, Shefer G et al (2016) Changes in newspaper coverage of mental illness from 2008 to 2014 in England. Acta Psychiatrica Scandinavica 134:45–52

Subramaniam M, Abdin E, Picco L et al (2017) Stigma towards people with mental disorders and its components – a perspective from multi-ethnic Singapore. Epidemiology and Psychiatric Sciences 26:371–382

Tan KA, Keng SL, Abu Talib M (2023) Editorial: mental health issues in Southeast Asia regions: looking back and moving forward. Front Psychiatry 14:1229079. https://doi.org/10.3389/fpsyt.2023.1229079

Vertovec S (1999) Conceiving and researching transnationalism. Ethnic and Racial Studies, 22(2), 447-462

Wahl OF (2003) News media portrayal of mental illness implications for public policy. American Behavioral Scientist 46(12):1594–1600

World Health Organisation (2022) Guidelines on mental health at work. https://www.who.int/publications/i/item/9789240053052. Accessed 1 October 2023

Sharifah Nurul Huda Alkaff is a Senior Assistant Professor in the English Studies Programme of the Faculty of Arts & Social Sciences at Universiti Brunei Darussalam (UBD). She has a PhD in Applied Linguistics from Curtin University of Technology, Australia. She teaches modules in linguistics and communication. Her research interests include media discourse analysis and investigating issues revolving around language, gender, and the media. She has published in a number of international journals in these areas. She is currently working on research investigating the intersections of Islam, gender and the media.

Chapter 18
The Instapoetry Community as a Transnational Imagined Community

Hana Ghani

Abstract Instapoetry is a poetic form that emerges from the intersection of poetry and Instagram. This research explores the transnational dynamics in the instapoetry community of Southeast Asia through an analysis of its poems and virtual interactions. Observation shows that while the instapoetry community in Southeast Asia is nascent, it holds potential for transnational collaboration in the region due to its shared poetic practice, themes, and cultural motifs. An adaptation of global instapoetic practices suggests a similar poetic practice that links nations within the Southeast Asian region. The similarity in approach to writing poetry on Instagram signals a potential for transnational creative collaboration in the region. Moreover, annual instapoetry events for Southeast Asian poets such as the Southeast Asia Poetry Writing Month or SEAPoWrimo substantiates the potential for cross-border collaboration. Despite this promising outlook, the region grapples with challenges rooted in the novelty of instapoetry as well as diverse sociopolitical configurations that may pose a hindrance to sustaining any cohesive community effort at a local and regional level. Ultimately, this research lays out the challenges, opportunities, and broader implications of the instapoetry community for transnational creative collaborations.

Keywords Imagined community · Instapoetry · Digital transnationalism · Instagram · Brunei Darussalam · Virtual community

18.1 Introduction

The instapoetry community, both globally and within Southeast Asia, can be seen as an imagined community. Despite geographical distances and sociocultural diversity, instapoets imagine themselves as part of a shared community united by a common interest in poetic expression. A historical overview of the Southeast Asian mediascape shows its key role in building imagined identities for the countries in the region. Following this thread, this study offers how digital media facilitates the formation of

H. Ghani (✉)
Faculty of Arts and Social Sciences, Universiti Brunei Darussalam, Gadong, Brunei Darussalam
e-mail: naqiyyah.hana@gmail.com

H. M. Y. Ho and Y. Chan (eds.), *Transnational Southeast Asia*, Asia in Transition 29,
https://doi.org/10.1007/978-981-96-3608-2_18

another imagined community—the instapoetry community. It addresses the question, how do instapoets in Southeast Asia navigate the instapoetic practice, and what are the implications, challenges, and opportunities that arise from this virtual community?

First, this study lays out an overview of Southeast Asian mediascape and its role in the formation of imagined community, before describing the instapoetry community in Southeast Asia. A close reading of Bruneian instapoems presents how the instapoets in the region are responding to particular tropes of a larger instapoetic movement shared by other poets in the region. This suggests an emergence of a collective identity emerges from ongoing cross-border exchanges. Through a comparative reading of instapoems from the region, the research shows how the instapoetry community in Southeast Asia influences collaborative creative efforts and shapes a sense of shared transnational identity beyond the rigid national state borders.

Exemplified by Sing Lit's Station's SEAPoWriMo—an annual virtual writing event for instapoets of Southeast Asia—the study investigates the long-term impact of regional initiatives for cross-border, transnational creative collaboration and the emergence of new Southeast Asian poets. Then the chapter continues by underlining potential threads for future research and concludes by highlighting opportunities and challenges for transnational collaboration brought upon by the instapoetry community in the Southeast Asian region.

18.2 Southeast Asia: Mediascape and Imagined Community

Benedict Anderson's *Imagined Communities* (2006) was initially an effort to work towards a more objective understanding of nationalism however the influence it has on further scholarly work has been immensely profound. Scholars find his proposition that a nation is imagined particularly useful as it offers a new critical understanding of modern nationalism and its relationship to the self. The statement alone has inspired a whole interdisciplinary body of academic research in sociology, anthropology, and spatial science (Cloke et al. 2013; Crang 2013; Oakes and Price 2008; Sidaway 2003) so much so that Spencer and Wollman (2002) claim that *Imagined Communities* has become "common sense" to nationalism that his "invocation has, in some cases, been a substitute for analysis" (p. 37). Following Anderson's theory, the instapoetry community is imagined based on a shared passion for writing and reading poetry on Instagram. The members of the community may never know each other but the collective passion for poetry becomes the point of unification platformised on Instagram.

An overview of the Southeast Asian mediascape may risk triggering crude generalisations. Modern Southeast Asia spans various historical, cultural, ethnic, religious, and political standings. Hamilton (1992) carefully proffers one generalisation as a foundation for her discussion of Southeast Asian mediascape and that is, "it is impossible to understand media and cultural expression in the region without recognising the central role of the state" (p. 81). Hamilton's statement reaffirms the relevance of Anderson's theory in that media plays a crucial role in facilitating an imagined identity for a community. Through media channels, symbols of a unified community are projected, and they promote a particular image that is aligned with the commitments of the state (Anderson 2006; Chatterjee 1993; Curran 2012; Greenfeld 1992).

However, Anderson's (2006) theory falls short in explaining the patterns of development in Southeast Asia as he only accounts for the "simple transposition of the European patterns of development" and fails to consider the complexities in forming imagined communities in postcolonial societies (p. 218). The printing press was first introduced in Southeast Asia by Western colonialists and missionaries and these prints are pivotal in disseminating their cultural and religious values throughout Southeast Asia. In other words, at the same, printed media facilitates an imagined community which were made available by dominant powers. The majority of Southeast Asian states were once subjected to a protectorate or colony. Hence for the majority of the region, creating a path to a modern nation—that is in terms of its ideologies, philosophies, and national goals—means moving forward as a postcolonial state and undeniably having to account for its colonial past.

Media policies become key to regulate the mediascape in Southeast Asia to aid in building a nationalist imagination. The independence of Malaysia, Indonesia, Philippines, and Singapore from the 1950s to 1960s signalled a "cross-cutting of movements: political and technological" in that nation-building is highly intertwined with media policies (Atkins 2013). The mediated image of a collective identity is built upon a difference from their colonial history, such as "the competing Western colonialism in the nineteenth century, the forms of decolonisation in the twentieth…and attitudes to the West" (Hamilton 1992, p. 82). Particularly referring to traditional media systems, Lent (1989) has expanded on what this entails, "Southeast Asian domestic media organisations were required to support and cooperate with the authorities by stressing positive news […] and by supporting governmental plans and ideologies" (p. 17). The media's role in consolidating an imagined community is most apparent in the reiteration of a single national language. For example, despite the multilingualism of the respective countries, state institutions such as Dewan Bahasa dan Pustaka (in Brunei and Malaysia) are designed to emphasise their national language, Bahasa Melayu through media ("Objektif Penubuhan" 2023; "Visi dan Misi" n.d.).

Although Anderson's theory of imagined community was originally prescribed to understand the formation of a modern nation, the theory is reconfigured to explain online ethnography following rapid digitalisation (Gruzd et al. 2011; Morimoto and Chin 2017). In order words, the notion of "imagined community" now expands and acts as a theoretical framework to understand the formation of virtual communities and their platform practices that shape the way community members participate and engage. This means that the term is readapted to describe formation of communities beyond the primary political boundaries. Rather, it explains the socially constructed, shared sense of identity and belonging such as the instapoetry community.

Critical observations conclude that new media has "dislocate[d], distantiat[ed] and refashion[ed] social connections over time and space" (Crang and Mohamed 2016, p. 280). It forces us to reconsider our former ideas of identity or community that are limited to the physical space. People from different parts of the world can connect, collaborate, and share their ideas, thus fostering transnational ties across regions as well as turning the world into a connected global village (McLuhan and Zingrone 1997, p. 121).

Generally online networks are seen as the supporting structure to maintain connections or establish reconnection between the country of origin and settlement (Bilecen and Cardona 2018; Kuşçu 2014; Shklovski 2011; Vorobeva et al. 2022), similar to how international calls were "the social glue of migrant transnationalism" (Vertovec 2004). Starikov et al. (2018) coined the term "transnationalism online" or "digital transnationalism" to explain the transnational ties and practices which are forged virtually. These terms effectively distinguish physical and virtual migration, making way for concepts such as virtual or digital diaspora, e-diaspora, or online migration (Lagurre 2005; Sahoo and De Kruijf 2016). Hence, it is tempting to study virtual communities as a phenomenon that is "other than" or oppositional to real-world communities. So much so that initial research on virtual communities is "focused on assessing whether these new formations were 'real' or 'authentic'" (Crang and Mohamed 2016, p. 126). What scholars began to realise is that communities formed online are as spatially complex as their offline counterparts and despite the perceived immateriality, virtual communities "are always materially practised" (Crang and Mohamed 2016, p. 126).

Digital transnationalism acknowledges a person's hybridity and multiplicity in inhabiting various imagined communities apart from "historical nationhood and state territorial imperatives" (Ong and Nonini 2003, p. 288). For example, Mohamad (2018) examines how Malay Malaysian international students in the UK self-disclose and perform their identity online to reaffirm their Malay Muslim identity and signal a connection to their home country, Malaysia. On Twitter, what started as an internet meme turned #MilkTeaAlliance into a symbol of regional and transnational solidarity among activists in Thailand, Hong Kong, and Taiwan fighting for democratic values (Huang and Svetanant 2022; Imawan 2023; Kreutz and Makrogianni 2024). Both examples highlight the role of digital platforms in expressing identity and fostering connections across geographical boundaries.

Although existing virtually, these virtual communities materialise through sustained, shared practices among members. Similar to how offline poetry communities are formed and sustained through physical interactions and shared practices, the instapoetry community may exist and interact through digital means. While the instapoets may not be aware of each other, each imagined themselves to belong in the community and thus understood that they are subject to material practices such as liking and commenting on posts, sharing poetry, or participating in online and offline poetry readings and writing challenges. The material practices make up the dynamics of the community. These shared digital or virtual actions are "material" because they hold real and tangible effects on the instapoetry community, as well as influencing the dynamics of the poetry landscape.

Virtual communities undeniably add new dimensions to the discourse on imagined communities and new media. The shift from print to digital as a medium of communication profoundly impacts the spatial and temporal dimensions of a community formed. In virtual and networked spaces, the community imagined is not limited to the fixity of space or time. The instapoetry community is transnational and transpatial, consisting of members from different countries with various cultures and national backgrounds, and yet all convene and interact in cyberspace. With physical mobility not being a limiting factor, the instapoets can convene based on shared interests transcending geographical boundaries. The interaction and feedback in this digital community happen in real time and can evolve quickly according to current events, trends, or discussions. Thanks to digital technology, an imagined community is more fluid and dynamic. Poets also can engage in various imagined communities and thus can adopt and evolve across multiple identities or belonging. Although digitally connected, as described formerly, the virtual community is still subjected to material practices and rituals which sustain the imagination of belonging to a community. The following section dissects the instapoetry community as a digitally mediated community and details its instapoetic practices that connect the virtual members.

18.3 Overview of the Instapoetry Community in Southeast Asia

Despite increasing steadily annually, 2023 data shows that the internet penetration for Southeast Asia is still relatively low at 63.7% of the total regional population (Kemp 2023). More than 95% of the population in Brunei, Singapore, and Malaysia are connected to the internet. It is no surprise that these three countries also hold the highest number of social media users. Meanwhile, the data shows that internet penetration in Timor-Leste and Burma is still below 50%. The digital divide can be attributed to levels of urbanisation, political unrest (Nanthakorn et al., 2023), and policy or regulatory variables (Lim 2018; Srinuan et al. 2010; UN.ESCAP 2019).

There is existing scholarly literature on digital expressions, literacy and language use on social media in Southeast Asia, although the majority of the studies focus on local and national contexts rather than regional cooperation. Instapoetry emerges as another intriguing dimension of creative digital expression contributing to the rich tapestry of creative interactions within the region.

The instapoetry community formed in the last decade, where its practice of poetry writing remains distinctive within the larger literary landscape. The poetic practice of an instapoet uniquely incorporates platform-specific labour such as social media engagements (likes, comments, or shares), building visibility on Instagram and developing an online community, with poetry writing. Apart from its close association with Instagram as a primary medium to share and consume poetry, instapoets often utilise visuals such as images (illustrations, photos, or artwork) and videos. The poems are short in length, adapted for mobile readers to quickly read and scroll

(Knox et al. 2023, p. 4; Lotman 2021, p. 72; Penke 2023, p. 2). Due to the poets' close association with Instagram, the instapoets also adapt to Instagram's technological changes. For example, with the implementation of reels, instapoet Tyler Knott Gregson began posting reels of poetry readings into his account. Themes explored within the instapoetry community often overlap with prevalent social justice topics.

While the practice of instapoetry is still largely polemic in the mainstream literary sphere, its transformative impact on poetic practice in a digital age is undeniable. The instapoetry community becomes a space for both professional and aspiring poets to share their work and connect with fellow writers and readers. The high interest and participation within the instapoetry community (whether in writing or reading) is apt as 92.3% of global users access the internet using mobile phones (Kemp 2023). The larger instapoetry community under the #instapoetry hashtag is now absorbed and expanded into sub-communities, most apparent by the iterations of instapoetry hashtags such as #instapuisi, #instapoesia, #instapoezja, and #thebruneiwriters. This not only indicates the worldwide phenomenon of instapoetry but also highlights transnational affiliations of instapoets and their readers. The extent of engagement and participation varies significantly across these sub-communities, with the majority of activity concentrated in the North American region, home to prominent instapoets like Rupi Kaur and Lang Leav.

Although Facebook reigns as the most popular social media platform across the region, the audience reach on Instagram remains notable with over 183 million active users in Southeast Asia (Kemp 2023). Despite the numbers, the usage and popularity of Instagram vary considerably according to the country. This also reflects or influence the level of presence of an instapoetry community for a particular country.

Brunei Darussalam has a nascent but growing writing community online, and among other social media platforms, Instagram seems to be a favourable choice for emerging and amateur Bruneian writers, particularly poets. This is perhaps unsurprising, as Instagram remains the most popular social media platform in the country with over 63% of the Bruneian internet population choosing to be active on Instagram. Instagram has arguably brought a positive impact on local small businesses, offering an expansive audience reach as well as a built-in digital infrastructure for professionals (Aji et al. 2023; Hashim et al. 2021). Similarly, the creative literary scene (whether amateur or professional) capitalises on Instagram for its reach and digital infrastructure. With a local literary landscape and creative book publishing (particularly Anglophone) that is budding, Instagram emerges as the preferred social media platform to network and connect with fellow writers and readers, and to share their work.

Bruneians post their poems under the common hashtag #TheBruneiWriters. There are also a few concentrated writing groups based on Instagram such as The Brunei Poets Community (@bruneipoets.community), and The Brunei Writers (@thebruneiwriters). The Brunei Writers is formed as a platform for local writers to connect and share their work. They have collaborated with local creative agencies such as 247 Studios and the Tiny Lit Fest, as well as regionally with the Sing Lit Station to create more opportunities for writers in Brunei to share their work.

Since their formation, they have opened space for poets of varied forms including performance poetry or spoken word.

Instapoetry has not only opened opportunities to Bruneian writers, it has also done the same in Malaysia. Jeyam (2018) argues that the integration of social media and poetry "has been wholly responsible in keeping the poetic word alive" in Malaysia (p. 126). While in the paper Jeyam particularly looks at the growth of spoken word in Malaysia thanks to social media platforms such as YouTube, general observation of Malaysian literary practices on Instagram shows a positive outcome as well. Charissa Ong, a Malaysian author of the best-selling poetry books *Midnight Monologues* (2016) and *Daylight Dialogues* (2018), utilises Instagram to reach an audience and readers of her poems. As aptly summarised by Jeyam (2018), social media channels allow new ways to write and publish which in turn "bodes well for the future of the Malaysian poetry in English" (p. 136). Instapoetry is not limited in language as poems written in Malay can be tagged under #instapuisi, a hashtag which hosts over 47 thousand instapoems to date.

Notably, countries like Timor-Leste and Burma do not exhibit a distinct instapoetry community. However, the poets still utilise social media platforms to make their poems accessible to a global audience. For example, Burmese poet Pandora (Pyait Hlaing Oo) started and raised to prominence through blogging (Wiles 2015). Poetry of protest has long withstanding association in cyberspace and the tradition continues to instapoetry. In Myanmar, digital poetry is part of activists' strategies for political commentary. In 2016, poet and activist Maung Saungkha used poetry as a mode of political commentary towards the ruling party. He posted a satirical poem on his personal Facebook page titled "Image", describing a tattoo of then-president Thein Sein (Ryan 2019). In Indonesia, visual digital poetry is recorded on YouTube (Mahyudi et al. 2017).

Admittedly, it is abundantly clear that instapoetry is comparably not as well-established as other parts of the world where the likes of Rupi Kaur and Lang Leav have topped book-selling lists alongside canonical writers such as Homer, Dante Alighieri, and Kahlil Gibran. Several factors can affect the prominence of instapoetry in a particular area such as social media usage patterns and the publishing infrastructures. As indicated previously, the social media usage pattern varies across Southeast Asia. There is also a markedly different extent of interest and presence of an instapoetry community among the Southeast Asian countries. Despite that, there is an online effort to galvanise regional ties between the instapoetry community in Southeast Asia.

18.4 Comparative Analyses of the Instapoetry Community in Southeast Asia

The prominence and accessibility of instapoetry makes the poetic genre attractive particularly to amateur poets. An instapoet from Southeast Asia negotiates a complex web of influences that are local, national, regional, and transnational. No longer are passive consumers of instapoetry, poets in Southeast Asia have begun to utilise Instagram as a platform for poetry. Instapoetry comes from the words "Instagram" and "Poetry" which not only alludes to the instantaneous nature of reading and writing poetry, it also represents Instagram as a platform for poetry. Hence, instapoets in the community perform instapoetic practices, which are practices that uniquely combine poetry writing and platform-specific practices. The emphasis on visuals or using hashtags are examples of these instapoetic practices. The practice is adopted by instapoets in Southeast Asia as well. The hashtags chosen encompass several motives. For example, Bruneian instapoets employ the hashtag #thebruneiwriters as a hub to connect with fellow writers in Brunei. Instapoems written in Malay are filed under the #instapuisi hashtag. Some instapoets also use hashtags that are related to the poem's theme. For instance, in an instapoem that describes the narrator's religiosity, an instapoet would employ hashtags such as #muslimpoet and #muslimah. In doing so, they are including themselves in the global collage of other Islamic-related posts on Instagram. Additionally, instapoets integrate elements of visual media such as digital illustrations and images in their poems. For example, in an untitled poem about past and repressed memories, a Bruneian instapoet included a distorted and obscured black and white photo to enhance and accompany with the theme. Another way an instapoet would emphasise a poem's visuality is by creating a video of a poem featuring the text and additional music. As observed, the shared practice across the instapoetry community demonstrates that the instapoets imagined themselves to belong in a community.

Instapoets from Southeast Asia simultaneously choose to feature cultural motifs or references, linguistic richness, and themes that are representative and distinctive to their country hence, indicating an engagement with the imagined border of the nation-state. For example, religion is a prominent theme and motif in Bruneian instapoetry. While not all poems expressing religiosity explicitly mention a particular faith, many allude to Islam. This aligns with Brunei's national religion, one component of the national tripartite known as the Malay Islamic Monarchy and as reiterated by His Majesty's titah or command (Ministry of Religious Affairs 2023). While a larger part of the instapoems by Bruneians is written in English, there are also poems written in Malay language, or mixed with the English language. Haji-Othman and McLellan (2014) studied the prevalence of "code-switching, mixing and meshing of English and Malay in both the spoken and keyboarded discourses of Bruneians" even though it is becoming more prevalent in digital conversations (p. 10). Brunei's multilingualism is reflected in the languages used in Bruneian instapoetry where the poems are written in English, Malay, code-mixed English and Malay and, in one instance, Dusun. In summary, the shared poetic practices such as the employment of hashtags and emphasis on visuals suggest the instapoet's collective consciousness as part of an imagined community. However, their instapoems are not entirely homogenous. They are also keen to integrate the rich literary and cultural traditions of their country.

While they are a part of a global poetic movement most indicative of the shared instapoetic practice with other instapoets globally, their work is distinctly shaped by transnational interactions within the region. These interactions facilitate a unique poetic identity beyond national borders, allowing the Southeast Asian instapoets to simultaneously assert their cultural particularities and engage in a broader, transnational dialogue that links their experiences across the region. The Southeast Asian Poetry Writing Month (SEAPoWriMo) is an initiative inspired by the National Poetry Writing Month (NaPoWriMo) (Ip et al. 2021). Sing Lit Station's iteration of the initiative for Southeast Asia started in 2019. SEAPoWriMo is "an annual poetry writing challenge that celebrates literary and cultural exchange through fun prompts, open mics, and video presentations" (Sing Lit Station n.d.). Every August, Sing Lit Station includes writers or writing collectives from Southeast Asia as moderators for a week. A writing prompt is released once a day on Instagram and Facebook by the moderators and the challenge is to write a poem every day in August.

An observation of instapoems from Southeast Asia that are posted under the SEAPoWriMo writing challenge in 2023 indicates a community of instapoets from Southeast Asia that are still in flux; navigating the influences and practice from the global instapoetry community, inserting unique cultural references and motifs to assert their national identity among regional poets or peers, and yet touching upon similar or common themes which suggests the attempt of creating a transnational cultural resonance that is beyond nation-state borders.

Instapoets in the region write about common themes that also suggest a shared transnational identity and cultural resonance that is beyond state borders and thus indicate a transnationally integrated region by a shared by a common creative cultural space, in this case, instapoetry. For a writing prompt about rice, three poets mention different types of rice dishes however, all four poems draw a connection between the rice dishes and themes of belonging. Rice, a staple and common food source in Southeast Asian cultures, becomes a recurring symbolic imagery of each poet's cultural roots. Using the symbol, each poet expresses the complexities of belonging while navigating forces such as tradition and globalisation.

For example, a Bruneian instapoem titled "sentimental rice" illustrates the speaker living in a foreign city. The speaker starts to reminisce the memories of their home after eating lunch with "crisply baked mentaiko sauce [...] hiding the rice underneath") made by her sister, "It tasted just like [...] the random picnics you have with family by the/ocean on school holidays back home" (Nadhir 2023). Despite the unfamiliar lunch in a foreign city, the hidden and ever-present rice represents the speaker's unwavering connection and sense of longing for their home. The rice dish itself, while it has no cultural significance to Brunei, has an enduring connection to her home due to a memory attached to the speaker's family. Through the symbolic imagery of the rice dish, the poet suggests that belonging is not merely forged through cultural affinity rather can also be maintained through personal memory. It is suggested that belonging is fluid and multifaceted, in that it can exist beyond cultural boundaries. In a poem titled *"Nasi Basi"*("Expired Rice"), the poet uses the symbol of stale rice to describe a language that has been long forgotten. The first two verses mention dishes that are not culturally linked to herself such as spaghetti ("at night eat spaghetti"), KFC ("in the morning eat KFC"), and sushi ("in the afternoon eat sushi"). The poem closes with the verse "Open the Tupperware the rice is expired/the language and culture that I love/please don't be erased from the era of today" (Efa 2023).[1] The use of stale rice symbolises the speaker's fear of losing her cultural and linguistic heritage, which are essential to her sense of belonging. The poem's imagery of culturally significant motif such as rice juxtaposed with global dishses such as spaghetti, KFC, and sushi, indicate how the speaker sees infiltration of foreign influences can corrode one's sense of belonging. The Tupperware itself conveys a need to protect or preserve cultural markers from decay, and against the tension of a globalised world. In short, the speaker emphasises the importance of maintaining markers of belonging in a changing world.

The majority of poets for SEAPoWriMo write in English. Despite the use of English, the poets still include cultural references or motifs unique to their identity. The blending of English and local experiences shows their desire to connect regionally through a common language while still staying true to their cultural authenticity. The use of English as a common language may also signal an ambition to forge transnational connection among other instapoets in the region. This is further amplified through efforts from SEAPoWriMo where its prompts stimulate cross-border creative exchanges and collaboration. This sheds light on the dynamic interplay that a Southeast Asian poet is subjected to, navigating the global influences from the broader instapoetry community, strengthening a transnational connection with poets in the region while also developing and safeguarding cultural references local to their nation. The instapoetry community in Southeast Asia involves a negotiation of identities where the instapoets are acknowledging both their local roots, maintaining regional ties by forging transnational commonality, and continuing their participation in a broader, global instapoetry community. Indeed, these overlapping, intricate dynamics contribute to the potential richness and depth of the instapoetry community in Southeast Asia.

18.5 Challenges and Opportunities

The nascent yet promising landscape of Southeast Asian instapoetry networks grapples with a host of formidable challenges rooted in the novelty of instapoetry as a poetic practice and limited engagement or participation rates across the region. The diverse sociocultural configurations especially social media usage patterns among countries in the region add another layer of complexity which hinders any centralisation or cohesive community effort at a local and regional level.

The lack of centralised effort to communise local instapoets is most apparent by the absence of standardised local hashtags for instapoets to use. This lack of virtual infrastructure makes it difficult for instapoets to share, collate, and digitally archive their work with other local instapoems. This fragmentation at a local level further impedes the facilitation of connection, collaboration, and visibility among countries in the region. Secondly, considering the linguistic diversity in Southeast Asia, the instapoems do not fully represent this variety. The practice of instapoetry remains highly influenced by the desire for global accessibility.

Regardless, the network of instapoetry community in the region still holds promise because it opens opportunities for collaborations among amateur poets on Instagram. The SEAPoWriMo writing challenge created by Sing Lit Station highlights instapoets in Southeast Asia to build and sustain regional creative cooperation both online and offline. The 2023 edition of the writing challenge brought about the creation of "Hot and Sour Magazine". This literary magazine "embodies the fruits of community" and features poetry from Southeast Asia which "speaks of our connection with our homes and how we endure the physicalities of our fraught lives and histories" (Sing Lit Station "Hot and Sour Mag" 2023, n.d.). The instapoems that are created by the members of the community establish a regional common ground. Thanks to SEAPoWriMo, the instapoetry community in Southeast Asia become a place of regional solidarity while also acknowledging the variety and diversity in creative cultural expression. Another example of transnational unity online is The Milk Tea Alliance borne out of a popular Twitter meme. It grows into a symbol for a digital political movement which advocates for democracy across various nations such as Thailand, Hong Kong, Taiwan, Indonesia, and India. While admittedly may not be as expansive or impactful, both exemplify how social media and shared interests can transcend national boundaries and foster cross-cultural connections.

18.6 Conclusion

Future inquiries may explore the evolving dynamics of Instagram and its impact on instapoetry in Southeast Asia. The limited and most importantly, disproportionate participation only allows for theoretical discourse, and limited close reading and comparative analyses. A focused and intensive close or comparative reading of instapoems within the region would shed more light on any narratives shared across the region and thus further enhance this research subject. Another area for research is to investigate how digital disparities, social media usage patterns, and internet policies (including freedom of net status) affect the level of engagement, participation, and visibility for the instapoetry community in the region.

Despite the geographical distance and sociocultural diversity, instapoets in Southeast Asia imagined themselves to be part of a community united by shared digital poetic practices and common interest in the poetic expression on Instagram. This allows ample opportunities for fellow amateur poets to collaborate creatively, turning the instapoetry community into a virtual place for regional solidarity and cooperation. However, the novelty of instapoetry as a poetry movement as well as the lack of centralised infrastructure on Instagram such as localised hashtags limit the participation and engagement of the instapoets. This is not to suggest absence or inactivity within the instapoetry community in Southeast Asia. Initiatives by Sing Lit Station such as SEAPoWriMo strengthen and invigorate creative regional collaboration between instapoets in the region resulting in an online literary magazine and a germination of new Southeast Asian poets. These creative literary initiatives denote how adaptation or integration of a global poetic movement on Instagram in the region encourage and facilitate cross-border collaborations which leads to the transformation of poetic practices on a local and regional level across Southeast Asia. While the instapoetry community in this region is still new, the active participation (despite its low number), cross-border collaborations, and shared transnational poetic themes such as belonging and preservation of cultural identity suggest a promising new channel for transnational collaboration, exchange of ideas, and cultivating a community that transcends rigid nation-state boundaries for the benefit of the Southeast Asia. The cultivation of transnational ties facilitated by the instapoetry community in the region enriches the poetic practice and contributes to a richer, dynamic, and interconnected poetic practice that extends beyond local or national boundaries.

Notes

1. English translation is poet's own.

References

Abigail (2023) In @avvigayilshin, Humble comfort (Photograph). [Instagram], 13 August. Available from https://www.instagram.com/p/Cv40rL-hNug/?igsh=d3JremVhcjY3OW91

Aji NS, Ibrahim NHR, Hamdan H et al (2023) Digitalisation of SMEs in the food and beverage industry: a case study of Brunei Darussalam. In: Almunawar MN, de Pablos PO, and Anshari M (eds) Digital Transformation for Business and Society. Routledge, pp 245–267

Anderson B (2006) Imagined communities: reflections on the origin and the spread of nationalism. Verso

Atkins W (2013) The politics of Southeast Asia's new media. Routledge

Bilecen B, Cardona A (2018) Do transnational brokers always win? a multilevel analysis of social support. Social Networks 53:90–100

Chatterjee P (1993) The nation and its fragments: colonial and postcolonial histories. Princeton University Press

Cloke P, Crang P, Goodwin M (2013) Introducing human geographies, 3rd edn. Routledge, London

Crang M (2013) Cultural geography. Routledge

Crang M, Mohamed SMH (2016) Researching virtual communities. In: Clifford N, Cope M, Gillespie T et al (eds) Key methods in geography. SAGE Publications, pp 270–284

Curran J (2012) Media and power. Routledge

Dewan Bahasa dan Pustaka Brunei (n.d.) Visi dan misi (Vision and mission). http://www.dbp.gov.bn/Tentang%20DBP/Visi%20dan%20Misi.aspx. Accessed 18 Jan 2024

Dewan Bahasa dan Pustaka Malaysia (2023) Objektif penubuhan (Objective of establishment). https://lamanweb.dbp.gov.my/objektif-penubuhan/. Accessed 18 Jan 2024

Efa (2023) In @crkdnghbrcrkdhrt, Expired rice (Photograph). [Instagram], 13 August. Available from https://www.instagram.com/p/Cv490DFx_Zx/?igsh=dnpweWVnNnBtNjdx
Fuchs C, Trottier D (2014) Theorising social media, politics and the state: an introduction. In: Trottier D, Fuchs C (eds) Social media, politics and the state. Routledge, pp 3–38
Greenfeld L (1992) Nationalism: five roads to modernity. Harvard University Press
Gruzd A, Wellman B, Takhteyev Y (2011) Imagining Twitter as an imagined community. American Behavioral Scientist 55(10):1294-1318
Haji-Othman NA, McLellan J (2014) English in Brunei. World Englishes 33(4):486–497
Hamilton A (1992) The mediascape of modern Southeast Asia. Screen 33(1):81–92. https://doi.org/10.1093/screen/33.1.81
Hashim NFA, Alek SK, Asmali KN et al (2021) Utilisation of social media by micro, small and medium enterprises in Brunei Darussalam. International Review of Management and Marketing 11(1):8–20
Huang RL, Svetanant C (2022) Challenging digital authoritarianism: milk tea alliance and transnational solidarity. In: Activism and Authoritarian Governance in Asia. Routledge, pp 130–142
Imawan R (2023) Milk tea alliance: the power of new democracy in ASEAN and efforts to increase the effectiveness of ASEAN norms. Strata Social and Humanities Studies 1(2):43–52. https://doi.org/10.59631/sshs.v1i2.96
Ip J, Ang ASL, Lee CJ (2021) Digital communities at work: Singapore poetry writing month. TEXT 25(64):1–9. https://doi.org/10.52086/001c.30984
Jeyam L (2018) "[T]he most underrated genre": Malaysian poetry in English in the 21st century. Asiatic: IIUM Journal of English Language and Literature 12(2):125–136
Kemp S (2023) Digital 2023: global overview report, DataReportal. https://datareportal.com/reports/digital-2023-global-overview-report. Accessed 18 Jan 2024
Knox J, Mackay J, Nacher A (2023) Global instapoetry. European Journal of English Studies 27(1):3–13
Kuşçu I (2014) The Uyghur diaspora in cyberspace: identity and homeland cause. Bilig: Journal of Social Sciences of the Turkish World 69:143–160
Kreutz J, Makrogianni AA (2024) Online repression and transnational social movements: Thailand and the #MilkTeaAlliance. Political Research Exchange. https://doi.org/10.1080/2474736X.2023.2299120
Laguerre MS (2005) Virtual diasporas and cyberspace. In: The digital city: the American metropolis and information technology. Palgrave Macmillan London, pp 125–142
Lent JA (1989) Mass communication in Asia and the Pacific: Recent trends and developments. Media Asia 16(1):16–24. https://doi.org/10.1080/01296612.1989.11726297
Lim M (2018) Disciplining dissent: freedom, control, and digital activism in Southeast Asia. In: Padawangi R (ed) Routledge handbook of urbanization in Southeast Asia. Routledge, pp 478–494
Lotman R (2021) The semiotics of new era poetry: Estonian instagram and rap poetry. Studia Metrica et Poetica 8(2):58–79
Mahyudi J, Saryono D, Siswanto W et al (2017) Construction of visual features of Indonesian digital poetry. International Journal of Linguistics, Literature and Culture 3(5):1–13
McLuhan E, Zingrone F (1997) Essential McLuhan. Routledge
Ministry of Religious Affairs (2023) Titah perutusan KDYMM sempena menyambut awal bulan Ramadhan 1444H/2023M. https://www.mora.gov.bn/Lists/Titah/NewDisplayItem.aspx?ID=369. Accessed 2 November 2023
Mohamad SMH (2018) Everyday lived Islam: Malaysian Muslim women's performance of religiosity online. Journal for Islamic Studies 37(1):74–100
Morimoto HL, Chin B (2017) Reimagining the imagined community: online media fandoms in the age of global convergence. In: Gray J, Sandvoss C, Harrington C (eds) Fandom: identities and communities in a mediated world, 2nd edn. New York University Press, New York, pp 174–188

Nadhir (2023) In @nadhirnadhirah, Sentimental rice [Photograph]. [Instagram], 26 August. Available from https://www.instagram.com/p/CwaBlrtxXYA/?igsh=MXFyMDF0bG9xYmV1Mw==
Nanthakorn B, Ractham P, Kaewkitipong L (2023) Double burden: exploring the digital divide in the Burmese educational system following the 2021 coup d'etat and the COVID-19 pandemic. Computers in Human Behavior Reports. https://doi.org/10.1016/j.chbr.2023.100310
Oakes T, Price P (2008) The cultural geography reader. Routledge
Ong A, Nonini D (2003) Ungrounded empires: the cultural politics of modern Chinese transnationalism. Routledge
Penke N (2023) Instapoetry: digital image texts. Springer Nature
Ryan JC (2019) Verses of resistance: The activist poetry of Myanmar. Southeast Asian Media Studies 1(2):19–40
Sahoo AK, De Kruijf JG (2016) Indian transnationalism online: new perspectives on diaspora. Routledge
Shklovski I (2011) Virtual transnationalism: technologically supported social network phenomenon. International workshop on transnational HCI: Humans, computers & interactions in global contexts at CHI 2011, 8 May 2011
Sidaway JD (2003) Imagined regional communities: integration and sovereignty in the global south. Routledge
Sing Lit Station (n.d.) Hot and sour magazine. https://www.singlitstation.com/hot-and-sour. Accessed 18 Jan 2024
Sing Lit Station (n.d.) SEAPoWriMo. https://www.singlitstation.com/seapowrimo. Accessed 24 Jan 2024
Spencer P, Wollman H (2002) Nationalism: A critical introduction. SAGE Publications
Srinuan C, Rohman IK, Srinuan P et al (2010) Digital divide in ASEAN countries: explaining the gap. In: Gugler P and Chaisse J (eds) Competitiveness of the ASEAN Countries. Edward Elgar Publishing, p 153–162
Starikov VS, Ivanova AA, Nee ML (2018) Transnationalism online: exploring migration processes with large data sets. Monitoring of Public Opinion: Economic and Social Changes, 147(5):213–232. https://doi.org/10.14515/monitoring.2018.5.17
UN.ESCAP (2019) Regulatory policies and ICT trends: insights from Timor-Leste. https://hdl.handle.net/20.500.12870/4120. Accessed 18 Jan 2024
Vertovec S (1999) Conceiving and researching transnationalism. Ethnic and racial studies 22(2):447–462
Vertovec S (2004) Cheap calls: the social glue of migrant transnationalism. Global Networks 4(2):219–224
Vorobeva E, Jauhiainen JS, Tammaru T (2022) Language, networks, and virtual transnationalism: The case of Russian speakers from Estonia living in Finland. International Migration 60(6):129–147
Wiles E (2015) Writers: the younger generation. In: Saffron shadows and salvaged scripts: literary life in Myanmar under censorship and in transition. Columbia University Press, pp 171–224

Hana Ghani is a PhD student in English Literature at Universiti Brunei Darussalam. Her doctoral thesis focuses on the treatment and representation of the female body in Rupi Kaur's poems in a new materialist lens. Her research interest encompasses the integration of social media and literature, Blue Humanities and Bruneian literature. She obtained her MA in English Literature from Universiti Malaya, where she dissects the perceived simplicity of instapoetry's writing style and poetic language. You may reach her at naqiyyah.hana@gmail.com

Chapter 19
Transnational Identity in the Shared Online Emotions of Environmental Activism

Meredian Alam and Muhammad Adil Iqbal

Abstract This article aims to delve deeper into the transnational identity shared in online emotions, which is embodied in the tweets in a youth-led environmental movement's organisation. Through case studies from the Twitter account of the Indonesian civic environmentalism Backsilmove, a Greenpeace's affiliated organisation based in Indonesia, this study explores the circulation of globalised transnational identity of activism, reflecting the meaning of grievance and contentions demonstrated in selected tweets from the Twitter handle of the organisation. This original study formulates, the environmental movement campaigns heralded by Backsilmove has resonated Greenpeace's transnational repertoire, which was presented in the effectiveness of its cyber-power by using Twitter to publicise its radical act of rescuing the urban forest and the capabilities of the activists to express their dissent as government plans to commercialise the forest unravelled.

Keywords Youth-led environmental organization · Twitter · Backsilmove · Spatial experiences

19.1 Introduction

The present article discusses the identity of transnationalism that Greenpeace young activists exert by mobilising social media to express their digital citizenship. First and foremost, Beneito-Montagut et al. (2018) define a transnational identity as a person's sense of belonging to more than one country or culture. Many influences, such as one's upbringing, social and political climate, and cultural heritage, all contribute to

---

M. Alam (✉)
Faculty of Arts and Social Sciences, Universiti Brunei Darussalam, Gadong, Brunei Darussalam
e-mail: meredian.alam@ubd.edu.bn

M. A. Iqbal
NUST Business School, National University of Sciences and Technology, Islamabad, Pakistan
e-mail: m.adiliqbal@nbs.nust.edu.pk

H. M. Y. Ho and Y. Chan (eds.), *Transnational Southeast Asia*, Asia in Transition 29,
https://doi.org/10.1007/978-981-96-3608-2_19

shaping one's values (Tedeschi et al. 2022). There is no single definition of transnational identity, and many particular externalities might shape its development (Ren and Liu 2022). To give just examples, one's transnational identity could be the result of having lived and worked in more than one country or using social media as part of globalisation products that corroborate transnationalism itself (Bilecen and Lubbers 2021). Globalisation and migration are two of the many social and political factors that might have an impact on transnational identities (Erdal 2020). Individuals in today's increasingly interconnected and multicultural world may have a sense of belonging to more than one cultural or national group (Chaudhary and Moss 2019). Much like migration can give rise to transnational identities, so too can the process of employing social media as the cognitive praxis of globalisation into a new culture of activism while maintaining ties to the original identity (Nowicka 2020; Sullivan 2019). On the one hand, following Hidalgo et al. (2015), it facilitates the establishment of international relationships and the acquisition of a richer appreciation for cultural diversity.

Social movement scholars in Southeast Asia have focused on transnational movements, such as those aimed at conserving the Mekong River, which flows through China, Myanmar, Cambodia, Thailand, and Laos. Since 2009, global environmental organisations have established subterranean forces utilising platforms such as Facebook, Twitter, and Instagram to amplify the demand for the liberation of the river from all forms of pollution, exploitation, and threats to its ecological integrity (Yong 2020; Young and Ear 2021). Furthermore, Greenpeace Southeast Asia (GPSEA) has taken the lead in bridging the activism aims of Southeast Asian nations and is actively engaged in efforts to safeguard the Mekong River (Kalland and Persoon 2013). This demonstrates that regional environmentalists continue to have an impact on the Southeast Asian environmental movement. Building on this, the following section describes the research's primary goals.

The present chapter unpacks the contentious thoughts of young citizen's environmentalist ideologies posted on the Backsilmove Twitter account in its role as an activist organisation. It is crucial to note that Backsilmove is arguably affiliated to Greenpeace Indonesia as all of the activists are the member of Greenpeace Indonesia (Alam and Nilan 2018). It is widely known that Twitter is an effective microblogging platform used to voice anxiety, anger, and other emotions in specific social settings (Torquati and Raffaelli 2004). Resistance activism worldwide has also fostered the different forms of moral sentiment and indignation experienced by various citizen groups. This phenomenon has been termed "online affective feedback" (Hidalgo et al. 2015, p. 364), where citizens' responses to certain events are aligned with the ideological interests of their group or organisation. Thus, there is urgency in revealing the meaning of tweets by factoring in both the context and relationship with the actor (Beneito-Montagut et al. 2018) as an interpretation of the social meaning of grievance. Moreover, through case studies from the environmental organisation Backsilmove's Twitter account, this chapter proposes to reconstruct the meaning of grievance based on tweets from this organisation.

In addition, we aim to observe and measure the sociocultural relevance of the environmental movement through the acts of protest staged by the Backsilmove activists.

Backsilmove is a youth-led organisation headquartered in metropolitan Bandung city, in West Java, Indonesia. Founded by younger members of the Indonesian subsidiary organisation Greenpeace (also known as Greenpeace Youth Bandung Indonesia), its primary activist goal was to protect the Citarum River from waste contamination. Although Backsilmove's does not put the Greenpeace logo on their banners, its actions nevertheless reflect characteristics of Greenpeace-style activism (Alam et al. 2019), marked by non-violent action and fearlessness during demonstrations. To this extent, Backsilmove's local action alludes to an extension to Greenpeace's ideology as a radical environmental organisation in that they herald direct protest against opposition to fight for environmental injustice at the local level. Manuel Castells describes this notion as a reflection of "glocalization"—a movement that incorporates both local and global repertoires to address a social injustice (Castells 2001, p. 23).

Some scholars investigating Indonesian politics argue that young Indonesian citizens nowadays benefit from being in the era of virtuality in which they exert transnational influence on social movements (Puspitasari and Ishii 2016; Suwana 2019; Tomsa and Setijadi 2018; Vidya et al. 2015). Meanwhile, Oberdabernig et al. (2018) regard this phenomenon as a cross-geographical democratisation of transparency, a consequence of Indonesia's disengagement from the grip of the New Order (Lim 2017). Social media tends to encourage democratisation in cyberspace, as well as transnational political participation involving various groups without geographical limits (Aduloju 2016). As such, a new citizenship phenomenon arose and became more widespread, undoubtedly facilitated by upholding the freedom of opinion in transnational arena.

19.2 Shared Emotions as Transnational Identity

Shared emotion as a transnational identity refers to the ability of individuals from many geographies and cultural backgrounds to come together and connect emotionally over a shared experience or sentiment (Célestine et al. 2022). Alam et al. (2019) argue, this shared emotion has the potential to foster a sense of identity that crosses national and cultural boundaries and brings people together. People from several countries may, for example, gather together to celebrate a global event such as the Earth Day and Black Lives Matter movement. As in Nwakanma's (2022) studies, while they speak different languages, have different cultures, and have different views, they share the same feelings of excitement, anticipation, and pride in their own countries' activists (see also Dache 2021). In this way, the shared emotion of national pride and sporting success becomes a global identity that brings people together across borders (Liu 2022). People may gather to lament a misfortune, such as a natural disaster or a terrorist assault (Avenell 2022; Guglielmi 2022; Tajuddin 2023). Regardless of nationality or cultural background, they may perpetuate feelings of loss, astonishment, and unity.

As Sorce and Dumitrica (2022) and Célestine et al. (2022) put in their study into collective emotions in transnational movement, this common emotion has the power to establish a sense of transnational identity, in which people come together as humans to support and console one another. Environmental movements can benefit greatly from shared feelings as a transnational identity (Huynh 2022). Alrmizan (2022) advances the idea of the constructive effects of transnational identity, that it has the potential to help individuals break down borders and build a sense of global community and interconnectedness. We can create a more compassionate and empathetic environment by acknowledging and respecting our shared emotional experiences.

The study of emotions in social movements was used to garner acceptance from scholars in this field because emotion is construed as irrational, raising issues over panic, and does not contribute to achieving movements' goals (Jasper and Owens 2014; Polletta and Jasper 2001). Rationalism recognises "sensible protestors as devoid of emotions" (Goodwin et al. 2001, p. 19). However, this gradually secured a place in the discourse of social movements alongside emerging evidence based on its effects of strengthening citizens' identity, collective solidarity, and the formation of new institutions through the course of its trajectory. For example, at the membership level, bonds between members are strong when there is positivity within the organisation, characterised by a sense of belonging and struggle (Jasper 1998, 2011; Polletta 1998). Moreover, the ability to share negative emotions that emerge from anger and disappointment towards the actions contradicting their ideology facilitates the mobilisation of activists. Goodwin and Jasper (2006, p. 260) explained that activists exert power in the public space to showcase their "moral outrage and anger to the world." Through this discourse, it is important to note that the structure of emotions serves as a "cognitive means" (Goodwin et al. 2000, p. 79) by pushing contentious feelings, urgency, fear, and deep concern. In the last 10 years of social media were chosen to express this contentiousness, as shared emotion is commonly verbalised by activists via these means (Al-Saggaf et al. 2016; Brummette and Sisco 2015; Evolvi 2019; Persson 2017).

Shared emotions are directed towards opponents, outsiders, and other groups (Creasap 2012). However, within a social movement, activists collaborate to create an identity, which involves both the community and target groups. This represents the virtual element of mobilisation which demands its expression through texts and creates common causes and collective narratives across social groups (Jasper 1998; Juris, 2008; Mansbridge and Morris 2001). According to Robnett (2004), it is a form of "emotional alignment" (p. 23) defined as a wholeness of sentimental resistance between groups to achieve certain goals. Meanwhile, within the environmental movement, the purpose of conflict is to stop destructive actions perpetrated by certain groups. These are usually government alliances that collaborate with private entities, colonising feelings and generating collective anger necessary to increase the resistance. In addition, Schrock et al. (2004) assert that this results in the occurrence of "emotional mobilisation," which refers to the "contentious feeling," (Schrock et al. 2004, p. 62) that is resurrected in order to foster and support activism. Social media also plays an important role in the experimental space. According to Jasper

(2011), words, sentences, and images are used by the activists to attempt to identify enemies. Furthermore, shared emotions have been a significant platform for engaging in activism as the activist becomes aware of their own pre-planned social activities in order to put pressure on the opponents (Jasper 2011).

The tweets from the Backsilmove activists are considered citizens' "shared emotions" and an expression of disappointment and hope towards a government that showed no consideration of the Babakan Siliwangi forest when developing the urban economy of Bandung city. In addition, being a public space, this medium is used to convince audiences that the natural environment is overshadowed by capitalism. Therefore, this medium is also used to call for sympathetic and citizen's consciousness with the use of selected commentary that puts political pressure on the government and private parties, and also rejects existing development policy. As environmentalists condemning the obvious destruction of nature, the Backsilmove activists' expressions on Twitter may be labelled as "feeling rules" (Hochschild 1979, p. 552), being authentic reactions to different situations that reflect their long-standing ideology of protecting the environment. Furthermore, the effective use of Twitter includes the imperialistic capitalism and neoliberalism, as a discourse to add the political arena to the environment. The effective involvement of several political debates by movement actors is in line with arguments by Bröer and Duyvendak (2009) that shared emotions through Twitter can produce reactions that vary from the public and allied movement organisations. Such responses arising from different parties evoke "moral shocks" (Jasper and Poulsen 1995, p. 500) that awaken a government's sense of crisis and leads it to realise that its actions are unethical. Jasper and Poulsen (1995) consider the emotionally expressive texts on Backsilmove's Twitter page as a means of protesting the morally reprehensible actions undertaken by governments and private institutions.

Success in environmental collectivism may be reached when groups of citizens with a similar political ideology set unbreakable resistance groups. Through Twitter, Backsilmove gained positive feedback from their followers and captured their feelings of concern about the future of Babakan Siliwangi forest. In the same vein, a Backsilmove activist, Pipin, re-tweets with this firm statement: "For the city forests, I am ready to defend!" The standing declaration publicised via Twitter implies a heart-breaking feeling from someone who was born, raised, and educated at one of the state colleges within Bandung. In addition, it was a rational act to actualize their disappointment about the decision of Bandung city government, which arbitrarily handed over the forest to a private company. Backsilmove tweets: "it could be impossible for us to push the government directly to rescind their decisions. Protesting in front of their offices might be dangerous because we don't want to be questioned by police or sent to jail. So, we use various resources from our repertoires such as drama, pantomime, and music concerts to engage and disseminate meaningful messages to the public.". The engagement was showcased by holding up a globe and performing pantomime while rallying towards Gedung Sate building. As for the Backsilmove on-the-street repertoire effects, one Backsilmove activist, Wirdan, asserted: "numerous journalists covered the activists, and Backsilmove was on national TV."

This young activist's response above marked a new chapter of "contentious politics" (Peterson et al. 2018, p. 377) model, moving away from conservative patterns (i.e. bureaucratic and parliamentary negotiations), through political oration in public places and demonstrations towards a post-modern way, which involves a wide variety of actions. This includes art concerning environmental issues, followed by disseminating and mediating into blogosphere publications. Moreover, the statements by Tio and Wirdan imply a new social movement to the ameliorating of the "creative city politics" (Mayer 2013, p. 5) of young Indonesian citizens, referring to activists who acknowledge contemporary social media as a means to mobilise their political powers.

The older generations heavily criticise the millennial generation as always positioned around marginal politics. They also regard them as mischievous and troublesome, as they act with scepticism towards social norms and obedience and display ignorance towards mainstream social norms (Anderson et al. 2016, p. 294). In the following tweet: "Young is an essential element of change," Backsilmove firmly states their fundamental ideology in regard to changing Indonesia socially, particularly in regard to the environment crises that are due to anthropocentric development. This kind of self-reflection constitutes power-evoking abilities and the Backsilmove's opposition to the dominating old senior activists in Bandung. Therefore, there is a need to take a closer look at their contentious feelings during the initial meetings of the Babakan Siliwangi Citizens Forum.

"Social distancing" (Krumer-Nevo and Benjamin 2010, p. 693) from the old activist generation implies that the movement cultivates great disappointment. This happens due to an inability to distribute and integrate ideologies into the Save Babakan Siliwangi movement dominated by senior environmentalists and Bandung activists. Apart from that, through the Backsilmove activists' involvement in the Babakan Siliwangi Citizens Forum, they crave autonomy, which is related to new normative values in the urban social movement. Hence, there is an emergence of social segregation between young and more mature activists—an issue quite apparent to Backsilmove activists. Meanwhile, such categories become a reason for activists to disassociate from circles of mature activists who obviously discriminate against them. Keeping the young activists outside of the core Babakan Siliwangi movement organiser group as the senior activists did to Backsilmove activists, according to Jensen (2011), has disempowered young people to articulate their concerns over the future of physical space for youth socialisation, which is the urban forest itself.

A year after its establishment in February 2013, Backsilmove chose a space-based repertoire of social actions held in public spaces which were enjoyed by the general public. Furthermore, integrated art exhibits revealed the emerging dynamics of the youthful energy. Capelli dubbed this nature of creative performativity as "artful resistance" (Capelli 2018, p. 1). Through this same reflection, there was an outpouring on Twitter to commemorate the one-year birthday of Backsilmove. They posted on their Twitter account: "The Spirit's Course, termed Backsilmove #1yearbacksilmove http://backsilmove.org/backsilmove-day/." These tweets included the hashtag #1yearbacksilmove as a public affirmation in order to frame the duration of the struggle. Conversely, the time-bound posts represented valuable memorabilia for activists and movement constituents.

That memorable event also illuminates the activists‘ joy when they reached the finish line of the protest in front of the parliament State Building. This action reminisces "place-based action frames" (Martin 2003, p. 733), explained as actions of people who evoke a strong feeling according to the social meaning of the places they previously occupied. In the conception of neighbourhood activism, Martin (2003) argues that it is the "material experiences" of those places that serve to cognitively bind the activists‘ solidarity (p. 735). From this standpoint, the collective dining event of the Backsilmove activists marks a celebration of the established social cohesiveness, memories of their struggle space, and ecological engagement with the Babakan Siliwangi forest.

Anticipative emotions are also shown in other tweets: "Let's prevent the representation of our voices by incompetents! #BandungJuarahttp://fb.me/141JmkU2Z." This was seen when the process of fighting over the forests of the Siliwangi Babakan city also coincided with a political period in 2013 when Bandung conducted a major election. This involved three pairs of candidates, including Ridwan Kamil, a leader among the young generation. In addition, Twitter and Instagram mobilisation were essential during the campaign in order to capture the hearts of the people and bring about populism. Awareness was also raised through social media by communicating using the *gahul* (colloquial) language style of the youths, in combination with Sundanese. Furthermore, attention towards the revitalisation of urban green space included convening discussions with artists in the Babakan Siliwangi urban forest, and routinely visiting Arts Workshop. Ridwan Kamil had already established himself by signing the agreement after his successful election as mayor of Bandung. Babakan Siliwangi prioritised his stylish, people-centred and intimate approach to society, and also the mass media, making it easier for him to win the hearts of the people. Thus, in the tweet, "Let's prevent the representation of our voices by incompetents!" #BandungJuarahttp://fb.me/141JmkU2Z Backsilmove was working very hard to present a political reminder to the public and to endorse Ridwan Kamil. This political alignment was shown through #ChampioningBandung, the jargon of the campaign by Ridwan Kamil. The expression in the hashtag implies support for the newly elected, and the only competent mayoral candidate, with the goal of preventing commercialisation of the forests. This demand is a manifestation of the anger of the youth towards the greediness of the previous Bandung mayor, Dada Rosada, and the sacrifice of their social space for this process. Dada Rosada was blamed for the destruction of the natural environment during his 10-year regime by developing a pro-profit development-oriented government regime (Martokusumo and Zulkaidi 2015; Soedirgo 2018). Rosada's poor reputation led activists to exert effective action to expel him from the mayoral position. Senior environmental and social activists in Bandung subsequently labelled him a corrupt and non-transparent leader after a Bandung local activists‘ protest uncovered embezzlement. His corruption was presented as *Kartu Merah Laporan Pembangunan* (Red Report Card Development), critiquing the practice of abuse of social funds. The protest coordinator of the Backsilmove movement argued: "Then, who else can be counted on, besides Ridwan

Kamil? We and the other partners from the government sector have already observed the weakness." This movement organisation was aware of this form of social process and the need for the power of activism was required to remain aligned. Therefore, the unwavering commitments tweeted included: "Strength does not come from physical capacity, but an indomitable will." -Mahatma Gandhi #1The Year Backsilmove. This was essential in maintaining the fire-panel resistance, as well as this spirit-incandescence, and reinforced through informal meetings, including *ngaliwet,* to keep his activists aligned with the values of the struggle for the environment. "Sometimes a decline is observed in the spirit of activists, indicating the need to take a break from routine", said Hadi, the activist. Some subculture social activities such as guerrilla planting at midnight on the street corners of Bandung, were also secretly but actively run by participants. Although the name is no longer Backsilmove, this subculture movement reflects the consistency of environmental action, and more importantly, it is able to channel outrage towards the government. A Backsilmove activist, Rayi, stated zealously: "we maintain this kind of movement, because it is more fun and engaging, and the contentious ambience is still there. If no similar movements emerge, perhaps someday I'll join up with a bolder action like Sea Shepherd." This statement points to the particular form of emotions which is sociologically referred to as "ilusio" (Bourdieu 1979, p. 80), which is a term alluding to the activists who tend to engage continuously in the movement organisation because they believe in the significance of their action. In the "ilusio" of Bourdieu's viewpoint, Rayi's key message is that the protest movement is precious in building his social identity as an activist, while his future involvement in other environmental organisations would be a social space to funnel and preserve his activists' identity.

Mahatma Gandhi was cited on Backsilmove's Twitter and revered by Backsilmove as a role model of an ideal peace activist. His name appeared 20 times within the 26 interviews conducted, although each was framed in a variant in the activism range. Furthermore, one individual, also a director of pantomimes of the State Buildings, defined him as a pro-environmental figure, and a daring social activist empowered by silence. Backsilmove launched a citizens' peaceful environmental movement inspired by Greenpeace with all activists undertaking non-violent direct-action training which featured preparatory action before involvement in an actual protest. In addition, the *Satyagraha* sparked by Mahatma Gandhi adopted Greenpeace practices as a resistance strategy, which was not tactical (Zelko 2004, p. 200). This was geared towards crippling the enemy, as participating in non-violent action without damaging the opponent is also effective.

The community is a part of the environmental transformation, as posted on the Backsilmove tweets: "It is certainly the optimistic mission of your community, for the sake of real and better changes for this beloved nation." Furthermore, the public is placed at the centre of social change, where individuals are ready to move and have a combative spirit. This is organised by young people, devising an agenda of movement actions or repertoire in order to counter the abuse of governmental power. Conversely, there is an incarnation known as "space of experiences" (Pleyers 2010), where something that happens inside Backsilmove implies a model of convergence, which is applied to counter globalisation, though only partially. Therefore, the community, in the eyes of Backsilmove, reflects on the phenomena of activism as an "alternative mode of (sub-) cultural practice," (Juris and Pleyers 2009, p. 15), from which emerges a form of citizenship among young people that prefigures wider social changes related to political commitment, cultural expression, and collaborative practice. Moreover, Twitter was also used to communicate with the public on related repertoire (i.e. pantomime, direct action), in order to assert the politico-ecological values, of which there is qualified alter-activism in itself. As Juris and Pleyers (2009, p. 57) stated:

> Alter-activism specifically involves the emphasis on lived experience and process, a commitment to horizontal, networked organization, creative direct action, the use of new information and communication technologies (ICTs), and the organisation of physical spaces and action camps, including laboratories for developing alternative values and practices.

Shared emotions are a crucial component in Instagram's application to draw more activists and public support from different national boundaries. The following conversation expresses the features of transnationalism in Backsilmove's activism:

> Meredian: Why is Instagram used so often in campaigns?
>
> Lisa (Backsilmove activist): Greenpeace is a multinational environmental advocacy organisation. Thailand, the Philippines, Indonesia, and Malaysia are home to the regional branch office of Greenpeace in Southeast Asia. These countries are home to the majority of Twitter and Instagram users.
>
> Author 1: Then why was Twitter used instead of Instagram?
>
> Meredian: It's because Twitter can swiftly assemble people, simply by applying a hashtag and indicating the location of the protest. Furthermore, the phrases employed there must be mind blowing and inspiring to activists outside of Bandung. In reality, we received numerous direct messages from activists in the Philippines and Thailand, who supported our cause.
>
> (Personal communication, 1 April 2015)

The Backsilmove movement, as a transnational activism, is well portrayed in this discussion. Although they do not represent Greenpeace when resisting the privatisation of the urban forest, the repertoire of using Twitter is an important component of the radical habitus they are developing as Greenpeace Indonesia activists. They use Twitter, a transnational platform pioneered by Greenpeace for sharing key environmental problems throughout environmental protests worldwide. This Backsilmove's global movement can thus be described as a "cognitive praxis" (Jamison 2006, p. 45), which refers to an action in activism that employs pre-existing techniques of communication learnt from and derived from the social movement organisations to which its activists belong (Jamison 2006). Similarly, the Backsilmove's use of Twitter reflects their revitalisation of Greenpeace's environmentalism. Riese (2015) claims this as a movement widely recognised as a mind-bombing by generating and posting catchphrases of activism that can emotionally captivate the public and compel them to support the cause online. In this conversation, transnationalism of shared emotions is well evinced.

19.3 Conclusion

With globalisation becomes pervasively evident as argued above, twitter is instrumental for the activists to circulate their transnational identity such as their discontent about encroachment to urban forest, their environmental responsibilities, and nationalism. As mentioned above, the research studies activists‘ claims and dissatisfaction conveyed via Twitter. Shared emotions in Backsilmove tweets also resonate with the youth's struggle and clearly demonstrate their environmental citizenship to the public. Through this chapter, it is learnt that transnational paradigm of activism is achievable when the respective activists spread out their contentious messages on Instagram so that the key values of such environmental movement can be exposed to other community members living in different geographical areas.

Author Contribution Meredian Alam was accountable for data collection, analysis, and writing the theories, and interviewing the research participants. Muhammad Adil Iqbal provided assistance in adding conclusions.

References

Aduloju AA (2016) Youth networks on Facebook and Twitter during the 2015 General Elections in Nigeria. J Afr Elections 15(2):28–49. https://doi.org/10.20940/jae/2016/v15i2a2

Alrmizan M (2022) The Arabs and the Muslims: between diaspora and transnationalism. Diaspora Stud 15(3):252–270

Avenell S (2022) Transnationalism, activism and civil society in Japan. In: Sahoo AK (ed) Routledge Handbook of Asian Transnationalism. Routledge, pp 17–29

Alam M, Nilan P (2018) The campaign to save the Bandung city forest in Indonesia: a cognitive praxis analysis of protest repertoires. Indonesia and the Malay World 46(136):343–359

Alam M, Nilan P, Leahy T (2019) Learning from Greenpeace: activist habitus in a local struggle. Electron Green J 1(42):1–18. https://doi.org/10.5070/g314237888 https://doi.org/10.1080/13639811.2018.1496623

Al-Saggaf Y, Utz S, Lin R (2016) Venting negative emotions on Twitter and the number of followers and followees. Int J Sociotechnol Knowledge Dev 8(1):44–55. https://doi.org/10.4018/ijskd.2016010103

Anderson E, Buchko AA, Buchko KJ (2016) Giving negative feedback to millennials: how can managers criticize the "most praised" generation. Manage Res Rev 39(6):692–705. https://doi.org/10.1108/mrr-05-2015-0118

Bilecen B, Lubbers MJ (2021) The networked character of migration and transnationalism. Global Net 21(4):837–852

Beneito-Montagut R, Cassián-Yde N, Begueria A (2018) What do we know about the relationship between internet-mediated interaction and social isolation and loneliness in later life? Quality Ageing Older Adults 19(1):14–30. https://doi.org/10.1108/qaoa-03-2017-0008

Bourdieu P (1979) Symbolic power. Critique Anthropol 4(13–14):77–85

Bröer C, Duyvendak J (2009) Discursive opportunities, feeling rules, and the rise of protests against aircraft noise. Mobilization 14(3):337–356. https://doi.org/10.17813/maiq.14.3.kx87723664558jm4

Brummette J, Sisco HF (2015) Using Twitter as a means of coping with emotions and uncontrollable crises. Public Relat Rev 41(1):89–96. https://doi.org/10.1016/j.pubrev.2014.10.009

Cappelli ML (2018) Standing with Standing Rock: affective alignment and artful resistance at the Native Nations Rise March. SAGE Open 8(3):1–13. https://doi.org/10.1177/2158244018785703

Castells M (2001) The new global economy. In: Muller J, Cloete N, Badat S (eds) Challenges of globalization: South African debates with Manuel Castells. Maskew Miller Longman, pp 2–21

Célestine A, Martin-Breteau N, Recoquillon C (2022) Introduction - Black Lives Matter: un mouvement transnational? (Introduction - Black Lives Matter: a transnational movement?) Esclavages & Post-esclavages. Slaveries Post-Slaveries 6

Chaudhary AR, Moss DM (2019) Suppressing transnationalism: bringing constraints into the study of transnational political action. Comparative Mig Stud 7(1):1–22

Creasap K (2012) Social movement scenes: place-based politics and everyday resistance. Sociol Compass 6(2):182–191. https://doi.org/10.1111/j.1751-9020.2011.00441.x

Dache A (2021) Teaching a transnational ethic of Black Lives Matter: an AfroCubana Americana's theory of Calle. In: Haynes C, Castillo-Montoya M, Hailu MF, Stewart S (eds) Black Liberation in Higher Education. Routledge, pp 28–41

Erdal MB (2020) Theorizing interactions of migrant transnationalism and integration through a multiscalar approach. Comparative Mig Stud 8(1):31

Evolvi G (2019) #Islamexit: inter-group antagonism on Twitter. Information, Com Society 22(3):386–401. https://doi.org/10.1080/1369118x.2017.1388427

Guglielmi M (2022) Entanglement and forms of orthodox transnationalism in Italy and beyond. In Guglielmi M (ed) The Romanian orthodox diaspora in Italy: eastern orthodoxy in a western European country. Springer International Publishing, pp 111–141

Goodwin J, Jasper JM (2006) Emotions and social movements. In: Stets JE, Turner JH (eds) Handbook of the sociology of emotions. Springer, pp 611–635

Goodwin J, Jasper JM, Polletta F (2000) The return of the repressed: the fall and rise of emotions in social movement theory. Mobilization 5(1):65–83. https://doi.org/10.17813/maiq.5.1.74u39102m107g748

Goodwin J, Jasper JM, Polletta F (2001) Why emotions matter. In: Goodwin J, Jasper JM, Polletta F (eds) Passionate politics: Emotions and social movements. University of Chicago Press, pp 1–24

Hidalgo CR, Tan ESH, Verlegh PW (2015) The social sharing of emotion (SSE) in online social networks: a case study in Live Journal. Computers Human Behav 52:364–372. https://doi.org/10.1016/j.chb.2015.05.009

Hochschild AR (1979) Emotion work, feeling rules, and social structure. Am J Sociol 85(3):551–575. https://doi.org/10.1086/227049

Huynh JA (2022) Assimilation through transnationalism: second-generation refugees and Vietnamese transnational organizing. Ethnic Racial Stud 45(14):2631–2650

Jamison A (2006) Social movements and science: cultural appropriations of cognitive praxis. Science as Culture 15(01):45–59. https://doi.org/10.1080/09505430500529722

Jasper JM (1998) The emotions of protest: affective and reactive emotions in and around social movements. In: Sociological forum. Kluwer Academic Publishers-Plenum Publishers, pp 397–424

Jasper JM (2011) Emotions and social movements: twenty years of theory and research. Ann Rev Sociol 37:285–303. https://doi.org/10.1146/annurev-soc-081309-150015

Jasper JM, Owens L (2014) Social movements and emotions. In: Stets JE, Turner JH (eds) Handbook of the Sociology of Emotions, vol II. Springer, pp 529–548

Jasper JM, Poulsen JD (1995) Recruiting strangers and friends: moral shocks and social networks in animal rights and anti-nuclear protests. Social Prob 42(4):493–512. https://doi.org/10.1525/sp.1995.42.4.03x0129y

Jensen SQ (2011) Othering, identity formation and agency. Qual Stud 2(2):63–78

Juris JS (2008) Performing politics: image, embodiment, and affective solidarity during anti-corporate globalization protests. Ethnography 9(1):61–97. https://doi.org/10.1177/1466138108088949

Juris JS, Pleyers GH (2009) Alter-activism: emerging cultures of participation among young global justice activists. J Youth Stud 12(1):57–75. https://doi.org/10.1080/13676260802345765
Kalland A, Persoon G (2013) Environmental movements in Asia. Routledge, London
Krumer-Nevo M, Benjamin O (2010) Critical poverty knowledge: contesting othering and social distancing. Current Sociol 58(5):693–714. https://doi.org/10.1177/0011392110372729
Lim M (2017) Freedom to hate: social media, algorithmic enclaves, and the rise of tribal nationalism in Indonesia. Critical Asian Stud 49(3):411–427. https://doi.org/10.1080/14672715.2017.1341188
Liu H (2022) Identity, Politics, and Transnationalism: deciphering New Chinese Diaspora in Singapore, 2010–2020. In: Santasombat Y (ed) Transnational Chinese Diaspora in Southeast Asia: Case Studies from Thailand, Malaysia, and Singapore. Springer Nature, Singapore, pp 125–152
Mansbridge JJ, Morris A (2001) Oppositional consciousness: the subjective roots of social protest. University of Chicago Press
Martin DG (2003) "Place-framing" as place-making: Constituting a neighborhood for organizing and activism. Ann Assoc Am Geographers 93(3):730–750. https://doi.org/10.1111/1467-8306.9303011
Martokusumo W, Zulkaidi D (2015) Some notes on area-based conservation: lesson learned from Bandung. Int J Built Environ Sustain 2(2):125–130. https://doi.org/10.11113/ijbes.v2.n2.67
Mayer M (2013) First world urban activism: beyond austerity urbanism and creative city politics. City 17(1):5–19. https://doi.org/10.1080/13604813.2013.757417
Nowicka M (2020) (Dis)connecting migration: transnationalism and nationalism beyond connectivity. Comparative Mig Stud 8(1):1–13
Nwakanma AP (2022) From black lives matter to EndSARS: women's socio-political power and the transnational movement for black lives. Perspectives Pol 20(4):1246–1259
Oberdabernig DA, Humer S, Cuaresma JC (2018) Democracy, geography and model uncertainty. Scottish J Pol Econ 65(2):154–185. https://doi.org/10.1111/sjpe.12140
Persson G (2017) Love, affiliation, and emotional recognition in #kämpamalmö: the social role of emotional language in Twitter discourse. Social Media+ Society 3(1):1–11. https://doi.org/10.1177/2056305117696522
Peterson A, Thörn H, Wahlström M (2018) Sweden 1950–2015: contentious politics and social movements between confrontation and conditioned cooperation. In: Mikkelsen F, Kjedstadli K, Nyzell S (eds) Popular struggle and democracy in Scandinavia. Palgrave Macmillan, pp 377–432
Pleyers G (2010) The global justice movement. Globality Stud J 19(1):1–14
Polletta F (1998) Contending stories: narrative in social movements. Qual Sociol 21(4):419–446
Polletta F, Jasper JM (2001) Collective identity and social movements. Ann Rev Sociol 27(1):283–305. https://doi.org/10.1146/annurev.soc.27.1.283
Puspitasari L, Ishii K (2016) Digital divides and mobile internet in Indonesia: impact of smartphones. Telematics Informatics 33(2):472–483. https://doi.org/10.1016/j.tele.2015.11.001
Ren N, Liu H (2022) Southeast Asian Chinese engage a rising China: business associations, institutionalised transnationalism, and the networked state. J Ethnic Mig Stud 48(4):873–893
Riese J (2015) Taking arms against a sea of troubles: the greenpeace whale campaign in Norway. J Appl Behav Sci 51(1):94–128. https://doi.org/10.1177/002188631453
Robnett B (2004) Emotional resonance, social location, and strategic framing. Sociol Focus 37(3):195–212. https://doi.org/10.1080/00380237.2004.10571242
Schrock D, Holden D, Reid L (2004) Creating emotional resonance: Interpersonal emotion work and motivational framing in a transgender community. Social Prob 51(1):61–81. https://doi.org/10.1525/sp.2004.51.1.61
Soedirgo J (2018) Informal networks and religious intolerance: how Clientelism incentivizes the discrimination of the Ahmadiyah in Indonesia. Citizenship Stud 22(2):191–207. https://doi.org/10.1080/13621025.2018.1445490
Sorce G, Dumitrica D (2022) Transnational dimensions in digital activism and protest. Rev Com 22(3):157–174

Sullivan MP (2019) Transnationalism, power politics, and the realities of the present system. In: Globalism Versus Realism: International Relations' Third Debate. Routledge, pp 195–221

Suwana F (2019) What motivates digital activism? the case of the Save KPK movement in Indonesia. Information Com Society:1–16. https://doi.org/10.1080/1369118x.2018.1563205

Tajuddin A (2023) Migration, transnationalism and development: the case of Indonesian diaspora. In: Sahoo AK (ed) Routledge Handbook of Asian Transnationalism. Routledge, pp 267–278

Tedeschi M, Vorobeva E, Jauhiainen JS (2022) Transnationalism: current debates and new perspectives. GeoJ 87(2):603–619

Tomsa D, Setijadi C (2018) New forms of political activism in Indonesia: redefining the nexus between electoral and movement politics. Asian Surv 58(3):557–581. https://doi.org/10.1525/as.2018.58.3.557

Torquati JC, Raffaelli M (2004) Daily experiences of emotions and social contexts of securely and insecurely attached young adults. J Adolescent Res 19(6):740–758. https://doi.org/10.1177/0743558403260023

Vidya NA, Fanany MI, Budi I (2015) Twitter sentiment to analyze net brand reputation of mobile phone providers. Proc Computer Sci 72:519–526. https://doi.org/10.1016/j.procs.2015.12.159

Yong ML (2020) Reclaiming community spaces in the Mekong River transboundary commons: Shifting territorialities in Chiang Khong, Thailand. Asia Pacific Viewpoint 61(2):203–218

Young S, Ear S (2021) Transnational political economic structures: explaining transnational environmental movements against dams in the lower Mekong region. Third World Quarterly 42(12):2993–3011

Zelko F (2004) Making greenpeace: the development of direct action environmentalism in British Columbia. BC Stud (142/3):197–239. https://doi.org/10.14288/bcs.v0i142/3.1725

Meredian Alam is Assistant Professor of Sociology at the Faculty of Arts and Social Sciences, UBD. He is currently undertaking research on the current trend of global environmental movement amongst young people. He earned a PhD in Sociology and Anthropology from the University of Newcastle Australia, an M.Phil in Culture and Environment from the University of Oslo, an MA in Development Sociology and a BSocSc from Gadjah Mada University.

Muhammad Adil Iqbal PhD is an Assistant Professor of History, International Relations and Pakistan Studies at NUST Business School (NBS), National University of Sciences and Technology (NUST), Islamabad, Pakistan. He is an active researcher and has established a very multidisciplinary educational background of Islamic Civilization, International Relations, Political Science, and Defence and Strategic Studies. He has worked in a variety of settings from industry to academia for over 15 years, learning about his subject matter from an array of perspectives and disciplines.

Chapter 20
Transnationalism of the Malay World: Evaluating Current Research Trends in Malay Linguistics

Najib Noorashid

Abstract Studies on transnational relations, practices, and contexts highlight the growing importance of interactions between people in different countries. These interactions involve negotiating cultural differences, communication, and social change in the contemporary world. In the Malay World—which includes Brunei Darussalam, Malaysia, Indonesia, and Singapore—discussions about the 'Malay vs. English' debate are central to contemporary sociolinguistic and identity studies due to national pride, persistence, and globalisation. This chapter discusses debates on language contact involving Malay and English in the transnational context of the Malay World. The discussion is based on academic experiences, fieldwork, and significant literature on Malay language and identity in the Malay World. This chapter focuses on three areas: (1) the concept of transnationalism in the Malay World; (2) current perspectives on 'Malay vs. English' language attitudes through a contemporary sociolinguistic lens; and (3) a reflective discourse and shifting paradigm of Malay linguistic research in the transnational Malay World. The chapter also sheds light on ongoing debates about Malay language, culture, and shared identity in relation to foreign influences such as English. It also explores research implications and prospects in the Malay World and contributions to the study of Malay linguistics and sociolinguistics in the context of globalisation.

Keywords Malay language · Malayness · English · Malay linguistics · Malay World

20.1 Introduction

In their seminal work on Malay contestations, Barnard and Maier (2021) delve into the intricate puzzle surrounding the origin and significance of the term 'Malay' which has been employed to describe 'a confusing variety of configurations of human

N. Noorashid (✉)
Faculty of Arts and Social Sciences, Universiti Brunei Darussalam, Gadong, Brunei Darussalam
e-mail: najib.noorashid@ubd.edu.bn

H. M. Y. Ho and Y. Chan (eds.), *Transnational Southeast Asia*, Asia in Transition 29,
https://doi.org/10.1007/978-981-96-3608-2_20

beings, locations, languages, customs, states, and objects' (p. ix) across various parts of the Southeast Asian region. Provocative and yet elusive, 'the primacy of [Malay] language has been the basis of further explorations into the boundaries of a Malay culture, a Malay nation, a Malay race, a Malay state, or a Malay identity' (p. x).

In essence, the term 'Malay' undergoes constant transformation, defying easy definition, remaining contested and never firmly established (Adelaar 2005; Milner 2008; Vickers 2021). Consequently, 'the thinking, writing, and talking about the Malay world becomes a daring endeavour, and the results will continue to be contested' (Barnard and Maier 2021, p. x), thus in return, offering various opportunities and perspectives in researching the possibilities surrounding the issues of Malays in the Malay World. Over the years, many researchers have remained steadfast in their interest in studying the development of the Malay language in the Malay World, which intersects with various factors in the region (Arka 2013; Aziz 2023; David et al. 2009; Noorashid, 2021b).

While emphasising the uniqueness of Malay studies in comparison to other areas such as Middle East and South Asian studies, Alatas (2021) writes that the development of Malay studies, including those of linguistics, has gone through a long tradition of complexities involving intellectual imperialism, colonial capitalism, and orientalist paradigms. He further asserts that methodological issues pertaining to collecting, interpreting, critiquing, demystifying, and deconstructing Malay issues have become matters that need 'a call for conscientious engagement' (p. 209). Alatas's critical essay resonates with a much earlier call by Collins (2002), who highlights a major issue concerning the lack of empiricism and inductivism in Malay linguistic research, which influences the attitudes of the Malays and the production of knowledge in the Malay world.

While conceptualising the notion of the Malay world, exploring the Malay language and identity perpetually involves examining the interrelations between the indigenisation of 'colonial knowledge' and the quest for a pan-Malay identity in the region (Soda 2020). However, fewer studies have explored the impacts of these aspects on the production of knowledge in Malay studies, particularly Malay linguistics. This is despite the fact that language has been used as a political tool to maintain the significance of Malay language, ideologies, and its vitality amid the hegemony of English across Malay countries (Aljunied 2011; Chee 1990; Jones 2015).

Focusing on sociolinguistic issues in the Malay World, and building upon earlier works of *A study on attitudes towards the Malay language and its vitality in Brunei, Malaysia, Indonesia, and Singapore* (Noorashid 2020) and *(Re)Negotiating 'Jati Diri Melayu': A researcher's reflection on Malay transnational identities and linguistic studies in the Malay World* (Noorashid 2022b), this chapter aims to discuss current debates on language contact involving Malay and English in the transnational context of the Malay World. Additionally, this chapter demonstrates that conducting research with the aim of contributing new knowledge aligns with the process of self-reflective discourse as a researcher.

20.2 The 'Transnational' Malay World

Today, research on transnational relations, practices, and contexts is marked by the increasing importance of interactions between actors across various national domains, shaped by cultural negotiations, communication dynamics, and societal transformations in the contemporary world (Boccagni 2011; Pries 2022). In essence, transnationalism encompasses the processes, movements, and exchanges of phenomena that transcend national borders, including those related to education, cultures, identities, social networks, politics, and the significance of language use (King 2023; Tedeschi et al. 2022; Ward et al. 2021).

While the notion of transnational studies is suggested in previous scholarly works of Malay linguistics in the Malay world (David et al. 2009; Jones 2000), fewer have correlated the current conceptualisation of transnationalism—the interactions of nations with similar vested interests and their symbiotic effects—in a definite way. In this section, this chapter argues that the conceptualisation and knowledge production in Malay linguistics and studies in the Malay world—whether it is intentional or organically developed—have intersected in various forms due to hybridity, shared attitudes, proximity in cultural and geographical existence, and similar socio-political histories and backgrounds.

The term 'Malay World' assumes diverse meanings contingent upon its contextual usage. Geographically and historically, it denotes a collective of nations in Southeast Asia, spanning the expanse between the Indian Ocean and the Pacific Ocean—the fourth largest archipelago globally, known as the Malay Archipelago. In geopolitical terms, the 'Malay World' is interchangeably used to describe the post-colonial region of the Malay Archipelago, recognised as *Alam Melayu* (Sweeney 2011), while in contemporary discourse, the concept of the 'Malay World' has evolved to encompass nations where Malay culture prevails and embraces the socio-cultural values of Islam (Mohd Shuhaimi and Osman 2012).

The Malay World emerges as the ancestral home of Malays, where Malay cultures and traditions thrive, and the Malay language serves as the daily mode of communication. Hence, the concept of the Malay World encapsulates the Southeast Asian region where the Malay race resides, upholds the Malay language, embraces Islam, and observes Malay cultures and traditions. The Malay people are identified as the '*Rumpun Bangsa Melayu*' or the descendants of the Malay race (Haji Mohammad and Haji Mahayudin, 2011, pp. 1–2).

Lukman and Bharuddin (2013) illuminate the Malay World through a linguistic lens, defining it as '…an area or district whose people speak the Malay language in Southeast Asia which includes Brunei Darussalam, Indonesia, Malaysia, Singapore as well as Southern Thailand […] which culturally refers to as the homeland for people of Malay descent historically led by the Malay Sultanate in South East Asia offshore' (p. 3).

The varieties of Malay spoken in Brunei and Malaysia (*Bahasa Malaysia*) have been influenced by British English due to the countries' histories as subjects of British colonialism. In contrast, *Bahasa Indonesia* has been shaped by Dutch colonialism.

The emergence of pluricentric Malay and its colonial legacy can be attributed to European colonialism and post-colonial experiences. This has influenced the rationale for recognising Malay as a national language in Malaysia, Singapore, and Indonesia. Colonised Malaysia (including Singapore, which was part of Malaya) experienced the intrusion of English in administration and education due to the need for communication with the British. However, the Dutch did not impose their language as strongly in Indonesia (Asmah 1992).

Different colonial experiences in Malaysia, Singapore, and Indonesia also influenced the use of the national language as a means to overcome ethnic divisions and promote regional integration and unity in these countries. Thus, colonialism affected the perceptions of the Malays regarding the coexistence of Malay and foreign languages, including the need to use Malay as an element of unity, power, and resistance to foreign languages (Lowenberg 1985). The relationships between colonial histories also influence the development of national language policies in these countries, including Brunei (Adelaar 2005; Jones 2015; Steinhauer 2004).

In the Malay World, focusing on Brunei, Indonesia, Malaysia, and Singapore (refer to Fig. 20.1 for the map of the Malay World), Malay (including *Bahasa Malaysia* and *Bahasa Indonesia*) is acknowledged as the official and national language. These nations collaboratively strive to preserve and uphold the significance of the Malay language through their respective governments and language bureaus.

Although these varieties of Malay exhibit a high degree of linguistic affinity, the four countries continue to 'converge' on standardising the orthography and terminology aspects through various national government broadcasting bodies (Asmah 1992). While there is a growing body of literature discussing on the contrastive development of terminologies and intelligibility between *Bahasa Melayu*, *Bahasa*

Fig. 20.1 The Malay World (Noorashid and Chin 2021)

Malaysia, and *Bahasa Indonesia* today, amid acknowledging that these varieties stem from the same Malayic root within the Austronesian language family (Adelaar 2005; Asmah 1992), this chapter refers to '*Bahasa Melayu*' or henceforth 'Malay language' as a reference to the varieties of Malay upheld by the Malays.

Additionally, efforts to safeguard Malay are extended through initiatives by *Dewan Bahasa dan Pustaka* (DBP) (Language and Literature Bureau) in the respective countries, or through the transnational organisation of *Majlis Bahasa Brunei, Indonesia and Malaysia* (MABBIM) (Language Council of Brunei, Indonesia, and Malaysia) and other regional associations. Another significant aspect of Malay nationalism is evident in the official national anthems practised by these countries, officially endorsed through national constitutions and/or federal laws (Reed and Bristow 1997). A comprehensive overview of these features is provided in Table 20.1.

While the conceptualisation of the Malay World is debatable in various principles and disciplines (Barnard and Maier 2021; Fee 2001; Milner 2008), the Malay world in linguistics and literature lens today is popularised by the Malays and the Malay countries through inter-regional endeavours, including MABBIM, DBP, and *Majlis Sastera Asia Tenggara* (MASTERA) (Council of Southeast Asia Literature), aimed at preserving the Malay language, culture, and identity, along with intellectual products supported by the governments of Brunei, Malaysia, Indonesia, and Singapore (refer to Table 20.1 for shared traits of the Malay World).

While these Malay countries project their specific Malay beings and traits through unique national philosophies (for instance, *Rukun Negara* for Malaysians, *Pancasila* for Indonesians, the National Pledge for Singaporeans, and *Melayu Islam Beraja* (Malay Islamic Monarchy) for Bruneians), the integration of these four countries through a shared Malay root, identity, cultures, and the commitment to embodying the essence of '*Jati Diri Melayu*' or being Malay or 'Malayness' are undeniable (Noorashid 2020).

Evidently, these shared identities are embraced by the Malays within '*daerah Melayu inti*' (the collective of Malay nations is commonly known as the core Malay

Table 20.1 Shared similarities of Malay in the four countries

Country	Malay population	Status of Malay	Malay language in education	Maintenance/ preservation efforts
Brunei	Majority	Official Language	One medium of instruction	DBP; MABBIM as a member country
Malaysia	Majority	National/Official Language	The main medium of instruction	DBP; MABBIM as a member country
Indonesia	Majority	National/Official Language	The main medium of instruction	*Balai Bahasa* (DBP); MABBIM as a member country
Singapore	Minority	National/Official Language; shared with English, Mandarin, and Tamil	One medium of instruction; 'mother–tongue' subject	DBP; MABBIM as an observer country

districts (Ismail 2000, p. 298)) affirming their status as Malay nations and ultimately fostering a collective sense of 'Malayness' among the people. This is amid complexities as the Malays in Brunei would prefer to be called 'Bruneians', whereas the Malays in Malaysia and Indonesian are usually referred to as 'Malaysians' and 'Indonesians' respectively due to different nationalities and geopolitical development which in turn created a sense of exclusivity (Bartlett 1954; Noorashid 2020). Milner (2008) states that defining 'Malays' or a '…real substance to a "Malay" essence—an "authentic", "mainstream" or "real Malay"—is clearly a challenge' (p. 196) as the 'character of 'Malay' communities can differ substantially from one place to another' (p. 10).

As 'the formation of the language [align with] the formation of the ethnic identity' (Vickers 2021, p. 53), in Malay scholarship, the celebrated embodiment of *Jati Diri Melayu* or shared 'pure Malayness,' serves as a symbol of unity and cohesion of ethnic identity within the Malay world (Barnard and Maier 2021; Fee 2001). *Jati Diri Melayu*, henceforth referred to as Malayness, encompasses individuals of Malay origin who speak the Malay language, adhere to Malay culture, practice Islam, and reside in core Malay districts (Abdullah 2009). Failure to uphold proficient Malay communication or embody ideal Malay characteristics and loyalty to the Malay monarchy system can often be viewed as sacrilegious (Noorashid 2020). Thus, over the years, the 'declining' of Malayness due to globalisation, mainstream culture, and neo-imperialism has sparked considerable discourse across various platforms, including politics (Ab Halim and Kamarul Azmi 2009; Zakaria et al. 2012).

The interrelationship between language, identity, and nationalism can become a powerful tool to represent people (Carter and Sealy 2007), where the relationship between transnational identities and linguistic practices has been a focus in academia (De Fina 2015; Duff 2015; Giampapa 2004; Warriner 2007). While language is a part of identity as: (1) a social and interactional construct/product; (2) influenced by macro-level demographic factors; (3) represented through labels and systems; (4) constructed through relationship among actors; and (5) formed intentionally or habitually (Bucholtz and Hall 2005), the notion of transnational identities is rather fluid as identities are not fixed but rather involves the result of processes and social practices.

While the contemporary concept of transnationalism within the Malay World has yet to receive widespread attention in academic discourse, there have been recurring suggestions regarding its essence and the perceived similarities and shared entities such as identity, language, culture, and socio-political elements. For instance, through an examination of language, sociology, and historical studies, it becomes apparent that past colonialism significantly influenced the perceptions of Malaysians and Indonesians regarding the coexistence of Malay and foreign languages, underscoring the necessity of utilising Malay as a unifying element, a source of power, and resistance against foreign linguistic influences (Lowenberg 1985; Steinhauer 2004). There is also an acknowledgement of shared attitudes towards language varieties in the Malay World because the Malays in the four countries share a similar language, culture, worldview, heritage, and political system (Milner 2008). Several literatures continue to fuel debates regarding linguistic actions and implementations of each Malay country, which consequently affect their counterparts within the Malay World

(Asmah 1992; Jones 2000; McLellan 2022; Noorashid 2022b, c; Roksana 2018). This is amidst a prevailing tendency of one Malay country to exemplify the other Malay counterpart in maintaining, revitalising, and interpreting attitudes and efforts towards the Malay language (Noorashid 2020).

20.3 Shifting Paradigm in Malay Linguistic Research

In the Malay world, the 'Malay vs. English' debate has become an ongoing topic among language experts, both Malay and international linguists. The hegemony of English as the primary lingua franca is evident worldwide, and other languages, including Malay, are assumed to be affected by this globalisation.

While there is a growing body of contemporary literature acknowledging the exclusivity of the national identity of each Malay country, the continuous efforts to maintain and revitalise Malay usage in the prevalence of English is one of the major language issues in the region. Malay speakers and experts are concerned that the globalisation of English may lead to the loss of Malay identity and culture. This can be seen in continued efforts to discuss the declining use and importance of Malay in various settings at the annual MABBIM and MASTERA meetings held in these Malay countries. Furthermore, there are alarming trends in scholarly works suggesting that English could replace Malay as the lingua franca of the Malay world, prompting several official and non-official associations to strive to address this situation.

Globalisation has created a world without boundaries, leading to the existence of multilingual societies where co-existing languages and dialects often compete to preserve their relevance to their speakers. Crystal (2000) claims that a language dies every two weeks from globalisation, while it is predicted that 90% of languages spoken in the world will face extinction by the end of the century (Krauss 1992). Consequently, language maintenance and the vitality of endangered languages have become major topics in sociolinguistics globally, with no exception for the Malay language.

However, in order to classify a language as endangered, one must not be misled by the superficial situation of the language; it should be analysed in its true context by taking various factors into account (Nelson 2007). Thus, in the Malay world, the opportunity to investigate the current situation of the Malay language due to the hegemony of English in a broader sociolinguistic perspective remains relevant, particularly because a plethora of literature has shown that the region is considerably 'polluted' by the proposition of linguistic purism (McLellan 1997).

In a critical essay on reviewing language vitality, Mufwene (2005) argues that the oversimplified yet well-received portrayal of 'English as the global language' in causing language endangerment, as suggested by mainstream studies, may not always be factual. This perspective tends to 'ignore the naturalness of egalitarian multilingualism as a way of life in many parts of the world' (p. 37). Similarly, Pennycook (2001) claims that the understanding of English as part of the colonial legacy and as a threat to other languages lacks an awareness of the 'social, historical, cultural, and political relationships' between English and the local languages (p. 78).

Thus, taking inspiration from these major studies in language contact, Noorashid (2020) sought to reassess issues surrounding the 'Malay vs. English' debate in the Malay World. The focus was on language usage and attitudes at both national and community levels, incorporating extensive assessments of language policies, landscapes, and usage. To achieve this, the research implemented a modified triangulation model that examined the natural use of and attitudes towards both languages through a series of observations and matched-guise tests. Additionally, the research analysed self-reported data, examination results, language policies, and incorporated interviews with over ten carefully selected Malay and international language experts, representing emic and etic perspectives on the 'Malay vs. English' debate as part of transnational language issues in the Malay world.

Noorashid's (2020) primary findings underscore that the attitudes towards the standardised varieties of Malay remain positive and are maintained through top-down approaches by the national governments of these Malay–speaking countries. However, this does not restrict the Malays' preference for English. Thus, language attitudes and usage among the Malays do not demonstrate extreme binary responses but are adaptive to contexts and audiences. This is due to the complementarity of both languages as they play different roles in the respective Malay societies. In conclusion, English will never replace Malay as the language of solidarity and of Malay culture and identity due to the awareness of the Malay speakers in differentiating instrumentalities of these languages. Therefore, as of today, Malay is far from being endangered and English does not threaten Malay, as previously claimed.

Nevertheless, Noorashid's (2020) research endeavour was undertaken in the face of a prevalent notion that the 'Malay language faced adverse effects due to the dominance of English,' as perpetuated by prominent Malay scholars in the region. This occurred amid the strong influence of Malay language missionaries, who perpetuated the recycling of knowledge production across generations of scholars in the Malay World (Collins 2002).

Multilingualism does not always contribute to language endangerment as the linguistic situation of one language (often seen in the West) should not be a fixed paradigm to other languages in different ecology (Mufwene 2002). In the past, several studies in international contexts have suggested a complementary role for English in a multilingual society as it provides more opportunities and empowerment to national/official and minority languages, as long as the speakers acknowledge its possibilities (House 2003; Johnson 2009; Majidi 2013).

Despite this reality in the Malay world, misconceptions persist regarding language contact between Malay and English, as well as the lingering impact of the period of Western colonialism, continuing to foster the belief that 'English has a negative effect on Malay,' even though this assertion may be an exaggeration. The pervasive politicisation and infusion of non-linguistic agendas into the discourse surrounding 'Malay vs English' debate has created a scenario where reporting and discussing linguistic issues through a neutral perspective is habitually perceived as 'going against the current' and may not align with popular sentiments among Malay language missionaries or the public. This is not unexpected, as numerous Malay scholarly works have widely acknowledged the view of English as causing a negative effect to the Malay language, identity, community, institutions, and nations (Chee 1990; Husni 2009; Milner 2008).

Convincing Malay experts and researchers to embrace an alternative truth poses an institutional and disciplinary challenge within the field of Malay linguistics. This resistance is further exemplified by a statement from prominent language policy scholar in Southeast Asia, Andy Kirkpatrick, who observes that '…people with prejudicial views are very difficult to change, and the purist argument sounds very attractive… it is very difficult to counter prejudice because people do not listen to fact' (Noorashid 2020, p. 217).

Furthermore, in the past, Malay purism has been connected to negative attitudes towards English, which is derived from Malay monolingualism. Language purism surrounding Malay development has been briefly discussed by Chee (1990), Gill (2002), Talib (2013), and Shahimah (2006), as well as by Mahmud (2000) who likens backward-minded purists to a '*wabak*' (virus) (p. 131) that can harm Malay development and survival. The experts of Malay language and policies in Noorashid's thesis (2020) further acknowledge the continued existence of these biassed attitudes among pro-Malay experts with vested non-linguistic interests. In his interview, Kirkpatrick emphasises the potential and imperative of expanding such investigations to incorporate aspects of religion and history within the region in understanding the issues behind language vitality vis-à-vis purism in the transnational context of the Malay world (Noorashid 2020).

In December 2022, the Southeast Asian Studies Regional Exchange Program (SEASREP) Foundation organised a regional workshop on 'The Making of Southeast Asianists in Southeast Asia' aimed at generating think pieces among contemporary scholars in Southeast Asian countries, with a specific focus on comparative work about countries or cultures across the region. Seeing the potential of researching Malay linguistics as part of transnational efforts, Noorashid's (2022d) think piece entitled *'Shifting selves' and 'against the current': Researching Malay linguistics, identities, and nationalism in the Malay World and beyond* discusses the shifting paradigm of being both the researched and the researcher in contemporary studies on Malay language contact with other languages. Recognising the potential for the advancement of Malay linguistics as part of Southeast Asian (SEA) studies, Noorashid (2022b) broadened the scope of investigation by incorporating transnational and nationalist perspectives on the 'Malay vs English' issue, culminating in the presentation of an early analysis titled, *(Re)Negotiating 'Jati Diri Melayu': A researcher's reflection on Malay transnational identities and linguistic studies in the Malay World* at the Global Citizenship and Transnational Identities in Southeast Asia conference held at Universiti Brunei Darussalam.

Both studies addressed a critical aspect that had not been previously discussed, emphasising the importance of re-evaluating the 'problematic' and traditional framework of '*Jati Diri Melayu*' (pure Malay identity or Malayness) intertwined with Malay nationalism (Noorashid 2022b, 2022d). The notion of 'Malayness,' while it has been used as an 'emblem of unification' and 'Malay maintenance' of the Malays across the Malay countries due to shared similarities in origin, culture, religion, and attitudes, has led to the oversimplification and modulation of the intricate interrelationships defining Malay identity across the Malay World. This, in turn, has shaped linguistic research and knowledge production in the region.

The challenges stemming from this traditional framework include difficulties in conducting empirical-based research, exacerbated by: (1) the politicisation of language issues; (2) the prevalence of post-colonialism and viral scholarship in linguistic research; (3) the misconceptions surrounding language phenomena, among other factors. These challenges, in turn, exert a profound influence on the attitudes and mentality of Malay researchers, policymakers, as well as the public. In fact, some scholarly works including Lee (2003), Noor Azam (2007), and Noorashid and McLellan (2021) have suggested on the flexibility of Malay identities amid multilingualism in Malay countries, challenging the pro-Malay scholars normalised projection of Malayness only through the Malay language.

Through a series of observations, encompassing both past and ongoing investigations into 'Malay vs. English' issues in the region, securing institutional support for various significant sociolinguistic endeavours may become a challenge. These extend beyond the preservation of the Malay language and identity to encompass studies on the safeguarding of ethnic and minority languages and identities, the potential exploration of translanguaging in pedagogy, the relevant discussions surrounding Malay diaspora and identity, and the imperative to incorporate technological advancements and rigorous quantitative analysis in Malay studies—a requirement aligned with the global demands of sociolinguistic research. Nevertheless, the focus on institutional support today remains towards safeguarding Malay maintenance and preservation (Noor Azam 2012; Noorashid 2020).

Nevertheless, despite the persisting challenges of insufficient institutional acknowledgement and lack of financial support, there remains the present opportunity to navigate towards resolutions and cultivate a new movement aimed at enhancing the comprehension of language issues in the Malay World. For instance, Noorashid's (2020, 2022a, 2022c) studies, which primarily aim to contribute novel insights to the extensive body of 'Malay vs. English' research, have revealed various aspects of Malay linguistic studies that need to be addressed. These aspects include the influence of contemporary changes and developments in socio-politics, religion (Islam), viral discourse on language use and policies, as well as the mentality and insecurities of the Malay community. The persistent challenges related to the exaggeration of the 'Malay vs. English' debate continue to arise, suggesting that discussions of 'language in contest' between Malay and English in the region as another language myth fuelled by non-linguistic agendas (Noorashid 2022c).

While Martin (1996) contends that 'Borneo is a veritable language laboratory,' in a transnational context, this chapter further recognised that '…the Malay World is an extended veritable language laboratory' (Noorashid, 2022a), offering abundant opportunities for linguistic research not only within a specific Malay country, but across boundaries in the Malay world and beyond. The trajectory of contemporary Malay studies, intricately linked with Malay linguistics, identity, and nationalism as integral components of transnationalism within the region and part of Southeast Asian studies, rests in the hands of a self-reflective, forward-thinking, and younger generation of academics within the Malay region. The field and the prospects of Malay linguistics and research can be significantly enhanced through increased exposure and training in contemporary knowledge and practices for the emerging cohort of researchers.

While institutional assistance from governments, language bodies, and research institutions—both local and international—may provide essential research funding and platforms for disseminating knowledge, it is also crucial to foster a conducive environment to academic and community. Such support facilitates the emergence of passionate and truth-seeking researchers who exhibit the courage to challenge prevailing norms and uphold scholarly principles, enabling the creation, presentation, and publication of academic works grounded in the tenets of empiricism and inductivism.

20.4 Reflective Discourse of a (Malay) Researcher

As research is known as a form of a 'translation process' through mutual engagement (Fook 1999, p. 14), involving the dynamics and negotiation of insiders and outsiders (Hoon 2021, p. 158), the importance of a reflexive approach to research has become central in social science and humanities research over the years. Adopting a reflexive approach in research remains imperative, emphasising its ability to scrutinise 'the epistemological dichotomies of researcher/researched, self/other, insider/outsider, and native/foreign' (Hoon 2006, p. 83).

Hertz (1996) and Olmos-Vega et al. (2023) emphasises the importance of examining the politics and ethics of a self-reflexive approach in qualitative research to acknowledge personal biases and power dynamics within the research process. Meanwhile, May and Perry (2017) and Jamieson et al. (2023) believe that incorporating a reflexive strategy in conducting and interpreting research will enhance the research process and its value. By understanding self-reflexivity and self-and-subject researcher, it can serve 'to challenge our understandings of the problems generated in accessing, gathering and presenting research materials, the nature of the subjects under investigation, and the interpretation of the societies, cultures, and histories of Southeast Asia' (Jammes and King 2021, p. 2).

In contemporary sociolinguistics, given the global significance of language endangerment and death, researchers often grapple with interpreting the impact of globalisation and English on language decline (Mufwene 2005; Nelson 2007). This challenge persists amid a pervasive misunderstanding of language contact dynamics between Malay and English in the Malay region, as discussed in the works of Noor Azam (2012), Noorashid (2020, 2021a, 2021b), and Noorashid and McLellan (2021). The prevailing influence of English has consistently been attributed to the diminishing use of Malay and the erosion of embedded cultural elements in the Malay countries. This perspective has permeated various channels, including academic scholarship, national news, social media, and parliamentary discussions, establishing itself as one of the paramount contemporary language issues in this domain.

Furthermore, the paucity of critical works in Malay linguistics within the Malay World poses a significant challenge for researchers, compounded by a tendency towards 'knowledge recycling' among them (Collins 2002; Collins and Zaharani 1999). Recent discussions in Malay studies have also highlighted the rise of viral scholarship characterised by non-linguistic agendas and publications based on hypothetical theories authored by influential pro-Malay language advocates (language missionaries) (Noorashid 2020). Furthermore, there are tendencies among scholars to exhibit bias and marginalise certain influential scholarly works due to 'intellectual imperialism and academic dependency' (Alatas 2021, p. 199). These phenomena have, to some extent, impacted the impartiality of Malay linguistic studies in the region (Noorashid 2022b) and further created a larger problem in knowledge production within Malay studies (Alatas 2021), affecting not only academia but also academics.

Delving into the limited body of critical Malay scholarship on 'Malay vs. English,' the 'misunderstanding' on language contact assertion by Mufwene (2005) and Nelson (2007) has been supported in sparse scholarly works by McLellan (1997) and McLellan and Noor Azam (2000). While Mufwene (2006, 2013) believes that the global discourse on language issues that often amplify concerns about language death due to English exists globally, further accentuating the need for empirical studies. Recognising the potential for simplification and misunderstanding in the discourse of 'Malay vs. English', conducting empirical-based studies, starting in Brunei and expanding to other Malay countries becomes imperative in Malay linguistic research (Collins 2002; Collins and Zaharani 1999; Noorashid 2020; 2022c).

In 2021, Noorashid (2021a) introduced the term 'methodological calamity' to describe methodological flaws that compromise the veracity of 'Malay vs. English' research and its reliability. These flaws, arising from non-linguistic agendas and political motivations, have created language myths and undermine the values of inductivism and empiricism in research, further challenging previous scholarship that pertains to oversimplify the linguistic issue. The issue also stemmed from the 'irresponsible act of recycling knowledge' by Malay researchers. While methodological errors are common in conducting research (Brewin 2023; Roulston 2011), they have become a taboo and an 'uncomfortable reality' in academic practices (Eckert 2020; Pillow 2003). However, addressing these issues and understanding errors as part of a self-reflexive approach in social science is crucial to avoid perpetuating misleading facts in knowledge production (Rigg 2021). This is amid several criticisms of methodological and interpretational flaws in Malay linguistic research raised by earlier works (Alatas 2021; Chee 1990; Collins 2002; Collins and Zaharani 1999; Ding 2011).

Following similar traditions of methodological and interpretation assessments in earlier Malay scholarly works, Noorashid (2021a) further discussed the politicisation of bilingual phenomena, simplistic methodological approaches and interpretations, and the recycling of viral scholarship by Malay language advocates were among the factors contributing to the misunderstanding of 'Malay vs. English' language issues in the Malay World. This, in turn, poses challenges in producing high-quality, neutral research papers within this field.

Nevertheless, guided by the principle that 'the more you read, the more you learn, and the better your understanding', Noorashid (2022a) further emphasises the need to be objective when approaching and interpreting issues surrounding 'Malay vs. English' in the region. He highlighted the importance of transitioning from a passive researcher to cultivating a sense of neutrality, which can become a turning point in advancing scientific research in Malay linguistics. This is amid grappling with the Malay dilemma of 'repeating' the sentiments of previous pro-Malay experts in the region. In the field of Malay linguistics, where politicisation and non-linguistic agendas are prevalent, it is imperative for a researcher to take a neutral yet scholarly stance against the existence of what Asmah (1990, p. 9) refers to as '*pelopor-pelopor bahasa ideal (sebagai bahasa standard)*' (pioneers of 'ideal' Standard Malay) who are reductionist and have been projecting '*sikap di luar tabii*' (p. 9) (attitudes of deviance) that hinder the development of Malay.

Good faith (Jammes and King 2021, p. 4) and commitment to scholarship and superior publication (Hasan 2023, p. 7) remain essential in conducting research. This includes promoting linguistic studies without political agendas or non-linguistic vested interests as part of advancing the dissemination of scholarly perspectives on understanding Malay linguistics and language research through academic approaches (Noorashid 2020).

Therefore, Crystal's (2011) acknowledgement of the importance of public speaking in language maintenance and preservation is crucial to addressing linguistic issues through a transnational lens when studying Malay within the wider Malay world. Thus, galvanising neutrality and critically examining the development of the Malay language in the region and beyond over time remains imperative in the contemporary lens of sociolinguistics and transnationalism. This is amid claims that, regardless of the efforts to maintain and preserve the Malay language while upholding Malay identity and hybridity across the Malay region, each of these Malay countries also needs to acknowledge their unique histories, cultural, and sociological development to understand the effects of the state and development of the Malay language across the Malay world today (Noorashid 2020; 2022b).

20.5 Conclusion

In 1990, Nelson Mandela reminds the idea of 'education as the most powerful weapon to change the world' (Ellis 2019). Undoubtedly, education signifies a transformative journey—a quest to unveil truths, address challenges, and reshape one's worldview in the relentless pursuit of knowledge and experience for the betterment of society. The role of education and research plays a symbiotic role as a life-learning approach to fulfil professional responsibilities while maintaining standard of performance to create new knowledge (Fattu 1960). Often times that research amid education reveals opportunities for individuals to evaluate and to innovate directions for the development of creative processes, attitudes, and reflections.

To date, the Malay world, consisting of Brunei, Indonesia, Malaysia, and Singapore, can be justified as a transnational concept through its historical and cultural ties rooted in the Malay language, customs, and traditions. Islam, as the predominant religion in the region, further serves as a unifying force that transcends national boundaries, contributing to the formation of a transnational Malay identity. Political ties, trade and economic networks, as well as diplomatic efforts among these countries continue to promote cooperation, dialogue, and collaboration on regional issues. As a unifying emblem and symbol of solidarity and shared identity, the Malay language serves as a lingua franca across the region, enabling communication and cultural exchange among diverse communities. This shared heritage creates a sense of belonging and connection among Malay communities across national borders, where the effort to maintain the language and identity remains vital to the people.

Seeing the potential and demand for reviewing and investigating sociolinguistic issues in the Malay world, while steadfastly adhering to the foundational principles of empiricism and inductivism in investigating language contact between Malay and English in the Malay region, this chapter also calls for enhancing cultural sensitivity to forge research that continues to recognise the transnational aspects of the Malay language. This endeavour can be a conscious effort to contribute seminal works to Malay linguistics studies and foster a movement that underscores the potential of bi–/multilingualism as an integral aspect of commodification and community practice, further raising an opportunity to change the course of academia in Malay linguistic studies.

Nevertheless, this chapter does not deny that the relentless pursuit of maintaining the importance of the Malay language, practices, and beliefs may have led to exaggerations in the 'Malay vs. English' claim, further fuelling misunderstandings about the actual situation of language contact between Malay and English or with any other languages. Collins's (2002) significant statement continues to serve as a poignant reminder to Malay researchers that '[l]inguistics is about presenting the facts and analyses in such a way that others can evaluate, criticize, and, if necessary, discard them' (2002, p. 2). Collins's critical remark on Malay knowledge production underscores the intricate conundrum of Malay language studies in the transnational context of the Malay World. Furthermore, it is also the responsibility of Malay researchers to present the facts of Malay linguistic research in the Malay world (Noorashid 2021a; 2022d).

Taking the perspective of a contemporary researcher in Malay sociolinguistic studies in the Malay world, and drawing inspiration from earlier reflexive articles by other scholars, this chapter illuminates the complexities and challenges inherent in the realms of Malay language studies, identity, and nationalism, contributing to the broader discourse of Southeast Asian studies. This chapter also marks a departure from traditional sociolinguistics by navigating against the currents of conventional scholarship, thereby contributing to the knowledge of contemporary Malay linguistic research within the Malay region and beyond. As language evolves and knowledge is dynamic, it is significant to integrate the discourse on 'Malay vs. English' into multidisciplinary research that spans sociolinguistics, media studies, and education, conducted in both Malay and English mediums.

References

Ab Halim Tamuri, Kamarul Azmi Jasmi (2009) Nilai, amalan, dan sistem pendidikan masyarakat Melayu Islam selepas penjajahan Barat hingga kini (Values, practices, and the education system of the Malay Islam civilisaton after Western colonisation until today). In: Peradaban Arab–Islam dan masyarakat Melayu (Arab–Muslim Culture and Malay Civilisation). Fakulti Pengajian Islam Bangi (Bangi Islamic Studies Faculty), pp 35–57

Abdullah Hassan (2009) Bahasa Melayu di persimpangan: antara jati diri dengan rempuhan globalisasi (The Malay language at crossroads: between identity and the onslaught of globalisation). Kemanusiaan (Humanity) 16:59–81

Adelaar KA (2005) The Austronesian languages of South East Asia and Madagascar: a historical perspective. In: Himmelman NP, Adelaar KA (eds) The Austronesian languages of Asia and Madagascar. Routledge, pp 1–41

Alatas SF (2021) Silencing as method: the case of Malay studies. In: Jammes J, King V (eds) Fieldwork and the self: changing research styles in Southeast Asia. Springer, pp 199–214

Aljunied SMK (2011) A theory of colonialism in the Malay World. Postcolonial Studies 14(1):7–21

Arka IW (2013) Language management and minority language maintenance in (Eastern) Indonesia: strategic issues. Language Documentation and Conservation 7:74–105

Asmah Haji Omar (1990) Attitudes in the learning of English among Malaysian students: a case study. Paper presented at the TESL seminar, Institut Aminuddin Baki (Aminuddin Baki Institute), Genting Highlands, 1990

Asmah Haji Omar (1992) Malay as a pluricentric language. In: Clyne M (ed) Plucentric languages. Mouton De Gruyter, pp 401–420

Aziz PG (2023) Keberdayaan bahasa Melayu: teori dan aplikasi (Empowering Malay language: theory and application). Dewan Bahasa dan Pustaka Malaysia (Malaysia Language and Literature Bureau)

Barnard TP, Maier HMJ (2021) Melayu, Malay, Maleis: journeys through the identity of a collection. In Barnard TP (ed) Contesting Malayness: Malay identity across boundaries. NUS Press, pp ix–xiii

Bartlett HH (1954) Malayan words in english. Michigan Quarterly Review 60:40–55

Boccagni P (2011) Rethinking transnational studies: transnational ties and the transnationalism of everyday life. European Journal of Social Theory 15(1):117–132. https://doi.org/10.1177/136843101142360

Brewin CR (2023) Inaccuracy in the scientific record and open post publication critique. Perspectives on Psychology Science 18(5):1244–1253. https://doi.org/10.1177/17456916221141357

Bucholtz M, Hall K (2005) Identity and interaction: a sociocultural linguistic approach. Discourse Studies, 7(4–5):585–614. https://doi.org/10.1177/1461445605054407

Carter B, Sealy A (2007) Languages, nations and identities. Methodological Innovations Online 2(2):20–31

Chee TS (1990) A study of evolution of the Malay language. NUS Press

Collins JT (2002) “Malay is fourth largest language” and other myths: language studies and ASEAN. Paper presented at the 2nd ASEAN linguistics conference: Bahasa pengukuh solidariti ASEAN (Language reinforces ASEAN solidarity), Universiti Brunei Darussalam, Tungku Link, 28–29 October 2002, pp 1–22

Collins JT, Ahmad Z (1999) The Malay language and ethnic identity in modern Malaysia. Akademika 55:33–148

Crystal D (2000) Language death. CUP

Crystal D (2011) Language diversity, endangerment, and public awareness. pp 12–20. https://www.davidcrystal.com/Files/BooksAndArticles/-4961.pdf. Accessed 14 July 2024

David MK, Cavallaro F, Coluzzi P (2009) Language policies – impact on language maintenance and teaching: focus on Malaysia, Singapore, Brunei and the Philippines. In Cavallaro F, Milde A, Sercombe P (eds) The Linguistics Journal – Special Edition:151–191. https://www.academia.edu/75543675/The_Linguistics_Journal_Special_Edition_Language_Culture_and_Identity_in_Asia_Special_Edition_Language_Culture_and_Identity_in_Asia. Accessed 13 July 2024

De Fina A (2015) Linguistic practices and transnational identities. In: The Routledge handbook of language and identity. Routledge, pp 163–178. https://doi.org/10.4324/9781315669816

Ding CM (2011) Cabaran bahasa Melayu di era ekonomi dan teknologi baru (Malay language challenges in the era of new economy and technology). In: Azurin Othman, Norati Bakar (eds) Kongres antarabangsa bahasa dan budaya (Congress of international language and culture), vol 1. Dewan Bahasa dan Pustaka Brunei (Brunei Language and Literature Bureau), pp 59–90

Duff PA (2015) Transnationalism, multilingualism, and identity. Annual Review of Applied Linguistics 35:57–80. https://doi.org/10.1017/S026719051400018X

Eckert J (2020) "Shoot! Can we restart the interview?": lesson from practicing "uncomfortable reflexivity". International Journal of Qualitative Methods. https://doi.org/10.1177/1609406920963810.

Ellis P (2019) Changing the world through education – how Nelson Mandela created conditions for success. https://blog.cambridgeinternational.org/nelson–mandela/. Accessed 13 July 2024

Fattu NA (1960) The role of research in education–present and future. Review of Educational Research, 30(5):409–421. https://doi.org/10.2307/1168922

Fee LK (2001) The construction of Malay identity across nations. Bijdragen tot de Taal–, Land– en Volkenkunde 157(4):861–879

Fook J (1999) Reflexivity as method. Annual Review of Health Social Sciences 9(1):11–20

Giampapa F (2004) The politics of identity, representation, and the discourses of self–identification: negotiating the periphery and the center. In: Pavlenko, A, Blackledge, A (eds) Negotiation of identities in multilingual contexts. Multilingual matters, pp 192–218

Gill SK (2002) International communication: English language challenges for Malaysia. UPM Press

Haji Mohammad Haji Abd Rahman, Haji Mahayudin Haji Yahaya (2011) Tamadun Melayu Islam di alam Melayu: kes Brunei Darussalam (Malay Islam Civilization in the Malay world: Brunei Darussalam case). Pusat Da'wah Islamiah Brunei (Islamic Da'wah Centre Brunei)

Hasan MM (2023) Good words, corrupt words, and publishing etiquette. Asiatic 17(1):1–8

Hertz R (1996) Introduction: Ethics, reflexivity and voice. Qualitative sociology 19:3–9. https://doi.org/10.1007/BF02393245

Hoon CY (2006) Defining (multiple) selves: Reflections on fieldwork in Jakarta. Life Writing 3(1):81–102

Hoon CY (2021) Shifting selves: Fieldwork, faith and reflexivity. In: Jammes J, King V (eds) Fieldwork and the self: changing research styles in Southeast Asia. Springer, pp 153–169

House J (2003). English as a lingua franca: a threat to multilingualism? Journal of Sociolinguistics 7(4):556–578

Husni Abu Bakar (2009) Code–switching in Kuala Lumpur Malay: the "Rojak" phenomenon. EXPLORATIONS 9:99–107

Ismail Hussein (2000) Pembakuan bahasa Melayu se–Nusantara (Standardization of the Malay language within the archipelago). Bahasa Jiwa Bangsa (Language as Soul of the Nation) 2:296–303. Dewan Bahasa dan Pustaka Brunei (Brunei Language and Literature Bureau)

Jamieson MK, Govaart GH, Pownall M (2023) Reflexivity in quantitative research: a rationale and beginner's guide. Soc Pers Psychol Compass 17(4):1–15. https://doi.org/10.1111/spc3.12735

Jammes J, King V (2021) Fieldwork and the self: Perspectives and debates. In: Jammes J, King V (eds) Fieldwork and the self: changing research styles in Southeast Asia. Springer, pp 1–30

Johnson A (2009) The rise of English: the language of globalization in China and the European Union. Macalester International 22(12):131–168

Jones GM (2000) Some language planning questions facing Brunei Darussalam, Singapore, Malaysia and the Philippines. In: Sibayan B, Bautista ML (eds) Fetschrift for Brother Andrew Gonzalez. Linguistics Society of the Philippines, pp 226–238
Jones GM (2015). Bilingual and multilingual education in Brunei and Malaysia: policies and practices. In: Wright WE, Boun S, Garcia O (eds) The handbook of bilingual and multilingual education. Wiley–Blackwell, pp 531–541
King HM (2023) Complex intersections of language and culture: the importance of an ethnographic lens for research within transnational communities. Journal of Multilingual and Multicultural Development 44(8):718–736 https://doi.org/10.1080/01434632.2023.2195385
Krauss M (1992) The world's languages in crisis. Language 68:4–10
Lee SK (2003) Multiple identities in a multicultural world: a Malaysian perspective. Journal of Language, Identity and Education 2(3):137–158
Lowenberg P (1985) Malay in Indonesia, Malaysia, and Singapore: three faces of a national language. In: Coulmas F (ed) What are national languages good for?. https://archive.org/details/ERIC_ED276272. Accessed 13 July 2024
Thaib L, Pa BC (2013) Regional cooperation: Malay world and the formation of ASEAN community. Global Journal of Human Social Science & Political Science 13(2):1–9
Mahmud Hj Bakry (2000) Unsur–unsur asing dalam bahasa (Foreign elements in language). Bahasa Jiwa Bangsa (Language as Soul of the Nation) 1:131–137. Dewan Bahasa dan Pustaka Brunei (Brunei Language and Literature Bureau)
Majidi A (2013) English as a global language; threat or opportunity for minority languages? Mediterranean Journal of Social Sciences 4(11):33–38
Martin PW (1996) Brunei Malay and bahasa Melayu: a sociolinguistic perspective. In: Martin PW, Ożóg C, Poedjosoedarmo G (eds) Language use & language change in Brunei Darussalam. Center of International Studies Ohio, pp 27–36
May T, Perry B (2017) Reflexivity: the essential guide. Sage
McLellan J (1997) Linguistic imperialism and the cultural politics of EIL. In: The context of Southeast Asia: outsiders' and insiders' perspectives on the roles of English. National Institute of Education Singapore, pp 156–165
McLellan J (2022) Malay and English language contact in social media texts in Brunei Darussalam and Malaysia. Frontiers in Communication. https://doi.org/10.3389/fcomm.2022.810838
McLellan J, Haji-Othman NA (2000) The myth of widespread English in Brunei Darussalam: a sociolinguistic investigation. Southeast Asia: A Multidisciplinary Journal 2(1&2):37–46
Milner A (2008) The Malays. Blackwell
Mohd Shuhaimi Haji Ishak, Osman Chuah Abdullah (2012) Islam and the Malay World: an insight into the assimilation of Islamic values. World Journal of Islamic History and Civilization 2(2):58–65
Mufwene SS (2002) Colonisation, globalisation and the future of languages in the twenty–first century. International Multicultural Societies 4(2):1–48
Mufwene SS (2005) Globalization and the myth of killer languages. In: Huggan G, Klasen S (eds) Perspectives on endangerment. Georg Olms Verlag, pp 19–48
Mufwene SS (2006) Language endangerment as embarrassment for linguistics. Bunting J, Desai S, Peachey R, Straughn C, Tomkova Z (eds) CLS 42: the Panels. Parasessions of the Forty–second Meeting of the Chicago Linguistic Society 42(2):111–140
Mufwene SS (2013) Globalization, Global English, and World English(es): myths and facts. In: Coupland N (ed) The handbook of language and globalization. Blackwell, pp 31–55
Nelson D (2007) Language death. In: Llamas C, Mullany L, Stockwell P (eds) The Routledge companion to sociolinguistics. Routledge, pp 199–204
Noor Azam Haji–Othman (2007) English and the bilingual Bruneian. In: Dunworth K (ed) English in South East Asia: challenges & changes. Curtin University of Technology Australia, pp 59–70
Noor Azam Haji–Othman (2012) Is it always English? 'Duelling aunties' in Brunei Darussalam. In: Rapatanaha C, Bunce P (eds) English language as hydra: its impact on non–English language cultures. Multilingual matters, pp 175–190

Noorashid N (2020) A study of attitudes towards Malay language and its vitality in Brunei, Malaysia, Indonesia and Singapore. Unpublished PhD thesis, Universiti Brunei Darussalam
Noorashid N (2021a) "Malay vs. English": a linguistic reality or a methodological calamity?. Paper presented at the 4th doing research in applied linguistic conference, King Mongkut's University of Technology, Thonburi, Thailand, 4 September 2021
Noorashid N (2021b) Vitaliti Bahasa Melayu di Negara Brunei Darussalam (Vitality of Bahasa Melayu in Brunei Darussalam). Kemanusiaan (Humanity) 28(2):159–187
Noorashid N (2022a) Linguistic research on Malay/Indonesian language in the Malay World: trend, issues, and implications. Paper presented at the seminar nasional kolaborasi akademik dosen–mahasiswa #2 (Seminar for lecturer–student national academic collaboration #2), Universitas Jenedral Soedirman (Jenedral Soedirman University), Jakarta, 22 October 2022
Noorashid N (2022b) (Re)Negotiating 'Jati diri melayu' (Malay identity): a researcher's reflection on Malay transnational identities and linguistic studies in the Malay world. Paper presented at the global citizenship and transnational identities in southeast asia, Universiti Brunei Darussalam, Tungku Link, 21–22 November 2022
Noorashid N (2022c) The myth of the Malay decline in the Malay World: a sociolinguistic perspective. Paper presented at the 4th Brunei–Malaysia forum, Universiti Brunei Darussalam, Tungku Link, 14–15 September 2022
Noorashid N (2022d) 'Shifting selves' and 'against the current': Researching Malay linguistics, identities, and nationalism in the Malay world and beyond. Paper presented at the making of southeast Asianists in Southeast Asia by the Southeast Asian Studies Regional Exchange Program, Bangkok, Thailand, 10–11 December 2022
Noorashid N, Chin WL (2021) Coping with COVID-19: the resilience and transformation of community-based tourism in Brunei Darussalam. Sustainability 13(15):8
Noorashid N, McLellan J (2021) "I speak English, but I'm still a Malay": language attitudes and identity amongst bilingual Bruneians living in London. rEFLections 28(1):121–143
Olmos–Vega FM, Stalmeijer RE, Varpio L et al (2023) A practical guide to reflexivity in qualitative research: AMEE Guide No. 149. Medical teacher 45(3):241–251. https://doi.org/10.1080/0142159X.2022.2057287
Pennycook A (2001) English in the world/the world in English. In: Burns A, Coffin C (eds) Analysing English in a global context. Routledge, pp 78–89
Pillow W (2003) Confession, catharsis, or cure? Rethinking the uses of reflexivity as methodological power in qualitative research. International Journal of Qualitative Studies in Education 16(2):175–196
Pries J (2022) Transnationalism. In: Scholten P (ed) Introduction to migration studies. Springer. pp 233–247.
Reed WL, Bristow MJ (eds) (1997) National anthems of the world, 9th edn. Cassell Publishers Limited
Rigg J (2021) The importance of being wrong: interpreting the roots of change in rural Thailand. In: Jammes J, King V (eds) Fieldwork and the self: changing research styles in Southeast Asia. Springer, pp 55–74
Roksana Bibi Abdullah (2018) The declining status of the Malay language and efforts in revitalizing its use in Singapore. SARJANA 33(2):1–10
Roulston K (2011) Working through challenges in doing interview research. International Journal of Qualitative Methods 10(4):348–366.
Shahimah Haja Mohideen (2006) A study of English loan words in selected bahasa Melayu newspaper articles. Unpublished masters dissertation, International Islamic University Malaysia
Soda N (2020) Conceptualizing the Malay world. Kyoto University Press
Steinhauer H (2004) Colonial history and language policy in insular Southeast Asia and Madagascar. In: Adelaar KA, Himmelmann NP (eds) The Austronesian languages of Asia and Madagascar. Routledge, pp 65–86

Sweeney A (2011) Pucok gunung es: lelisanan dan keberaksaraan dalam kebudayaan Melayu–Indonesia (Peak of the iceberg: oral and literary culture in Malay–Indonesia). Popular Gramedia (KPG)

Talib IS (2013) The encroachment of English in Malaysian cultural expression. In: Wee L, Goh R, Lim L (eds) The politics of English: South Asia, Southeast Asia and the Asia Pacific. John Benjamins, pp 145–166

Tedeschi M, Vorobeva E, Jauhiainen JS (2022) Transnationalism: current debates and new perspectives. GeoJournal 87:603–619. https://doi.org/10.1007/s10708-020-10271-8

Vickers A (2021) 'Malay identity': modernity, invented tradition and forms of knowledge. In: Barnard TP (ed) Contesting Malayness: Malay identity across boundaries. NUS Press, pp 25–55

Ward P, Devrilmez E, Ayvazo S et al (2021) A transnational research collaboration: a social network analysis and perspectives on our community of practice. Quest 73(4):342–356. https://doi.org/10.1080/00336297.2021.1965892

Warriner DS (2007) Transnational literacies: immigration, language learning, and identity. Linguistics and Education 18(3–4):201–214. https://doi.org/10.1016/j.linged.2007.10.003

Zakaria Stapa, Noranizah Yusuf, Abdul Fatah Shaharudin (2012) Islam asas pembentukan jati diri bangsa Melayu–Muslim (Islam as the foundation for the formation of the Malay–Muslim identity). Jurnal Hadhari (Hadhiri Journal) Special Edition:129–141

Najib Noorashid is a Lecturer in Malay Language and Linguistics at Universiti Brunei Darussalam. His research focuses on linguistic diversity, culture, and education in the Malay World and beyond. He has published a book and articles with Cambridge University Press, Elsevier, Springer, and Palgrave Macmillan. His previous roles include an Associate Consultant at i–Bayan Venture Brunei, Adjunct Professor of English at Seri Begawan Religious Teachers University College Brunei, and Editorial Journalist at Radio Television Brunei. He also serves as an editor and reviewer for several international journals.

Index

H. M. Y. Ho and Y. Chan (eds.), *Transnational Southeast Asia*, Asia in Transition 29,
https://doi.org/10.1007/978-981-96-3608-2

D

E

The manufacturer's authorised representative in the EU is Springer Nature Customer Service Centre GmbH, Europaplatz 3, 69115 Heidelberg, Germany. If you have any concerns regarding our products, please contact ProductSafety@springernature.com

Printed and bound by CPI Group (UK) Ltd, Croydon, CR0 4YY
01/10/2026
02225928-0011